Tracks & Sign
of Reptiles & Amphibians

Tracks & Sign
of Reptiles & Amphibians
A Guide to
North American Species

Filip Tkaczyk

STACKPOLE
BOOKS

Published by
STACKPOLE BOOKS
5067 Ritter Road
Mechanicsburg, PA 17055
www.stackpolebooks.com

Printed in the United States of America

10 9 8 7 6 5 4 3 2 1

First edition

Cover design by Caroline M. Stover
Photos by the author except where noted

Library of Congress Cataloging-in-Publication Data

Tkaczyk, Filip.
   Tracks & sign of reptiles & amphibians : a guide to North American species / Filip Tkaczyk. — First edition.
      pages cm
   Includes bibliographical references and index.
   ISBN 978-0-8117-1186-9
1. Reptiles—North America—Identification. 2. Amphibians—North America—Identification. 3. Animal tracks—North America—Identification. I. Title. II. Title: Tracks and sign of reptiles and amphibians.
   QL651.T55 2015
   591.47'9—dc23
                                            2015029091

This book is dedicated to my son,
Brayden Tkaczyk,
and to the future generation of trackers and
naturalists who will carry this vital
knowledge forward to our descendants.

CONTENTS

ACKNOWLEDGMENTS

First and foremost, I want to send out a huge thank-you to my family. My greatest supporter during the work on this project has been my wife, Jen. I am incredibly grateful for her unwavering encouragement, sacrifice, and incredible patience through the process of making this book. Great thanks also goes out to my mother, Wiesia, who has always supported my passion for the natural world and sacrificed a great deal to allow me to have access to wild places. Also, special thanks to our family for their love and support, especially Nancy Ikeda, Bob Ikeda, and Tamara Ikeda.

This book was made possible because of many helpful and generous contributors. These wonderful people put in a great deal of their personal time and energy to help me bring this book into the world. The first and greatest contributor is Joe Letsche, who shared his deep well of knowledge on many species of reptiles and amphibians. Joe's wonderful photos grace many pages of this book, and his words and considerable field experience heavily influenced the information on turtles. His friendship, incredibly generous spirit, and support leave me in his debt.

Another major contributor was wildlife tracker and nature enthusiast Kim Cabrera. Her keen observations, numerous wonderful photos, and continued support helped see this book through to completion. Her incredible willingness to give so much of her time and energy to this project also leaves me in her debt.

Jason Knight, my boss at Alderleaf Wilderness College and fellow wildlife tracker, was also a key player in this project. His contributions to this book, his advice, and his patience with this process have been immensely valuable.

It has been a real treat to get to know and work with Jonah Evans. He has a keen intellect, incredible tracking skills, and a giving nature, and he helped refine this book into the professional text you see before you. His continued support and shared words are always incredibly appreciated.

Special thanks also go out to those who adventured with me to the wildlife-rich landscapes of the United States in search of the tracks and sign left behind by scaly and slimy creatures. My heartfelt thanks go out to Joanna Wright and Robert Mellinger, Jeremy Williams, Ted Packard, and Preston Wolf. Thank you for the jokes, stories, and adventures, and for all being good travel companions. The times we shared in the field will shine as wonderful memories for a long time to come.

Thanks also to friends and fellow field herpers Travis Knudtson and Jeff Nordland, who both gave a great deal of their time in and out of the field to help support me in this work. Thanks for putting up with the guy who likes to pick apart lizard poop to see what's inside, and thanks for getting sucked in right along with me.

Also thanks to other generous people who have helped me by sending photos and making time to be out in the field with me. Especially thanks to Steve Fortin, Jim Bass, Mike Wolverton, George L. Heinrich, Nathan Smith, James Maughn, Owen Holt, and Cameron Rognan.

Many people contributed photos to this project that were vital in furthering my knowledge, and whether or not their photos appear in this book, I must acknowledge their significant contributions. They include Kyle Loucks, Chris Hyde, Trevor Ose, Jessica Moreno, Carl Koch, William Flaxington, Jeremy Westerman, John Vanek, Dave Scott, Mark Tavor, Dirk Stevenson, Dick Newell, Marcus Reynerson, Tim Borski, Joshua Jones, Connor O'Malley, Zachary Cava, Michael Cravens, Todd A. Hogan, Tiffany Pritchard, and Bethany Avilla.

Also my thanks to the many people who gave precious advice on when and where to go and seek various species of reptiles and amphibians, shared vital tracking information, or got me in touch with those who could help. They include Gary Nafis, Dr. Cameron Barrows, Jackson Shedd, Joshua Holbrook, Chad M. Lane, Joshua Wallace, Scott Gibson, James Holley, Michelle Van Naerssen, Richard F. Hoyer, Michael A. Peters, Chris Byrd, Rob Speiden, John Sullivan, Sara Viernum, Robert E. Weaver, and many others.

And, of course, none of this would be possible without the patience and hard work of Mark Allison and Tim Gahr at Stackpole Books. Thank you for taking my words and photos and making them into a beautiful book I can be proud of!

FOREWORD

We were born to track animals. Our ancestors studied animal tracks and sign since the very origins of humanity. At one time, an individual's ability to identify and follow animal tracks could mean the difference between finding food and going hungry. In *The Art of Tracking: The Origin of Science*, Louis Liebenberg posits that learning to track animals was fundamental to the development of the early human brain. He argues that we were wired, through millennia of natural selection, to track animals, and that this skill greatly influenced the way our brains function today. We may owe our capacity to recognize obscure patterns and solve complex puzzles, at least in part, to our history of tracking animals.

In today's modern society, we are bombarded with stimuli. Advertisements fight for our attention, colors are oversaturated, flavors are enhanced, and the more subtle stimuli are all but drowned out. You may be surprised, then, that despite all the noise and allure of these conveniences, interest in wildlife, the outdoors, and tracking animals has surged in recent decades.

As a teenager, I was drawn into animal tracking by some inexplicable force. The more I learned about the natural world and the better I became at solving natural mysteries, the more I wanted to know. Tracking felt natural to me, and we now know that this may in fact be the case for us all.

As I began to realize the value of tracking, I became determined to be able to identify every track I could find, and to find ways to apply the concepts and skills associated with animal tracking within modern society. I studied extensively with teachers and field-guide authors across the United States, and photographed everything I found. I also took numerous wildlife technician jobs around the country where I could apply tracking abilities. As my knowledge grew and as new resources became available, I learned how to recognize the tracks of many mice, birds, and even small insects.

It was my passion for animal tracking that eventually led me to become a CyberTracker Track and Sign Evaluator, coauthor *Animal Tracks and Scat of California*, and write *iTrack Wildlife*—a guide to animal tracks for the iPhone. It is also what led me to become a wildlife biologist and eventually to my current position as Mammalogist for Texas Parks and Wildlife.

Despite the depth of my interest and passion for animal tracking, reptile and amphibian tracks remained stubbornly out of reach. Even the best trackers tended to identify them to only the coarsest level. My knowledge rarely allowed me to identify beyond the basic salamander, lizard, frog, turtle, snake, or alligator tracks.

Then one day, I asked Filip Tkaczyk to review some of my photos—his answers forever changed how I look at reptile and amphibian tracks. He was able to easily identify many tracks to a specificity that I never thought possible. Filip taught me how to differentiate the tracks of softshell turtles from snapping turtles, bullfrogs from leopard frogs, and even between different lizards.

The publication of *Tracks & Sign of Reptiles & Amphibians* is a landmark. Every major taxonomic group of animals now has an in-depth tracking guide. While many excellent field guides are available that make tracking easier to learn than ever before, reptiles and amphibians have received little attention. Prior to this book, Stackpole published excellent field guides to mammal, bird, and insect tracks, but never reptile and amphibian tracks.

Tracks & Sign of Reptiles & Amphibians is an incredible contribution to the field of tracking *and* to the field of herpetology. Filip's research took him to places no tracker has been before. As he explored new ways to identify tracks, he had to invent terminology, test hypotheses, and reach far into unexplored and difficult territory.

This book will open up a new chapter in my journey to becoming a better tracker and learning more about reptiles and amphibians. I hope you too will accept Filip's invitation to learn more about the animals that surround us. He has painstakingly researched and compiled this book in order to show us the fascinating world we live in—as we've never seen it before. Perhaps you will discover that tracking is what you were *born* to do as well.

—Jonah Evans
Texas Parks and Wildlife Mammalogist
Track and Sign Evaluator for CyberTracker Conservation

Introduction

This book is an invitation to stop, look more closely, and learn more about the wildlife that surrounds us wherever we go. The natural world is rich in stories, small and grand events that occurred in the very distant or recent past. When an animal moves, it cannot do so without influencing and impacting the elements and lives of others around. These impacts and the connections they represent are the focus of the craft of tracking.

Look around a natural landscape and you'll see evidence of animals everywhere. The holes in the leaves of a tree may have been created by feeding caterpillars. The torn buds and young growth of a shrub may be the sign of a browsing deer. The scat at a trail intersection could be the scent mark of a coyote. Those sets of four tracks decorating that mud puddle may be the tracks of a frog or toad. We live in, and move through, rich tapestries of stories made by these animals and told by their tracks.

Even a seemingly barren landscape, such as the downtown area of a major city, is covered in the tracks and sign of animals. That

A set of parallel trenches formed by an old turtle trail in the drying clay of a shallow pond. DAVID SCOTT

The rich tapestry of tracks on a sand dune in the Mojave Desert, including the trail of a sidewinder, a bounding pocket mouse, and the bipedal hopping trail of a kangaroo rat.

loose mess of grass stuck in the crevice of a concrete wall that you may have ignored as random flotsam might be the nest of a house sparrow. The white, runny stains high along a ledge of a skyscraper could be the droppings of a nest of peregrine falcons. And those pale, gray feathers scattered about and blowing along the sidewalk? They may be part of the remains of a pigeon that was the falcon family's previous meal. Anywhere you go, there is wildlife around you.

It is my sincere hope that this book helps you to slow down and look more closely at your wild neighbors, and at the sign they leave behind them. In reading the stories of their lives, we also enrich our own.

The skills of tracking are also incredibly useful to help us gather information about our rapidly changing world. They allow us to study the animal life around us, to look into their lives and see their relationships with other living things. Through tracking we can see what animals eat, where they seek water, shelter, and mates, and what parts of the terrain hold the greatest importance to them. In order to properly preserve or conserve natural landscapes we must first understand how they function, and tracking is a vital window through which we can survey the complex relationships that make up the natural world.

Why Track Reptiles and Amphibians?

Reptiles and amphibians have enriched the lives of countless people around the world, especially children. For many of us, our first interactions with wild animals may have been with frogs or toads, snakes, turtles, or lizards. These early meetings can have a powerful effect on how we view the importance of environmental conservation and preservation, continuing to influence us as adults. Though not all of us may become naturalists or herpetologists, we might still look back fondly on those first meetings with these wild creatures.

Observing reptiles and amphibians in their natural environments is a richly rewarding and fascinating pastime. The study of reptiles and amphibians is known as herpetology, a term from which we get the commonly used idiom "herps." Though different from humans, these animals are worthy of our close scrutiny and respect. Reptiles and especially amphibians have become vital indicator species because they are sensitive to environmental changes such as pollutants in the air, soil, and water. Herps have also been the focus of studies in biomimicry, modern medicines, healing abilities of vertebrates, and animal anatomy, and have even given us a better understanding of how some animals can survive the process of physically freezing in the winter.

So why look at their tracks and sign? Consider that just like other animals, reptiles and amphibians leave evidence of their activities behind them as they travel about the landscape. The details of tracks and sign can be observed, recorded, and interpreted to a surprisingly accurate degree. Wildlife tracking has been used to study the lives of many other animals, especially mammals and birds. The potential for the skills of wildlife tracking to be used by both the layperson and the professional field researcher is incredibly broad.

Learning to Track Wildlife

Learning to track means first noticing things that are irregular within the landscape. These might be dramatic, such as shadowed footprints in an otherwise pristine sand dune. They might also be subtle, such as sign left by the feeding of a chuckwalla (*Sauromalus ater*) on a flowering desert shrub. To notice such details, you have to look very closely at the landscape.

Furthermore, tracking is the practice of observing evidence left by an animal and gleaning information from it. Wildlife tracking includes looking at such evidence as tracks, trails, burrows, digs, feeding sign, scat, shed skins, nests, eggs, bones, kill sites, and anything else that might carry information about that animal. It requires slowing down, inspecting the clues with all of your senses, and interpreting the information. And although the use of field guides such as this one can be of great benefit, it is still essential to learn and to continually practice these skills in the field.

Tracking can not only be used to study the lives of animals, it can also help you locate them. Once learned, tracking can greatly increase your success rate of seeing animals in the field. This is especially true for smaller, more cryptic wildlife species, such as reptiles and amphibians, that are often hidden in plain sight. Tracking is a pathway of exploration on which to follow your curiosity and your sense of wonder at the life around you.

By its nature, wildlife tracking includes both a hard scientific element as well as a softer, more subjective approach. Both aspects have value, and are complementary to a complete study of the wildlife you're tracking. When tracking an animal, you must follow a logical protocol of careful observation of what is present. At some point, however, you will also have to engage your skills of imagination in order to interpret what information the bits of evidence are communicating. The accuracy of your interpretation increases according to the time you spend in the field. This is precisely why wildlife tracking takes both persistence and a great deal of practice.

In studying the wildlife you're tracking, it also helps to research the most up-to-date literature on the natural history and behavior of the animals. This can really help you make the most accurate interpretations of what you observe. This is why I have chosen to include the natural history for each species I cover in this book.

Tracking in and of itself can be a personal learning journey with no end point. There is always more to study and to understand.

Tracking skills are attainable by anyone who has the interest and is willing to put in the field time it takes them to learn. Remember to remain humble as you learn the skills of wildlife tracking, as admitting you are wrong is an essential part of the process. Even seasoned trackers make mistakes and will easily admit that they are sometimes wrong or that they don't have the answer. Allow your natural curiosity to guide your learning process. Remember to treat wildlife tracking as a skill, not as an identity or an inborn talent. We were born to track, to look at the signs left behind by the passing bodies of others, but skilled trackers are made over time and with focused effort. Masterful trackers dedicate years of time, and make countless mistakes to gain the knowledge of wildlife tracking they come to earn.

Also consider joining a field-based herpetological group or club, and look into joining wildlife tracking and naturalist classes in your area. For many, learning with others will help you to become skilled at a more accelerated rate. It also allows you to engage in conversation about your interpretation, and helps you see things you might otherwise miss.

This Book's Format

This book is designed to be as user-friendly, simple, and accessible as possible. Many of the terms used to describe different kinds of tracks and sign are the same ones used in previous tracking books published by Stackpole Books, especially those by Mark Elbroch such as *Field Guide to Animal Tracks and Scat of California*, and *Mammal Tracks & Sign*. Where necessary, terms that strictly refer to reptiles or amphibians will be used and defined within the book.

I have focused on the tracks, scat, and other types of sign that I felt were most likely to be observed in the field. In no way does this text cover every species of reptile and amphibian found north of the Mexican border. It does, however, attempt to describe the tracks and sign of the most commonly encountered species in several major regions of the United States.

As the first guide purely dedicated to the tracks and sign of reptiles and amphibians, this book attempts to break a tremendous amount of new ground. Many of the tracks and sign recorded for and presented in this book came from wild specimens, while some came from captive animals. I will denote examples of the latter as "captive" next to the photos and text.

A major challenge in preparing this book was acquiring the tracks of the smallest species that I cover. Some are too small and lightweight, and live in areas where substrates are too coarse, to leave any clear tracks at all. Others were larger and heavier, but were mainly aquatic and hardly ever left tracks or sign above water. That said, tracks and sign found in puddles or streambeds are some of the more exciting due to their ephemeral nature.

The tracks and sign you will see in this book are those you'll most likely encounter in the field. Admittedly, these are biased by my and the various contributors' experience, luck, and field time. As this area of the study grows, observers will likely find tracks or types of animal sign that are not discussed here.

Indeed, it is my desire that the information found here will inspire others to continue to study the tracks and sign of reptiles and amphibians in even greater depth. As with any text available to scientific critique, this book will need to be updated in the future as new information is discovered and our understanding grows. You have an opportunity to not only to study this area of tracking, but also to contribute to it. If you want to contact me with questions or observations that add to our shared body of growing tracking knowledge, send an e-mail to: herptrackers@gmail.com

What Are Tracks?

Tracks are not the lifeless stamps of an animal foot on an inert piece of ground. They are actually a dynamic expression of an animal's movements through the push and pull of its body against a movable and variable substrate. They are, generally speaking, made by the feet of the animal, although the marks made by the belly or tail are also considered tracks.

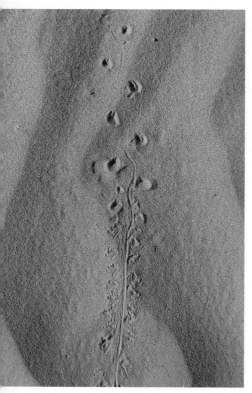

A single track of an individual animal—for instance a fence lizard—will look completely different depending on how quickly it's moving, what it's responding to, and the actual material across which it's traveling. The semifluid surface of deep and loose sand might quickly obscure the fence lizard's details. But a substrate like a thin layer of fine silt might capture the details of every toe, even the tiny scales, and hold them there for a long time.

In-field practice is required for you to identify the tracks of even a single species under various conditions and at different speeds. Take a look at this trail (a series of tracks) of a western fence lizard to see how the appearance of the tracks changes with the speed of its movement.

Because many herps are small and lightweight, even the tiniest variation in the quality of substrate may dramatically affect the clarity of the tracks you will see. Keep this in mind as you follow a trail and record the tracks you see either mentally, on paper, or with a camera.

Notice how these tracks from a western fence lizard change as it goes from a walk to a very fast overstep trotting gait. The animal started moving slowly at the bottom of the frame, and sped up as it moved away from the camera. JOE LETSCHE

Tracks can reveal a wealth of information about an animal, including its size, species, approximate current body temperature, speed, and even its gender. They may also reveal something about the health of an animal, for instance whether it is missing toes or even an entire foot or leg.

Observing closely the behavior of an animal through the tracks and sign it leaves can tell you a rich story about its life. Whether it is out foraging, fighting off a rival, or pursuing a potential mate, it leaves behind a story that can be read and understood by the skilled observer.

Tracking as a Discipline

Tracking requires you to make educated assumptions about what you observe. With time, you can reliably learn to glean all kinds of information from a set of tracks or sign. I must stress, however, that you should always apply some scientific scrutiny to the process.

When looking at a set of tracks, always seek to see what is actually there in the landscape. Look with your eyes, and keep an open mind. Resist the temptation to get stuck in the idea that there is only one possibility. Look for supporting evidence that might help you find the most likely and accurate interpretation of what has happened in the landscape. Use the evidence to lead you to a possible conclusion, rather than allowing your immediate assumption to cloud the evidence.

One mental exercise to help you see more clearly is to temporarily set aside the immediate need to identify the maker of the tracks. If possible, focus in on a small area that includes some clear tracks. Next, ask questions about what you can see to draw out the details of what is in front of you. Here are some useful questions to start with:

- What is the context of the tracks; where are they? What is the general habitat? What is the microhabitat? Are the tracks near cover or out in the open?
- What part of the body made these tracks?
- If the tracks were made by feet, can you see toes? If so, how many? Are they of equal length? Are there tubercles (round nodules) on the toes or feet in general?
- What is the overall pattern the tracks make?
- Are there claws present? Are they present on all the toes? Are they of equal length on all the toes? Is webbing present?
- What size (length and width) are the tracks?
- How far apart are the tracks? How wide is the trail overall?
- What direction was this animal traveling?
- How fast was this animal traveling? What details indicate that?
- Are there any other kinds of associated tracks or sign (such as tail drags, scat, digs, etc.)?
- Do the tracks appear to change in response to this associated sign? If so, how?
- Do the tracks appear to change in response to other elements in the environment?
- Finally, what group or species might have made these tracks?

As you look at the tracks or sign in front of you, other questions might arise. Try to answer as many as you can with what you can see and observe with your senses.

It is possible in many cases to identify the tracks or sign of a reptile or amphibian down to a single species. If you wish to learn more about its life, however, you will need to look deeper than just the tracks or sign. Reptiles and amphibians live in complex and dynamic interrelations with many other forms of life.

You may discover that the animal you are tracking has a particular liking for a certain kind of plant as cover. You might notice that certain species of invertebrates continue to show up in the animal's scat. This will inevitably lead you to study the greater environment around you, and lead you down the path of the naturalist.

Tracking as a Naturalist Pathway

Tracking is best practiced through the lens of broader ecological under-standing. No reptiles or amphibians or even humans exist in a bubble, and so to understand them you must also attempt to understand at least some of their relationships with the natural world. Anyone who wishes to master wildlife tracking must also study through the perspective of the naturalist. That means gaining an understanding about the other

The tracks of a killdeer that predated these trapped Fowler's toad tadpoles. The circular depressions in the puddle are made by the swimming action of the tad-poles. JOE LETSCHE

animals, plants, geological features, and weather patterns that influence the lives of the reptiles or amphibians you are tracking.

For example, if you want to study the life and behavior of the coast horned lizard (*Phrynosoma blainvillii*), you must understand the relationships it has with its preferred prey species. In southern California for instance, coast horned lizards feed heavily on harvester ants of the genus *Pogonomyrmex* and *Messor* (Jones and Lovich 2009). Those same ants act as seed dispersers in the native environment. It is helpful to also know the preferred foods of those ant species, as the ants will tend to concentrate where those foods are abundant. In turn, coast horned lizards are often observed feeding along busy ant trails or near their nests. Therefore, knowing something about the lives of the ants and the plants they depend on in the coastal chaparral habitats of California will help you to better understand the coast horned lizards.

Thus, through the eyes of a naturalist, studying tracks and sign also invites us to see the world more holistically. Animals are, in many ways, an expression of the greater landscapes in which they live, and from which they come. To better understand the landscape is to have deeper insight into that particular species, and vice versa.

Recording Track and Sign Information

Make a record of your findings to encourage your learning process as well as to have something to share with others.

You can use a variety of methods to record tracks or sign in the field. You should certainly carry a notepad or notebook and several writing implements. Log such general details as date, time of day, and location. You may want to note other valuable details, such as air or ground temperature, moon phase, or a description of the habitat and microhabitat in which you found the tracks. For example, bright moonlight can affect the behavior of nocturnal herps and may even limit when they are surface active.

While drawing tracks and sign is another time-tested method for recording naturalist observations, the ubiquity of cell phones and the quality of their cameras make photography an incredibly valuable and easy aid in the field. You can also use a traditional camera. Tracks photographed properly not only make an excellent record, but will also allow you to facilitate the process of sharing and collaborating with others, and often prove to be sources of learning in the future. I still continue to glean useful information from quality photos of tracks and sign I took years ago.

The best times to photograph tracks are when the sunlight is enhancing the shadows within them, which is generally in the early morning close to sunrise and in the evening close to sunset while the sun is lower in the sky.

Consider what kind of information you want to capture in the photo before you start taking pictures. The following details are some of the most valuable to record:

- An example of the overall trail pattern
- Clear, close-up series of a few tracks
- Clear, close-up images of individual tracks, including good examples of both fronts and hinds
- Tracks of the same species under different conditions
- Tracks or sign that are recorded with some kind of measuring device for scale, such as a ruler, tape measure, or other object of known length
- Include photos of the general area that show the habitat and microhabitat structures

The best angle to photograph tracks for the purpose of scientific study is straight on (directly overhead). Make sure the camera lens glass surface and tracking surface are parallel. If the camera is tipped at an angle toward the tracks or away from them, it tends to distort their shape and size. Artistic shots of tracks taken at various other angles can be beautiful, but do not present the most scientifically accurate records.

I recommend to always take multiple photographs of the tracks or sign without manipulating them in anyway. Once you've done so, photograph everything again with a measuring device included in your photo. That way, if you leave any marks when placing the measuring device, or in the future you want to see the images without additional elements in the frame, then you have records of both versions.

It is also sometimes necessary to remove distracting elements from the area you intend to photograph, such as pieces of garbage that can be distracting or make the photo unpleasant to look at. This kind of cleanup should be done with great care so as not to significantly alter the surface around the tracks. Remember, though, to take some photos that include everything as it was found—objects you wish to remove could actually be part of the story that sent the animal down that particular pathway in the first place.

Understanding Herp Feet

The feet of both reptiles and amphibians are generally arranged according to the same physical plan as human fingers and toes. Many have five toes on both front and hind feet. Some groups, such as frogs, toads, and

Hind   **Front**

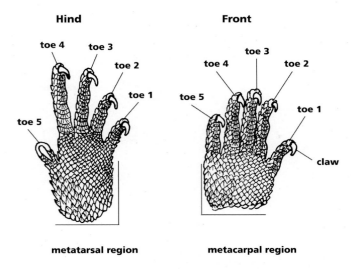

metatarsal region   **metacarpal region**

The right hind and right front of a chuckwalla.

salamanders, have only four toes on their front feet, with five on their hinds. Other species, such as slender salamanders (*Batrachoseps* sp.) and four-toed salamanders (*Hemidactylium scutatum*) have only four toes on both front and hind feet.

For ease of communication, I will refer to specific toes on the feet of various herps according to numbers. Consistent with previously established systems of numbering already in use for the toes of mammals and birds, I will number them from 1 to 5. Toe 1 is the toe observed nearest to the body while the foot is facing forward and parallel to the body. From here we number outward, so that toe 1 is followed by toe 2, and toe 2 is followed by toe 3, etc. Generally speaking, the toe found furthest from the body of the reptile or amphibian will be toe 4 on a front foot or toe 5 on a hind foot. Please take a close look at the diagram for a visual representation.

Using Foot and Track Morphology

One of the most valuable methods for studying the tracks of any animal is to learn to understand and identify differences in the foot and track morphology. This is one of the main keys to identification used in this book. This means looking at the size, shape, and arrangement of features on an animal's foot. For each species, I will describe these features in detail to help clearly identify its tracks in the field. Here are some general examples for the different groups of reptiles and amphibians.

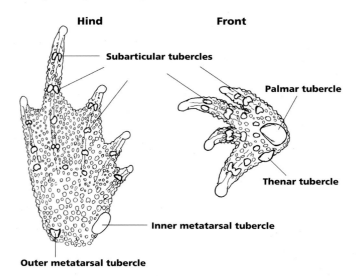

The right hind and left front of an *Anaxyrus* toad.

It is helpful to use the following terms when describing the different parts of the feet. For reptiles, features include toes, claws, heels or palms, and for some species, webbing. In amphibians, the feet generally include toes, heels or palms, tubercles, toe pads, and webbing. Tubercles are rounded outgrowths of skin, especially on the underside of the feet, that give frogs and toads additional grip.

Many reptiles and amphibians have unique features to their feet. For instance, members of the genus *Spea* and *Scaphiopus* (commonly called spadefoots) have uniquely smooth surfaces on the underside of their feet. Unlike many other frogs and toads, spadefoots do not have numerous, bumpy tubercles on their toes or pads. They do, however, all possess a single hardened, spade-like appendage (modified tubercle) on their hind foot that they use for digging.

Another example of unique morphology are the hind feet of members of the genus *Coelonyx* (banded geckos), which all have a toe on their rear foot that is angled backward and to the outside of the trail. They also have very fine, granular scales all over their bodies—including the undersides of their feet—so they do not register any noticeable scale patterns in their tracks.

The preceding examples are some of the more easily observable, but there are many others described in this book.

Movement Patterns

Reptiles and amphibians move in ways that are most efficient for their body type and lifestyle. Lizards, snakes, salamanders, and crocodilians move with mostly side-to-side flexion and extension of their spines. Mammal spines, by contrast, have a majority of movement in an up-and-down, forward-and-backward direction during forward motion. Turtles and tortoises have spinal columns that are fused to their shells and can't move either laterally or side to side; they move only their legs, head, and tail during locomotion.

Hop

In general, frogs and toads move by jumping or by using a modified form of walking. More specifically, frogs and toads jump using a gait called a "hop." This means that when they jump, their front feet always land ahead of their hind feet, so they leave a group of four tracks together. Some lizards also hop, as do a few salamander species. The stride of a hop is measured from the anterior end of one group of tracks to the posterior end of the next group of tracks.

Walks

Walking is a common gait used by all legged reptiles and many amphibians. When frogs or toads walk, they leave behind a set of paired tracks. These pairs include one front and one hind track from the same side of the body, with the front track ahead of the hind. This type of walk is more specifically known as an "understep walk." A walking stride is measured by taking

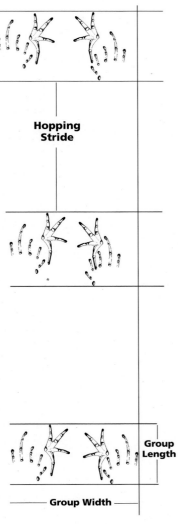

Hopping Stride

Group Length

Group Width

A typical pattern of a hopping frog.

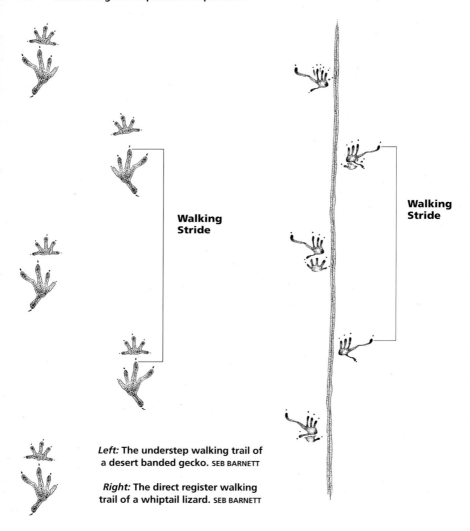

Walking Stride

Walking Stride

Left: The understep walking trail of a desert banded gecko. SEB BARNETT

Right: The direct register walking trail of a whiptail lizard. SEB BARNETT

the distance from the anterior end of one track to the same spot of the next track made by that same foot.

All salamander species who regularly travel in terrestrial habitats also use this gait. Lizards that are moving very slowly, such as when they are stalking prey, are cold, or are trying not to be noticed by a nearby predator, may use this gait. This is a baseline gait for all species of turtles and tortoises.

At a slightly higher speed, frogs and toads may occasionally walk in a manner in which their hind tracks land on top or directly adjacent and

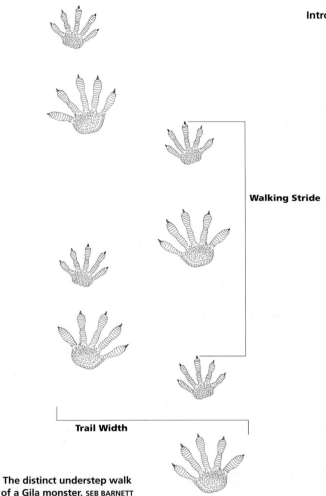

Walking Stride

Trail Width

**The distinct understep walk
of a Gila monster.** SEB BARNETT

overlapping with the fronts. This is called a "direct register walk." This is
rather unusual for frogs and toads, since they don't often move fast
enough to leave this type of track, but it's not out of place for salamanders moving at a fast walk.

For many reptile species, however, a direct register walk is common.
Many lizard species will use an understep or direct register walk at
slower speeds. Measuring this type of stride is the same as that of the
understep walk.

The next type of walk as speed increases is the "overstep walk," where
the animal moves in a manner so that the hind tracks end up ahead of
the front tracks in a sequence. This gait is common to many lizards,

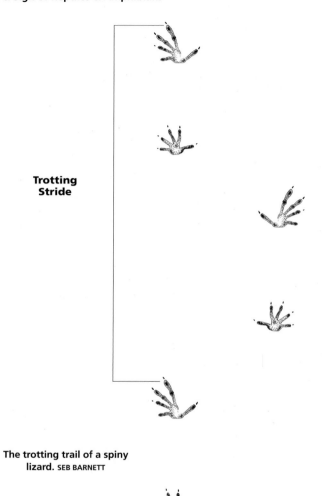

**Trotting
Stride**

**The trotting trail of a spiny
lizard.** SEB BARNETT

though it is physically impossible for turtles or tortoises to accomplish. This is because they can only move one limb at a time while the other three limbs support the body.

   Some lizard species—such as alligator lizards and Gila monsters—have a visually distinct form of this walk in which the tracks are very similar to the common walking gait of a raccoon. In this case, tracks in the overstep gait land so that the hind on one side of the body lands next to the front from the other side of the body.

For the sake of simplicity, I will use the same term used by CyberTracker Track and Sign evaluators to refer to the typical raccoon walk and call it a "2 x 2 walk" (2 by 2 walk).

Trots

Trots are faster gaits, and the next is a "direct register trot." This is a common gait used by many lizard species as they speed up from a walk to a high-speed trotting gait. For fast-moving species such as fence lizards and zebra-tailed lizards, this is a transitional gait from which they progress to even faster movements.

The next gait is a modified "overstep trot." Here, the animal moves forward at such a rapid pace that the front tracks are made and then the hind tracks overstep them and end up ahead as the momentum carries the animal's body forward. A trotting stride is measured by taking the distance from the anterior end of one track to the same spot of the next track made by that same foot.

Bipedal Run

The overstep trot is the fastest gait for many lizards. Some

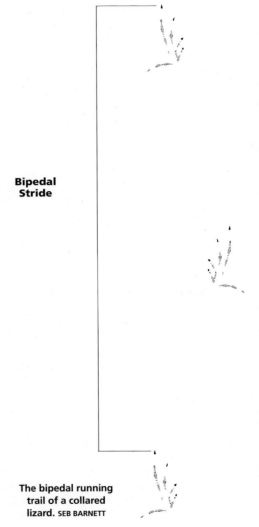

Bipedal Stride

The bipedal running trail of a collared lizard. SEB BARNETT

can go even faster by running on only their hind legs in a motion called a "bipedal run." In this case, only the hind feet register any tracks. Bipedal running is most often used in pursuit of prey or when escaping potential danger. Some species of lizards have noticeably larger and more muscular hind legs, which aid in bipedal locomotion. This stride is measured from the anterior end of one hind track to the anterior end of the next track made by the same foot.

Tracking Measurements

The measurements given in this book are helpful guidelines to identification and interpretation. Track length and width, trail width and stride length, and width and length of scats and other signs are given in a numeric range. I've gathered these measurements from as many specimens as possible, but I certainly haven't included the entire possible spectrum for track or sign measurements. Such measurements are useful for differentiating similar species from one another as well as informing us on some of the behavior of the animal in question.

Never rely only on measurements to determine the identity of a species. Remember to use all of the available evidence, especially tracks that clearly show the foot morphology, scat, sheds, digs, and whatever other evidence you can find nearby, including the context of the surrounding environment. Measurements are meant to support the other evidence, but generally do not stand alone.

Basics of Herp Biology

It is very helpful to have at least a basic understanding of the natural history of reptiles and amphibians when seeking to learn their tracks and sign. Reptiles and amphibians are known as *ectotherms*, formerly "cold-blooded." This outmoded label is deceptive and inaccurate, since the active body temperatures of some herps can be above 100 degrees Fahrenheit. The word "ectotherm" describes how these animals acquire their bodily temperatures from their surrounding environment. This is in contrast to *endotherms*, like mammals and birds, which acquire and maintain their body temperatures through internally generated heat created mainly through their metabolic processes.

Although it may seem simplistic to rely on the temperature of surroundings to warm one's own body, ectotherms like reptiles and amphibians can be incredibly energy efficient, considerably more so than endothermic animals. This is because reptiles and amphibians use only a fraction of the energy they derive from their food to maintain their body temperatures and metabolic processes. This allows many of them to survive on considerably less food per day as compared to mammals or birds of a similar size. This characteristic has allowed reptiles and amphibians to live in abundance even in areas that are incredibly harsh and relatively poor in food, such as the most arid portions of the deserts in the American Southwest.

Many have special physiological and behavioral adaptations to dealing with extreme conditions. For example, some lizard species are active at very high temperatures. Others limit their activity periods to cooler times of day during the summer months. Some desert-dwelling amphibians, such as spadefoots, might remain underground in a state of suspended animation for months at a time as they await the return of the sporadic desert rains.

On the other hand, reptiles and amphibians that live in colder parts of North America must adapt to the more temperate seasonal conditions. Generally speaking, cold weather tends to suppress or completely inhibit activity in herps.

Knowing this is vital to understanding the greater context of the lives of reptiles and amphibians. Their periods of activity—and, therefore, when they are most likely to leave tracks or sign—are determined by their temperature needs. Some reptiles are surprisingly cold tolerant, such as some garter snake species (*Thamnophis* ssp.) or rubber boas (*Charina bottae*), which I have observed being active at air temperatures down into the lower sixties Fahrenheit. On the other end of the spectrum, you have reptile species such the desert iguana (*Dipsosaurus dorsalis*), which can be active at air temperatures of 115 degrees Fahrenheit (Jones and Lovich 2009).

Most amphibians prefer cooler and wetter conditions. For example, a cool, rainy fall or spring day with temperatures in the fifties is a great time for many salamander species to be surface active. The need for amphibians to maintain moist skin precludes those except for the most drought tolerant from dwelling in the particularly arid parts of North America.

An adult male chuckwalla uses the morning sun to warm its body before becoming more active.

Field Code of Ethics

To gain a deeper understanding of reptiles and amphibians, and how their behaviors are reflected in their tracks and sign, it is important to spend time watching the animals themselves. Though you can accumulate knowledge from observing these animals in captivity, doing so presents at most a small fragment of their life story. In order to find and to study them closely, you must enter into and interact with their environments. You cannot do this without having an effect on the environment, but you can choose how to affect it. It is absolutely vital to take full responsibility by being conscious of your effects.

Practice care in how you interact with the animals and the landscape. Here are some helpful things to keep in mind:

- Replace any microhabitat features you've disturbed while seeking a herp. Cover objects often create unique moisture-holding shelters that are vital to reptiles and amphibians. If you can't move a cover object without being able to replace it just as it was or destroying it, then leave it alone.

- Be careful about what you carry with you on your clothes and equipment. If you have been in a wetland environment recently, carefully clean all of your gear so as to help reduce or even prevent the spread of Chytrid fungus. This invasive fungus introduced to the United States by humans has devastated many populations of amphibians.

- Do not harm any animals in your attempt to study them.

- Do not handle herps unnecessarily, as this stress can affect their health. For instance, causing a lizard to lose its tail costs that animal a great deal of resources, to the point where its ability to successfully reproduce that year may be reduced, and, more obviously, puts it at a disadvantage when trying to avoid real predators.

- If you feel the need to collect an animal from the wild, please respect local and federal laws. Limit your impact by collecting from healthy populations with an abundance of animals. Remember, every animal you remove from a population is no longer able to reproduce within that population.

- Do not introduce any species where they do not belong. Nonnative and invasive species compete with native animals and often spread new diseases into wild populations. Also, do not return long-term captive animals back into the wild as this may also bring in disease elements and will likely cost that animal its life.

- Whenever possible, observe wild animals from a distance that they are comfortable with. This allows them to behave in a relaxed, nat-

This canyon tree frog is a well camouflaged, streamside frog of the desert South-west. Its color and pattern closely match the rock crevices where it rests and hides during the day.

ural manner, and gives you a chance to see a deeper, more intimate view of that animal's life. Animals going about their daily tasks are far more interesting than those simply showing anxious or fearful behaviors such as hiding or running away from a human being that is approaching too closely.

• Watch for reptiles and amphibians on roadways and busy trails, and whenever possible and safe to do so, encourage them off of roads.

Follow these simple suggestions and enjoy using this book as a guide to enrich your time studying reptiles and amphibians in the field.

Illustrated Tracks

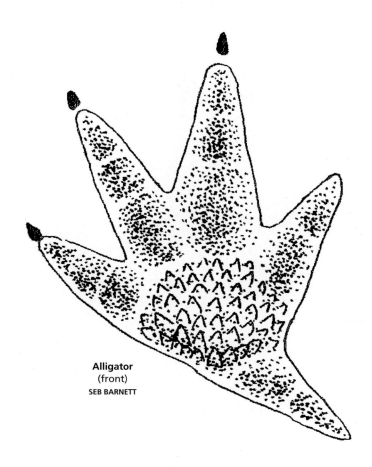

Alligator
(front)
SEB BARNETT

Tracks shown at actual size.

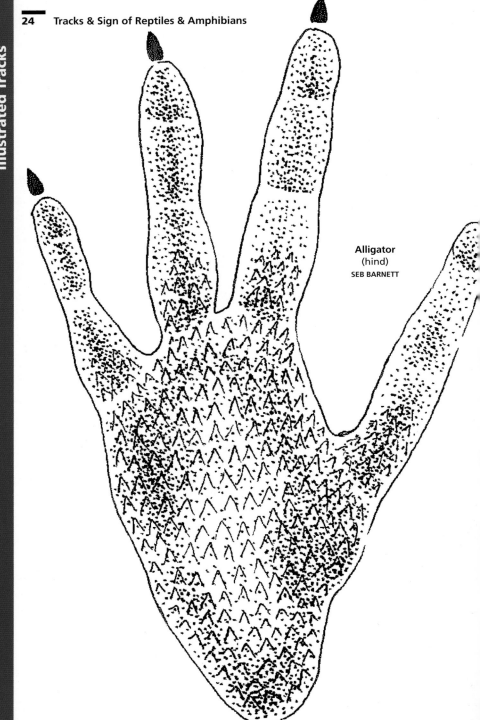

1

Alligator
(hind)
SEB BARNETT

Blanchard's cricket frog
SEB BARNETT

American toad
SEB BARNETT

Boreal toad SEB BARNETT

California slender salamander
SEB BARNETT

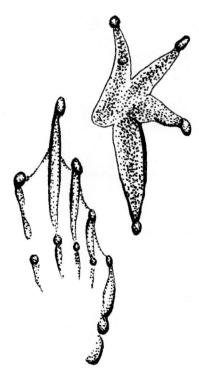

Bullfrog SEB BARNETT

Chuckwalla
SEB BARNETT

Common gray tree frog

Coastal giant salamander
SEB BARNETT

Cuban tree frog

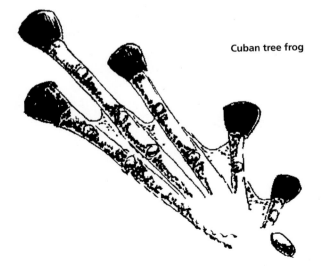

Desert banded gecko
SEB BARNETT

Desert night lizard

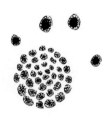

Eastern collared lizard
SEB BARNETT

Desert tortoise SEB BARNETT

Eastern newt

**Eastern spadefoot
toad** SEB BARNETT

**Foothills yellow-
legged frog**
SEB BARNETT

Gila monster
SEB BARNETT

Granite spiny lizard

Great Plains toad SEB BARNETT

Green anole
SEB BARNETT

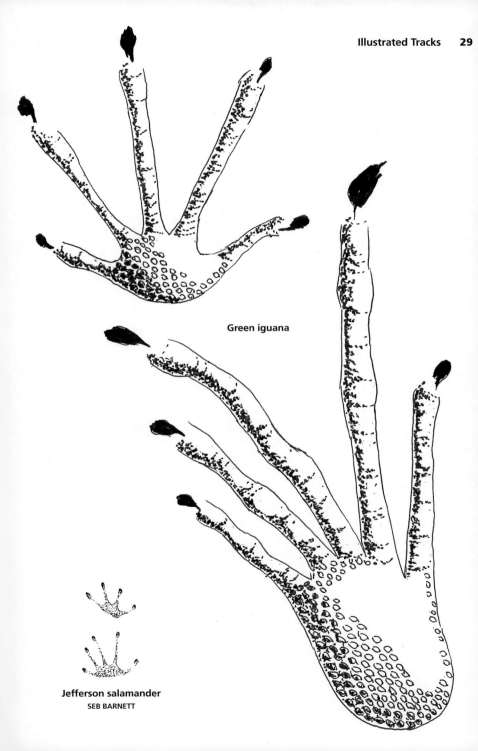

Green iguana

Jefferson salamander
SEB BARNETT

Illustrated Tracks

Leopard frog

Long-nosed
leopard lizard
SEB BARNETT

Long-toed salamander
SEB BARNETT

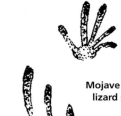

Mojave fringe-toed
lizard SEB BARNETT

Northern dusky salamander
SEB BARNETT

Northern Pacific chorus frog
SEB BARNETT

Ornate box turtle
SEB BARNETT

Painted turtle SEB BARNETT

Pacific pond turtle

Pickerel frog

Pig frog

Red-legged frog
SEB BARNETT

Red-eared slider

Red-spotted toad
SEB BARNETT

Rough-skinned newt
SEB BARNETT

Side-blotched lizard
SEB BARNETT

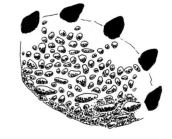

Snapping turtle

Southern alligator lizard
SEB BARNETT

Southern toad

Spotted salamander
SEB BARNETT

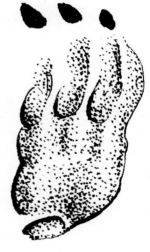

Spiny softshell turtle
SEB BARNETT

Texas horned lizard SEB BARNETT

Tiger salamander SEB BARNETT   **Tiger whiptail** SEB BARNETT

Western fence lizard SEB BARNETT

Woodhouse's toad SEB BARNETT

Salamanders

2

Salamanders are members of the order Caudata. There are approximately 127 species of salamanders in the United States. There are 9 families of salamanders in the United States, including: mole salamanders (Ambystomatidae), amphiumas (Amphiumidae), hellbenders and giant salamanders (Cryptobranchidae), Pacific giant salamanders (Dicamptodontidae), lungless salamanders (Plethodontidae), waterdogs and mudpuppies (Proteidae), torrent salamanders (Rhyacotritonidae), newts (Salamandridae), and sirens (Sirenidae). The order Caudata includes aquatic, semiaquatic, biphasic, and fully terrestrial species. The species covered in this book will be terrestrial, biphasic, or semiaquatic.

Generally speaking, salamanders are long, cylindrical amphibians with two to four legs and long tails. Though widespread, diverse, and in some places abundant, salamanders are not as well known as other herps. This is likely due to their tendency to live under the cover of rocks, leaves, or soil, or in water. The only time they are regularly seen by the casual observer is when they travel to and from their breeding sites.

Salamanders are generally slow, methodical walkers. As the front foot on one side of the body is placed, the hind foot on the opposite side of the body is placed. Then, with the next step, the other front foot is placed, while nearly simultaneously, the hind foot on the opposite side of the body is placed. Each step is associated with a sideways flexing of the spine. When sped up, this movement is not unlike a fish swimming.

When looking at the tracks, you'll generally see one of several gait patterns. Scientific studies have demonstrated that salamanders use only symmetrical gaits (walks and trots; Petranka 1998). The most commonly observed gait is the understep walk. This is where a front track is made, and then a hind track falls directly behind it on the same side of the trail.

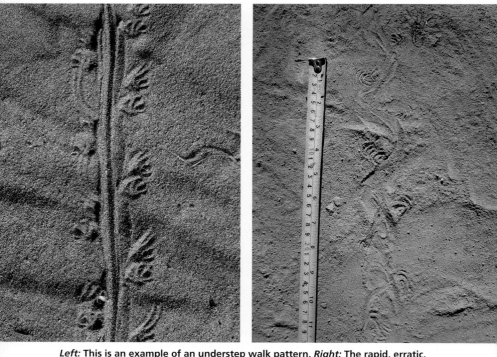

Left: This is an example of an understep walk pattern. *Right:* The rapid, erratic, direct register trotting trail of an adult northwestern salamander.

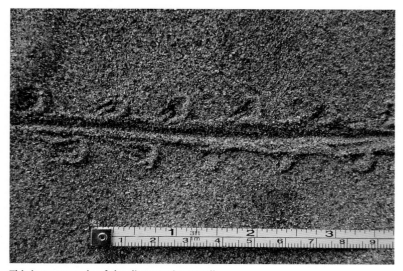

This is an example of the direct register walk.

At a slightly faster walking pace, a salamander may create a direct register walk pattern. This is where the hind track falls directly on top of the front track. This is the top speed for many species.

A few salamanders can speed up to an overstep trot, where the hind track actually ends up falling past the front track. Although a common gait for lizards, this is rather unusual for salamanders. The trail on page 36 made by a northwestern salamander (*Ambystoma gracile*) demonstrates an overstep trot. This animal made these tracks in a wild attempt to escape perceived danger and get out of an exposed, open area. Salamanders use quite a bit of energy when moving this way, and as such, they only use an overstep trot over short distances.

A small number of salamanders can actively hop when being pursued or when traveling over rough terrain. They don't hop often, because the action is very energy intensive. When hopping, they leap and then land with their front feet in front and their hind feet behind. Unlike many mammals, salamanders are unable to bound—this action requires that the hind feet land ahead of the front feet when the animal jumps. Because of their long torso and relatively short limbs, salamanders cannot do this.

Foot Morphology

The feet of salamanders vary depending on where they live and what kinds of landscapes they need to travel in. Generally speaking, most terrestrial salamanders have four toes on their front feet and five toes on their hind feet. There are some exceptions to this rule, including slender salamanders (*Batrachoseps* ssp.) and the aptly named four-toed salamander (*Hemidactylium scutatum*), which possess four toes on both front and hind feet. Some fully aquatic species, such as sirens, only possess the forelimbs. A few species, such as the amphiumas, have greatly reduced forelimbs, lack hind limbs, and may only possess one or two toes on their front feet. These limbs may be largely ineffective for locomotion when the animals are forced to struggle across terrestrial environments, but likely help guide these salamanders through vegetated portions of their aquatic habitat.

Looking at the undersides of the feet of a variety of salamander species, you might see differences in the texture of the feet, the location, length, and shape of toes, and the presence of other physical adaptations. Unlike the feet of frogs and toads, many salamander species don't have obvious and well-developed tubercles.

Two groups of salamanders that do possess some tubercles are tiger salamander species and some members of the genus *Taricha*. Tiger salamanders (including both *Ambystoma tigrinum* and *Ambystoma*

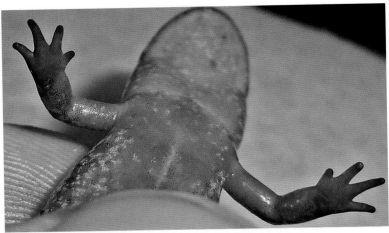

The front feet of the northern dusky salamander (*Desmognathus fuscus*).
TIFFANY PRITCHARD

californica) possess two tubercles on their hind feet. *Taricha* newts (specifically *Taricha torosa*) possess a single tubercle at the base of toe 5 on their hind foot, and one at the base of toe 4 on the front.

Most members of the climbing salamanders (genus *Aneides*) have feet adapted for climbing. The tips of their toes are expanded into rounded or squared-off pads that aid in climbing, similar to those of tree frogs. The tails of some *Aneides* species are also prehensile, and are used as a climbing aid.

Members of the slender salamander family have the smallest feet in proportion to their bodies. Their tracks are distinct in their appearance because their tail drag is very wide, while their tracks are incredibly small. Without the rows of dots made by their feet, the trails of slender salamanders could be mistaken for the trails of large earthworms.

Some species have distinct lengths of toes. For instance, both of the long-toed salamander species have unique feet in that toe 4 on their hind foot is significantly longer than the rest. This helps distinguish them from most other western salamander species.

The Importance of Range and Habitat

Some species of salamander have very similar foot morphology, so it's vital to use different methods to distinguish one from another. Understanding the animals' habitat preferences and range is vital to determine what you're observing. Closely related species include the coast range newt (*Taricha torosa*) and rough-skinned newt (*Taricha granulosa*), or Jefferson's salamander (*Ambystoma jeffersonianum*) and blue-spotted salamander (*Ambystoma laterale*).

Some parts of the United States have a richly diverse salamander population, such as the Pacific Northwest region, stretching from northern California to southern coastal British Columbia, and the Appalachian Mountain range from Georgia to Maine. The arid regions, such as the desert regions of the Southwest, contain very few salamanders. In the Chihuahuan Desert of southeastern Arizona, there is only a single species, the barred tiger salamander (*Ambystoma mavortium*).

Some salamanders, such as the tiger salamander complex (*Ambystoma tigrinum*), range throughout the United States. Some species are much more limited. Both the Shasta salamander (*Hydromantes shastae*) and the Scott Bar salamander (*Plethodon asupak*) are limited to a single county in the state of California. One subspecies—the desert slender salamander (*Batrachoseps major aridus*)—has a range so limited it is only found in one or two canyons in the arid eastern foothills of the Peninsular Range in southern California.

This deep pond in the Pacific Northwest is home to a variety of amphibians, including three different salamander species.

Seasonal Patterns

Salamanders tend to follow relatively cyclical patterns of activity based around temperature, moisture, time of day, and time of year. Species of salamander that require standing or flowing water for reproduction usually breed sometime in the spring. A few may even start breeding before ice disappears from their ponds and streams.

Many species may be surface active during periods of rain, and may be found under cover objects or even observed actively exploring in the open in a slow, plodding manner. During drier times of the year, salamanders will seek sheltered spots in moist or wet locations. Some species, such as tiger salamanders (*Ambystoma tigrinum* and *californica*) and northwestern salamanders (*Ambystoma gracile*) may spend much of their adult lives underground. Despite how common these species can be in prime locations, little is known about their lives while they are concealed.

Biphasic species, such as the coastal giant salamander (*Dicamptodon tenebrosus*) and California giant salamander (*Dicamptodon ensatus*) spend their entire larval stage as stream-dwelling creatures. As they mature, they lose their external gills and transform into terrestrial animals. The adults, however, do not generally wander far from water.

Climbing Salamanders (*Aneides* spp.)

These salamanders are known for their ability to climb, which is aided by their particular foot structure. All members of this group have mesial webbing (reaching midway to the toe tips), and toe tips that end in enlarged pads similar to those of tree frogs.

One of these, the wandering salamander (*Aneides vagrans*), sometimes lives high in the forest canopy as an adult. Some individuals have been found over 150 feet off the ground in the mossy mats and crevices in the canopy of living redwood trees.

Arboreal Salamander (*Aneides lugubris*)

This is the largest of the *Aneides* genus with a total length between 4 and 7 inches. It has a brown color on top, with small yellowish to cream-colored spots. Males have impressively muscular heads and can give a surprising defensive bite—they have rather long and sharp teeth for a salamander.

This species is a capable climber, like many members of the *Aneides* genus. It has a rounded tail and long toes with expanded toe tips that help with climbing. Although it can climb, this species is also often found on the ground under cover objects that hold moisture.

In its range in California, the arboreal salamander tends to be most frequently found in coastal oak woodlands, yellow pine and black oak forests, moist areas along rock faces, and some areas in urban yards. It is also relatively tolerant of dry conditions compared to other salamanders in the same range, and can even be found in coastal dune systems. This species shows a dependence on live oaks both as aestivation sites and as places for nesting. In southern California, these salamanders are often associated with sycamores along seasonal

The understep walk of an adult arboreal salamander. Notice the way the rear of the larger hind tracks slant downward toward the outside of the trail.

A detailed image of tracks from an adult arboreal salamander. Notice the enlarged toe tips common to *Aneides* salamanders.

stream corridors (observations by Gary Nafis).

Aneides lugubris is generally known to be a sit-and-wait predator that mostly ambushes invertebrates wandering past its hiding spot. This species will also eat slender salamanders. Unlike many other salamanders, it's aggressively territorial, and both sexes will bite other adult members of their own species. Many arboreal salamanders—especially males—show pale bite scars on their heads or other parts of their bodies.

This is one of the few salamander species that is known to vocalize, and may squeak when handled by humans. The sound might serve to scare off potential predators. It's a surprising sound to hear, especially from a member of a group known as "lungless salamanders." The sound is actually more likely made by downward pressure of the eyeballs as they are pressed into the mouth cavity, forcing air out through the nostrils or the salamander's closed lips (Stebbins 2012).

Track: Front: 0.3–0.5 in (0.76–1.27 cm) x 0.2–0.4 in (0.5–1 cm)
The front feet have four toes, all of which tend to register in the track. The tip of each toe is expanded into a pad. The bases of all four toes are fused together via a thin webbing of skin.
Hind: 0.3–0.6 in (0.76–1.52 cm) x 0.4–0.6 in (1–1.52 cm)
The hind feet have five toes, of which four or five toes tend to register. If a toe does not register, it tends to be toe 5. The bases of all five toes are fused together via a thin webbing of skin. This webbing appears further up on the toes than on the front feet.

Trail: Understep Walk   Trail Width: 1.4–2.1 in (3.56–5.33 cm)
Stride: 1.5–2 in (3.81–5.08 cm)
Tail Drag: 0.2–0.3 in (0.5–0.76 cm)
Tail drag shows a rather wide trail created by the thick, rounded tail. This helps distinguish the species from others with acute laterally compressed tails.

Notes:
• Watch for the tracks of these salamanders in fine silt or small patches of mud in woodland habitats where they are typically found.
• Be careful to not disturb microhabitat cover objects where this salamander may be found, especially in the drier periods of the year when such moist havens are a matter of life or death to these animals.

Dusky Salamanders—Genus *Desmognathus*

This is a moderately large group of plethodontid salamanders, including around twenty-one different species found in the United States. Among this number is the pygmy salamander (*Desmognathus wrighti*), one of the smallest salamander species in North America, with adults growing to between 1.5 and 2 inches in total length.

The greatest diversity of these salamanders exists in the Appalachian Mountains. Their legs are short to moderate in length. The tails have a thick, triangular base that tapers to a laterally compressed tip.

Dusky Salamander (*Desmognathus fuscus*)

This small to medium-sized salamander is the most commonly encountered species in this genus. Adults grow to between 2.5 and 5.5 inches in total length. Adults are tan or dark grayish brown on top, uniform in color or mottled. A pale line runs from the eye to the corner of the mouth.

This is a salamander of rocky springs, seepages, and creeks in woodlands. You'll often find them by lifting rocks in this kind of habitat. The closely related southern dusky salamander (*Desmognathus conanti*) is found in floodplains, sloughs, and muddy streams near the edges of swamps.

This species deposits its eggs in moist spots inside of logs, under rocks in creeks, and in wet piles of leaves. The female will remain with the eggs until they hatch. She will defend her eggs from other dusky salamanders, which will eat them if given the chance. This species also eats earthworms, spiders, and water-dwelling insect larvae. Large adults sometimes eat smaller members of their own species.

Typically, this salamander moves rapidly when exposed and is able to trot as well as hop. Watch for its tracks in soft, fine silt along the edges of rocky creeks. The trail may show the animal understep walking, overstep trotting, or even hopping.

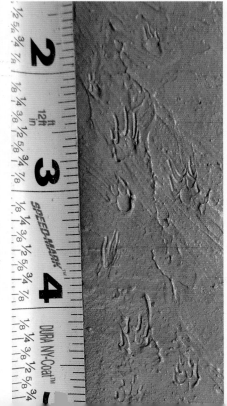

The messy overstep trotting trail of the northern dusky salamander.
TIFFANY PRITCHARD

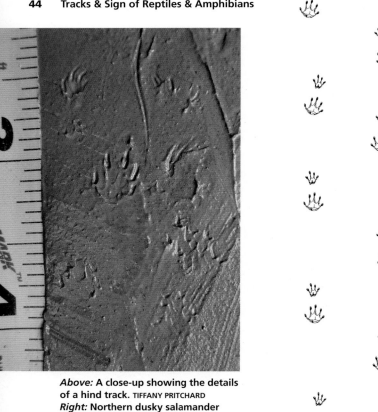

Above: A close-up showing the details of a hind track. TIFFANY PRITCHARD
Right: Northern dusky salamander understep walk trail.

Track: Front: 0.2–0.3 in (0.5–0.76 cm) x 0.2 in (0.5 cm)

Tracks are small. The front foot possesses four toes, and all four tend to register in the tracks. Toe 3 is the longest. Toe 2 and toe 4 are about equal in length. Toe 1 is the shortest, but the base is on the same plane as toe 4 on the opposite side of the foot. The tips of the toes often curve downward, creating distinct dots at the toe tips. In some tracks, the back of the metacarpal region may show one or two shallow clefts. The metacarpal region is slanted at the rear, with the area below the bases of toe 3 and toe 4 reaching further back than the area below the bases of toe 1 and toe 2. Front tracks tend to register forward or pitch slightly outward.

Hind: 0.2–0.4 in (0.5–1 cm) x 0.2–0.4 in (0.5–1 cm)

Tracks are small and fan shaped. The hind foot possesses five toes, and all five tend to register in the tracks. Toe 3 is the longest, and toe 4 is only slightly shorter. Toe 2 and toe 4 are about the same length. Toe 1 is the shortest and positioned slightly lower on the foot than toe 5. A shallow, upside-down crescent is found on the metacarpal region of the foot. This may or may not register in the track, and this area may also show two shallow clefts near the middle of the metatarsal area. The hind tracks tend to register directly forward or pitch in slightly toward the center line of the trail.

Trail: Overstep Trot   Trail Width: 0.7–1.2 in (1.78–3 cm)
                               Stride: 0.8–1.7 in (2–4.32 cm)
         Tail Drag: 0.1–0.2 in (0.25–0.5 cm)
         The tail drag is inconsistent, but when present tends to be very thin. It is V-shaped in cross section and is made by the strongly laterally compressed portion of the tail near the tip.

Notes:

- When this salamander is found, it is often able to move at surprising speed. It throws itself from side to side in a rapid trot that appears almost lizard-like.
- Look carefully at the tracks of this species, and use habitat and microhabitat cues when trying to distinguish it from other salamander species.

Eastern Newts—Genus *Notophthalmus*

This is a small group of salamanders found from the Atlantic coast to as far west as central Texas. These newts have toxic skin secretions, especially in their terrestrial stage.

Eastern Newt (*Notophthalmus viridescens*)

This small salamander is one of the most familiar salamander species, especially in the northeast United States, due to its bright red terrestrial stage called the "red eft." These pale to brilliant red subadults advertise their toxic skin secretions during this stage in their life cycle, when they are up to ten times more toxic than the larval or adult stage (Watkins 1968). Adults measure between 2.5 and 5.5 inches in total length.

Adults of this species are largely aquatic, and are yellowish, pale olive green, or brown with yellow ventral surface. The back and belly are covered in sporadic black dots. Adult and subadult eastern newts also often have a single row of widely spaced, black-bordered red dots on each side of their body.

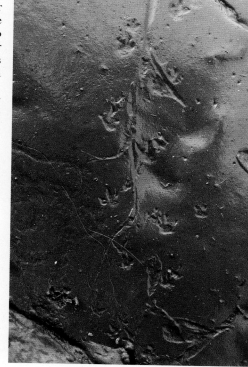

The understep walking trail of an adult eastern newt. DAN GARDOQUI

**Eastern newt
understep walk trail.**

Adult eastern newts are found in farm ponds, lakes, reservoirs, swamps, canals, and slow-flowing streams. The red efts are found the damp leaf litter of woodlands and forested landscapes.

These small salamanders feed largely on the invertebrates they encounter under leaf litter as subadults, such as spiders, springtails, and mites. Adults feed on water-dwelling invertebrates, small tadpoles, and amphibian eggs.

This salamander has feet that have distinctly different toes from those of any of the salamanders that share its range, as described below. The tracks of the red eft stage are most likely to be encountered in small puddles of mud on forested trails or on dirt roads. The tracks of adults are found in mud at the edges of wetlands.

Track: Front: 0.3–0.4 in (0.76–1 cm) x 0.2–0.3 in (0.5–0.76 cm)

Small tracks. The front foot possesses four toes, and all four tend to register in the track. Toe 3 is the longest, and toe 2 the second longest, projecting significantly further than the other two toes. Toe 1 is so short it appears as just a dot. Toe 4 is longer, closer to toe 2 than to toe 1 in length. There are no tubercles present. When the metacarpal region registers, the back of it slants downward toward toe 4. The tracks generally point directly forward in the direction of travel or pitch slightly toward the center line of the trail.

Hind: 0.3–0.4 in (0.76–1 cm) x 0.3–0.4 in (0.76–1 cm)

Tracks are small. The hind foot possesses five toes, and all five tend to register in the track. Toe 3 is the longest. Toe 2 and toe 4 are the same length, and pitch away from the central toe at about 45 degrees. This gives the track a trident-like appearance. Toe 1 and toe 5 are very short, and appear as dots on either side of the back of the track. Breeding males develop enlarged, hardened toe tips, a line of raised tubercles on the outer edge of their metatarsal region, and dramatically swollen vents. The hind track points directly forward in the direction of travel.

Trail: Understep Walk   Trail Width: 0.8–1.25 in (2–3.17 cm)
                            Stride: 0.6–1.25 in (1.52–3.17 cm)
            Tail Drag: 0.1–0.2 in (0.25–0.5 cm)
            The tail drag of this species is typically V-shaped in cross section, because
            it is made by the lower edge of the strongly laterally compressed tail.
Notes:

• This small, beautifully colored salamander leaves some of the most easily
  recognizable tracks.

• Be careful when handling these animals, as the skin secretions can be irritating
  if rubbed into mucus membranes.

Black-spotted Newt (*Notophthalmus meridionalis*)

This small salamander is limited to the southern tip of Texas, near the coast-
line of the Gulf of Mexico. Adults measure between 2.75 and 4.25 inches in
total length. Adults are olive green on top, with scattered black spots. The
ventral surface is yellowish and also spotted. Like the previous species, these
newts produce toxic skin secretions.

This species is found in both permanent and seasonal wetland habitats,
such as ponds, ditches, and swamps. They prefer well-vegetated bodies
of water.

Black-spotted newts appear to lack a well-defined terrestrial (eft) stage.
Adults generally move overland only when seeking new breeding habitat
or when their pond dries up. Look
closely at the edge of such receding
wetlands for the trails of these small
salamanders.

Track: Front: 0.3–0.4 in (0.76–1 cm)
            x 0.2–0.3 in (0.5–0.76 cm)

Tracks are small. The front foot
possesses four toes, and all four
tend to register in the track.
Toe 3 is the longest, and toe 2
the second longest. They project
significantly further than the
other two toes. Toe 1 is so short
it appears as just a dot. Toe 4 is
longer, closer to toe 2 than to
toe 1 in length. There are no
tubercles present. When the
metacarpal region registers,
the back of it slants downward
toward toe 4. The tracks generally
point directly forward in the
direction of travel or pitch
slightly toward the center line
of the trail.

**The walking trail of an adult black-
spotted newt.** JASON KNIGHT

Hind: 0.3–0.4 in (0.76–1 cm) x 0.3–0.4 in (0.76–1 cm)

Tracks are small. The hind foot possesses five toes, and all five tend to register in the track. Toe 3 is the longest. Toe 2 and toe 4 are the same length, and pitch away from the central toe at about 45 degrees. This gives the track a trident-like appearance. Toe 1 and toe 5 are very short, and appear as dots on either side of the back of the track. Breeding males develop enlarged, hardened toe tips, a line of raised tubercles on the outer edge of their metatarsal region, and dramatically swollen vents. The hind track points directly forward in the direction of travel.

Trail: Understep Walk   Trail Width: 0.8–1.5 in (2–3.81 cm)
                             Stride: 0.6–1.5 in (1.52–3.81 cm)

Tail Drag: 0.1–0.2 in (0.25–0.5 cm)

The tail drag of this species is typically V-shaped in cross section, because it is made by the lower edge of the strongly laterally compressed tail.

Notes:
- This species is still rather understudied in the field. Basic aspects of its life cycle are not well known. Tracking these salamanders might help reveal more about their terrestrial habits.

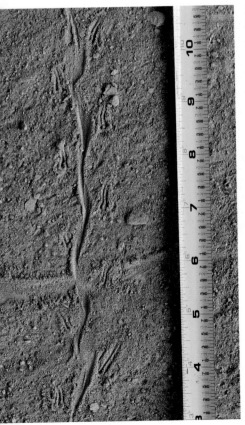

Ensatina (*Ensatina eschscholtzii*)

This West Coast salamander complex includes a diverse assortment of sub-species, a total of seven known at this time. These subspecies show dramati-cally different color patterns, and one subspecies, the aptly named yellow-eyed ensatina (*Ensatina eschscholtzii xanthopicta*) might be mimicking the toxic Taricha newts with its appear-ance. One of the subspecies, the large-blotched ensatina (*Ensatina esch-scholtzii klauberii*) is considered by some as a separate species (Stebbins 2012).

This medium-sized plethodontid salamander is found from southern British Columbia, down along the

The rapid direct register walk of an adult ensatina.

Left: The slow understep walk typical of this species. *Right:* An understep walking trail of an ensatina in fine dust showing a detailed hind track.

coast and just into Baja California. It has a relatively short body, but rather long legs. It also has a round tail with an obvious constriction that encircles the tail base. This constriction allows the tail to be autotomized when it is grabbed by a predator. Tail regeneration is slow, taking about two years to fully form.

This species is sexually dimorphic, and adult males can be distinguished from females by their swollen cloacal glands and enlarged, overhanging upper lip. Adults measure between 3 and 6.5 inches in total length.

This species lives in a variety of habitats, including forests, woodlands, chaparral, and moist areas of grasslands. Like most salamander species, this one seeks out moist microhabitats under cover objects. It can be seen being surface active on cool, rainy days.

This is one of the few species known to vocalize. The salamanders squeak or hiss when they are picked up.

Look for their tracks crossing drying mud puddles in woodland trails or dirt roads, as well as in dust under bridges.

Track: Front: 0.22–0.3 in (0.55–0.76 cm) x 0.27–0.35 in (0.68–0.89 cm)

Front tracks have four toes. Toe 1 is shortest. Toes 2 and 4 are similar lengths. Toe 3 is the longest toe. The front foot tends to face almost directly forward. The metacarpal region is reduced, but has two small tubercles. There is also a small, round tubercle at the widest point of each toe. Toe tips are rounded.

Hind: 0.24–0.33 in (0.6–0.83 cm) x 0.3–0.45 in (0.76–1.14 cm)

Hind tracks have five toes. The hinds are similar to other plethodontids. Toe 1 is the shortest. Toes 2 and 4 are similar in length. Toe 5 is longer than toe 1. Toe 3 is the longest. Hind tracks often pitch toward the middle of the trail. The widest part of each toe has a small, round tubercle. The largest tubercle is at the base of the longest toe, toe 3. The other tubercles are proportionally smaller. Toe tips are rounded.

Trail: Understep Walk   Trail Width: 0.9–1.14 in (2.286–2.89 cm)

Tail Drag: 0.1–0.26 in (0.254–0.66 cm)

Tail drag is U-shaped in cross section from the round tail.

Notes:

• This species is commonly encountered, even in suburban gardens with the appropriate moisture-holding habitats of cover objects.

• The U-shaped tail drag helps distinguish the ensatina from other medium-sized salamanders such as newts, coastal giant salamanders, and the mole salamander species, all of which have a V-shaped tail drag because of their laterally compressed tail.

Pacific Giant Salamanders

This group of salamanders belongs to the genus *Dicamptodon*, which means "double curving teeth." Only four species are known in this genus: Cope's giant salamander (*Dicamptodon copei*), Idaho giant salamander (*Dicamptodon aterrimus*), California giant salamander (*Dicamptodon ensatus*), and coastal giant salamander (*Dicamptodon tenebrosus*). These thick-bodied and large-headed salamanders are unique to the western United States. They have well-developed legs and feet, as well as large laterally flattened tails.

These large, rather charismatic salamanders are delightful to find and observe in the wild. They tend to be found in and near cold, rocky streams and rivers, mountain lakes, and sometimes ponds. Transformed adults are generally found within fifty meters of a stream.

The large size and impressive heads of the adult giant salamanders allow them to eat a variety of prey items, including an assortment of invertebrates such as large insects, spiders, and slugs. They also include some vertebrate prey in their diets, such as small fish, frogs, other salamanders, small snakes, and even shrews and mice. They are generally understood to be ambush predators, and capture prey with short lunges and snaps of their jaws.

Be especially careful if you are searching for these salamanders, as they often have strong site fidelity and will often be found under the same cover objects over time. Pacific giant salamanders can also live quite long.

Coastal Giant Salamander (*Dicamptodon tenebrosus*)

This robust species grows the largest and also has the widest range, from southern British Columbia down to northern California. It is arguably one of the largest terrestrial salamanders in the world—transformed adults can grow over 12 inches in total length.

The color of adults is dark brown overlaid with light spots, blotches, and marbling. Very old adults may show little or no markings on the body, and the clearest markings are on their heads.

This giant species feeds on almost anything small enough to fit in its mouth. Most often this includes slugs, earthworms, other invertebrates, smaller salamanders, small lizards, and even small mammals such as mice or shrews.

Pacific giant salamander species are most likely to leave tracks in areas on the edge of waterways, especially in small pockets of fine silt along creeks. As reflects their size, their tracks are impressively large for a salamander.

Left: The understep trail of an adult coastal giant salamander leaving a shallow wetland. KIM CABRERA *Right:* The front and hind tracks of a large, adult coastal giant salamander in firm silt along the edge of a small stream.

The front and hind tracks of an adult coastal giant salamander. Notice that toe 5 is barely visible on the hind track. KIM CABRERA

Track: Front: 0.26–0.54 in (0.66–1.37 cm) x 0.36–0.63 in (0.91–1.6 cm)

Coastal giant salamander walking trail.

Tracks are medium in size. The features of the foot are relatively robust. Four toes are present, and generally all will register. All toes are medium in length. Toe 3 is the only slightly longer than the rest. Overall track aligns forward or pitches slightly to the inside of the trail. Metacarpals are moderate and may register in track.

Hind: 0.43–0.8 in (1.09–2 cm) x 0.49–0.725 in (1.24–1.84 cm)

Tracks are medium in size and fan shaped. The features of the foot are relatively robust. Five toes present, and four or all five may register in the track. If only four toes register, toe 1 is generally missing. Toe 3 is the longest, but sometimes toes 3 and 4 are equal length. On some substrates, toe tips will be the most pronounced parts of the track. The metatarsal region is moderate and may show up in the track. Track may align forward or pitch slightly toward the middle of the trail.

Trail: Understep Walk   Trail Width: 1.45–2 in (3.68–5.08 cm)
                              Stride: 1.37–3 in (3.47–7.62 cm)
          Direct Register Trot Trail Width: 1.75–2.25 in (4.4–5.71 cm)
                              Stride: 3.3–4.2 in (8.38–10.67 cm)
          Tail Drag: 0.1–0.2 in (0.25–0.5 cm)
          Tail strongly laterally compressed. If tail drag is present, it is made mostly with the thin lower edge of the tail. Line resembles that which may be left behind by dragging a finely edged tool through the substrate. Tail drag may be straight, or undulating.

Notes:

- This potentially very large salamander species is generally found near cold, or at least cool, and relatively fast-flowing streams. Watch for its tracks especially in the silty pockets at the edges of streams that have very low water flow. Tracks are most often found with the return of the first rains after a long dry spell or when that particular pool of the creek has mostly dried up and forced the salamander to move overland.

- This salamander species can also vocalize. It may growl, hiss, or squeak when picked up or molested by a predator. It can also bite surprisingly hard, though is generally mild mannered.

- The coastal giant salamander can be a long-lived species, with some individuals recorded living close to thirty years in the wild. Because of their strong site fidelity, adults can often be found again and again under cover objects in the same small area.

- This is one of the few salamander species that has been recorded to trot.

Jefferson Salamander (*Ambystoma jeffersonianum*)

The Jefferson salamander is a long, slender salamander with a dark gray or brownish body, with light blue-gray or silvery flecks on its lower body, tail, and limbs. This species has a long tail that is round at the base and tapers to a laterally compressed tip. Breeding males have more dramatically compressed tails and swollen vents. The

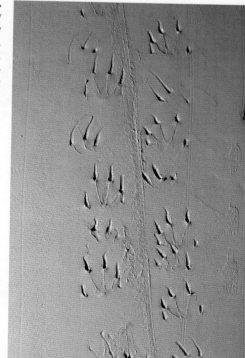

The perfect tracks of a Jefferson sala-mander made under controlled condi-tions. Notice the wide drag mark created by the swollen vent of this adult male.
JOE LETSCHE

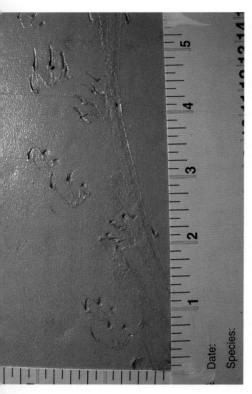

Another perfect understep walking trail of a Jefferson salamander, also made under controlled conditions. JOE LETSCHE

toes are very long and thin on both front and hind feet.

This northeastern dweller of decid-uous forests relies heavily on vernal pools as breeding grounds. Jefferson salamanders (*Ambystoma jeffersoni-anum*) seek these pools during warm spells in mid-winter and have been observed walking on patches of snow to get to these breeding sites. This species often shares its breeding pools with spotted salamanders (*Amby-stoma maculatum*), though Jefferson salamanders seek out these pools earlier. Often this species is leaving these breeding sites as spotted sala-manders arrive.

Males deposit spermatophores that are picked up by females to fertilize their eggs. The eggs are generally deposited on sticks or leaves sub-merged in the pools. Multiple females often lay egg masses close together and may arrange the oblong masses lengthwise along the same stick. The egg masses of this species develop along with a green symbiotic algae (*Oophila amblystomatis*) that covers them.

Outside of the breeding season, little is still known about this species' ecol-ogy because like many other Ambystoma salamanders, it spends much of its adult life underground. Seek their tracks in and around breeding sites during early spring.

Track: Front: 0.3–0.45 in (0.76–1.14 cm) x 0.4–0.55 in (1–1.49 cm)

The front feet have four toes, all of which tend to register in the tracks. The toes are very long and taper only slightly. Toe 3 is the longest, and only slightly longer than toe 2. Toes 2 and 4 are similar in length. Toe 1 is the shortest. Overall, the front foot tends to point directly forward relative to the center of the trail or may pitch slightly toward the center of the trail.

Hind: 0.45–0.65 in (1.14–1.65 cm) x 0.5–0.65 in (1.27–1.65 cm)

The hind feet have five toes, all of which tend to register in the tracks. The toes are relatively long. Toe 1 is the shortest. From there, toes 2 through 4 are increasingly longer. Toe 4 is distinctly the longest and noticeably longer than even toe 3, which is the most similar in length to toe 4. Toes 2 and 5 are about the same length. Hind tracks are as wide as long, or are slightly wider than long, and tend to face directly forward or may pitch slightly outward away from the center of the trail. This species

may register less metatarsal surface area than the related spotted
salamander (*Ambystoma maculatum*).

Trail: Understep Walk   Trail Width: 1.8–2.2 in (4.57–5.58 cm)
                           Stride: 1.4–2.5 in (3.55–6.35 cm)
        Direct Register Walk Trail Width: 0.8–1.6 in (2–4 cm)
                           Stride: 2.6–2.8 in (6.6–7.11 cm)

Notes:

- The toes of this species are very long, especially those on the hind feet, helping to differentiate it from many other salamander species. These long toes can also help differentiate a pure Jefferson salamander from hybrid salamanders in the Jefferson complex.
- The tail drags of this species, when observed, are similar to those made by the northwestern salamander (*Ambystoma gracile*).

Long-toed Salamander (*Ambystoma macrodactylum*)

This is a medium-sized salamander in the Ambystomatidae or "mole salamander" family. It has five recognized subspecies, one of which is critically endangered. The Santa Cruz long-toed salamander (*Ambystoma macrodactylum croceum*) has an incredibly restricted range, and it continues to be threatened by development. The other four subspecies are found farther north, and generally have much larger ranges where they are one of the most common salamander species.

Members of this species have a wide, dull- to bright-yellowish to greenish-yellow stripe starting at or near the head and running onto the tail partly or all the way to the tail tip. This dorsal stripe varies considerably from individual to individual, with some having only a partial stripe or one broken up into yellowish blotches. To varying degrees, yellowish blotches or spots may also be present on the head or dorsal surface of the legs. The sides of the body, head, and legs are dark charcoal gray. The belly tends to be somewhat paler gray, with a fine peppering of white to pale yellowish dots.

The toes of both the fronts and hinds are relatively long. It is the long

The perfect trail of a western long-toed salamander in fine river silt.

fourth toe on the hind foot that gives this species its name. Watch for this feature in helping to distinguish their tracks from those of other western salamander species.

The tail is relatively round in cross section at the base, but increasingly becomes laterally compressed toward the tip. Males have longer tails and longer limbs than females.

Their preferred habitats include coniferous forests, mixed woodlands, alpine meadows, and in sagebrush communities. They generally seek out ponds, lakes, and flooded fields as breeding sites in these same habitats. Long-toed salamanders are more difficult to observe breeding than other western salamanders because they tend to do so under the cover of night and often down in the leaf litter at the bottom of a pond.

This species is most obviously surface active during breeding season. At lower elevations, they may start breeding between October and February. Meanwhile, at high-elevation sites, they may not breed until July when the melting snow has freed up their preferred sites.

They can often be found during wet weather under the cover of large rocks, bark slabs, or logs. Their tracks are most likely to be encountered as they travel to and from breeding sites.

The foot morphology resembles that of other eastern, long-toed *Ambystoma* species that do not overlap with this species in range. It is most similar to the northwestern salamander (*Ambystoma gracile*), which shares part of its range. Adults of this related species are generally twice as large, and in most populations, toe 3 is the longest on the hind foot.

Track: Front: 0.19–0.25 in (0.48–0.63 cm) x 0.22–0.31 in (0.56–0.78 cm)
The front feet have four toes, all of which tend to register in the tracks. The toes are relatively long. Toes 1 and 4 are similar length. Toe 3 is

The rapid direct register walk of a western long-toed salamander. The lashing action of the tail created the wide, zigzagging pattern down the middle of the trail.

slightly shorter than toe 4. Front tracks are slightly wider than long, and tend to face forward or pitch slightly toward the center line of the trail.

Hind: 0.26–0.4 in (0.66–1 cm) x 0.28–0.45 in (0.71–1.14 cm)

The hind feet have five toes, all of which tend to register in the tracks. The toes are relatively long. Toe 1 is the shortest. From there, toes 2 through 4 are increasingly longer. Toe 4 is distinctly the longest and often angled somewhat outward at about 45 degrees from the center of the trail. Look for this as a distinct feature to help separate it from other western salamanders. Toes 2 and 5 are about the same length. Hind tracks are as wide as they are long, or slightly wider than long. Hind tracks tend to face directly forward or pitch slightly toward the center line of the trail.

Trail: Understep Walk   Trail Width: 1–1.8 in (2.54–4.57 cm)

Stride:1 in, 1.31 in, 1.22 in, 1.4 in, 1–1.4 in (2.54–3.55 cm)

Tail Drag: 0.1–0.48 in (0.25–1.21 cm)

Tail drag is generally V-shaped in cross section. Mainly the region near the tail tip drags, while the thicker and rounder portion of the tail closer to the base is kept clear of the substrate. At higher speeds, the middle of the tail will be swept back and forth as the animal swings its body more dramatically from side to side. This might create a relatively wide area of disturbance that is similar in width to the animal's body. The tail tip will create a secondary, thinner line that may create a zigzagging pattern within this wider line created by the mid-portion of the tail.

Notes:

- This is a common salamander in most of its range, and it can even be found in urban and suburban parks if the right habitat features are present. I was surprised to find this species present at a restored wetland on top of a capped landfill site in the city of Seattle. It was the only salamander species I observed at this location at the time.

- Outside of the breeding season, these salamanders may not be visible, as they spend a considerable part of their adult lives in burrows. Laboratory research done by Ducey in 1989 showed that this species will act aggressively toward members of its own species and that they tend to avoid each other when housed in the same tank. This suggests that they likely defend their burrows from intruders.

- Similar to the larvae of tiger salamanders (*Ambystoma tigrinum*), in some high-elevation sites long-toed salamander larvae (*Ambystoma macrodactylum*) may become cannibalistic. They may also have similar physical adaptations, including larger vomerine teeth and larger heads than other larvae in their breeding ponds.

Northwestern Salamander (*Ambystoma gracile*)

This large member of the Ambystomatid salamanders tends to be relatively drab in appearance, colored in subtly differing shades of brown. This species has particularly well-developed parotoid glands found behind the eyes as well as a ridgeline of glands on top and running the length of the tail.

This species has large, brown, frog-like eyes. It also possesses relatively short, but muscular, limbs. The front and hind feet have long, well-developed toes. Males' hind limbs become enlarged during the breeding season, which helps them hold onto females. The tails of both sexes are laterally compressed and are thickest at their base.

Despite being common within its range, adult northwestern salamanders are rarely observed outside of the breeding season. They spend the majority of

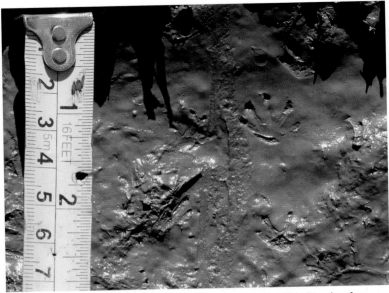

The clear front and hind tracks of an adult male northwestern salamander in fine mud.

their lives underground. Relatively little is still known about their subterranean existence.

The northwestern salamander is known to take on a rigid, defensive posture when touched by a potential predator. This posture involves the animal standing stiff-legged, with the head pointed down and the tail raised and waving. You can often observe sticky white liquid oozing out of the parotoid glands behind the eyes and from the glandular ridge on top of the tail. This liquid tastes foul and is somewhat poisonous, and can cause skin irritation in some people.

This species prefers temperate coniferous forest habitat, mixed woodlands,

The understep walking trail of a heavily gravid adult female northwestern salamander. The thin line is made by the tail drag, while the wide drag mark is made by the female's swollen belly.

and grasslands. They range west of the crest of the Cascade and Coast Range mountains from northern California to central coastal British Columbia, including all of Vancouver Island. There are two recognized subspecies: the brown salamander (*Ambystoma gracile gracile*) and the British Columbia salamander (*Ambystoma gracile decorticatum*). The first subspecies is found throughout the range, while the second is limited to the most northern portion along the central coast of British Columbia.

Northwestern salamanders seek out permanent ponds, lakes, wetlands, and, to a lesser degree, slow-moving rivers and streams during the breeding season. Breeding occurs between January and April at lower-elevation sites. Meanwhile, breeding does not occur until snowmelt at high-elevation sites, which can be as late as between June and August.

You can distinguish the egg masses of this species from other amphibian egg masses found in the same habitat because of their structure. The masses usually are relatively large, and attached to a branch, cattail, or similar support structure. The egg mass is also relatively firm, and will retain its shape if you pick it up out of the water. Older egg masses are often covered and interpenetrated by green algae. This algae (*Oophila amblystomatis*) is symbiotic and actually helps provide oxygen to the growing salamander embryos within the egg capsules (Petranka 1998).

In some permanent breeding habitats, a proportion of the adults never fully transform into terrestrial dwellers. Instead, they retain their external gills and other larval features in a state known as "neoteny." Such neotenic adults are less common in lowland breeding sites, but significantly more common at higher-elevation sites. These individuals are capable of reproducing with fellow neotenic adults or fully transformed adults.

The tracks of this species are most often observed near breeding sites. Watch for them in areas of fine silt or sand in small puddles or along pond edges.

Track: Front: 0.3–0.45 in (0.76–1.14 cm) x 0.4–0.52 in (1–1.32 cm)

Tracks are medium in size. The front feet have four long toes, all of which tend to register in the tracks. Toes 1 and 4 are similar length. Toes 2 and 3 are similar length, with toe 3 sometimes only slightly longer. Front tracks are as long as they are wide, or are slightly wider than long. No tubercles appear on the front feet. The front tracks tend to be angled in line directly forward.

Hind: 0.43–0.72 in (1.09–1.83 cm) x 0.5–0.83 in (1.27–2.1 cm)

Tracks are medium in size, and fan shaped. The hind feet have five long toes, all of which tend to register in the tracks. Toe 1 is the shortest. Toes 2 and 5 are similar lengths. Toe 3 is the longest, with toe 4 slightly shorter. Hind track is fan shaped overall. No tubercles appear on the front feet. Hind tracks tend to angle in line directly forward.

Trail: Understep Walk   Trail Width: 1.6–2.4 in (4–6.1 cm)
                          Stride: 1.82–2.8 in (4.57–7.1 cm)

Tail Drag: 0.1–0.6 in (0.25–1.52 cm)

This species has a V-shaped tail drag in cross section from its laterally compressed ventral tail edge. The tail is so long that as the salamander walks, the sides of the tail may drag on the ground, with part of the ventral edge creating a thin line down the middle, about 0.1 in (0.25 cm), with the wider marks of the undulating tail up to 0.6 in (1.52 cm) in width. Sometimes the body will also leave a drag in deeper substrates up to 0.6 in (1.52 cm) wide.

Notes:

- The northern subspecies (along the central coast of British Columbia) of the northwestern salamander (*Ambystoma gracile decorticatum*) has an additional bone in the fourth toe of the hind foot. This gives it a foot structure similar to that of the long-toed salamander (*Ambystoma macrodactylum*), and it is in this particular area that you may confuse the tracks of the two. The tracks of the northwestern salamander, though, will be considerably larger.
- This is one of the few salamanders known to vocalize. Generally, they emit a ticking sound when in a defensive posture. They have been observed making these sounds during encounters with potential predators or during aggressive encounters with their own species.
- The mildly poisonous secretions help this salamander species survive in lakes and streams where predatory fish and bullfrogs are present.

Pacific Newts (Genus *Taricha*)

This group of western newts all share several features: rough-textured skin, and a brightly colored ventral surface of orange-yellow or red that contrasts with the darker dorsal color and highly toxic skin secretions. Newts of this genus also have skin toxins that contain tetrodotoxin, which is a potent neurotoxin shared by some oceanic fish, such as the infamous Japanese delicacy of pufferfish "fugu."

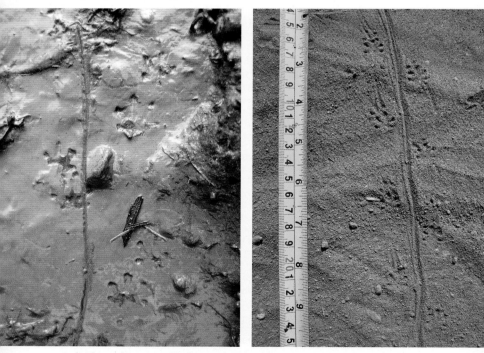

Left: The understep walk of a rough-skinned newt moving through a puddle of fine silt. KIM CABRERA *Right:* The direct register walk of a rough-skinned newt in fine sand. Notice the short, thick toes on both the front and hind tracks.

A close-up image of the direct register walking trail of a large coast range newt, a relative of the rough-skinned newt, moving in the direction toward the top of the photo. Notice the very thick toes.

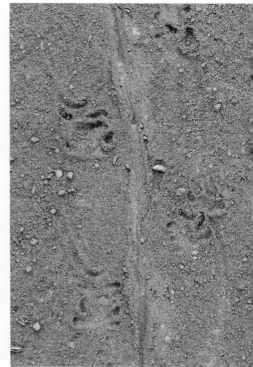

These toxins are dangerous if ingested, and have caused accidental human fatalities. The animals are, however, generally safe to handle, though their skin secretions should be kept clear of mucous membranes and areas of broken skin.

There are four main species: rough-skinned newt (*Taricha granulosa*), California newt (*Taricha torosa*), Sierra newt (*Taricha sierrae*), and red-bellied newt (*Taricha rivularis*). All four have similar foot morphology, with rather short and thick toes on both front and hind feet. Males develop swollen limbs and nuptial pads on their toe tips to help them grip females during breeding.

All Pacific newts will respond in a similar fashion when prodded by a potential predator. They will arch their spine down, lift their head so the ventral surface is exposed, and curl their tail over the back. This is all done to expose as much of the ventral surface as possible, likely as a means to warn of their highly toxic skin secretions.

The tracks of all four *Taricha* newts are similar. Use range and habitat as helpful clues to help separate them.

Rough-skinned Newt (*Taricha granulosa*)

The most wide-ranging Pacific newt lives in an area that includes coastal southeastern Alaska along the western coast down to the San Francisco Bay Area of California. Rough-skinned newts are frequently encountered, and are most visible when adults travel to and from breeding sites.

You can tell this species from the California newt (*Taricha torosa*) by several physical differences: when viewed from directly above, the eyes of *T. granulosa* do not meet the margin of the head, and the lower lids of *T. granulosa* are dark instead of pale as in *T. torosa*.

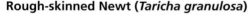

Rough-skinned newt adults measure between 3.5 and 7.75 inches in total length. This species typically has a reddish-brown, brown, or black dorsal area. On the ventral surface, it is bright orange or yellow.

Given the huge numbers of adults that migrate to breeding sites along the West Coast of the United States, the tracks of this newt are some of the most likely to be encountered. You may observe their tracks crossing coastal dunes during wet periods as they move toward or away from shallow wetlands where they breed.

Track: Front: 0.23–0.45 in (0.58–1.14 cm) x 0.3–0.5 in (0.76–1.27 cm)

Front feet have four toes, and all tend to register in the tracks. The toes are short and wide overall. Toe 1 is the shortest, followed by toe 4. Toe 3 is the longest, with toe 2 being slightly shorter. A small tubercle is found at the base of toe 4. The metacarpal area that registers is similar in size to the area covered by the toes in the tracks. The front tracks face directly forward along the line of travel. Fronts tracks are often difficult to see as the hind tracks tend to partially obscure them, or the front tracks may be obscured by toe drags.

Hind: 0.35–0.6 in (0.89–1.52 cm) x 0.4–0.6 in (1–1.53 cm)

The hind feet have five toes, and all tend to register in the tracks. The tracks are fan shaped overall. Toe 1 is the shortest, and is basically just a nubbin. Toes 2 and 3 are a similar length. Toe 3 is the longest, followed closely by toe 4. A small tubercle is found at the base of toe 5. The metatarsal area that registers is similar in size to the area covered by the toes in the tracks. The hind tracks face directly forward along the line of travel.

Trail: Understep Walk   Trail Width: 1.3–1.8 in (3.3–4.57 cm)
                                      Stride: 1.5–2.35 in (3.81–5.97 cm)

Tail Drag: 0.1–0.35 in (0.25–0.89 cm)

The tail drag is generally narrow and V-shaped in cross section due to the finely tapered edge of the laterally compressed tail.

Notes:

• Male rough-skinned newts develop rougher and drier skin on the ventral surface of the toes and across the metacarpal and metatarsal areas. This likely gives them a better grip on slippery females during mating.

• You can sometimes observe this species in large mating congregations called "mating balls." These might be composed of twenty to more than a hundred individuals underwater in a swirling mass of entangled bodies. These newts return to the same breeding sites annually. Watch for their tracks entering and exiting such breeding sites.

• Rough-skinned newts are known to eat the eggs of other amphibians.

Slender Salamanders (Genus *Bathrachoseps*)

With very long, thin bodies and long tails, these are sometimes referred to as "worm salamanders." They do somewhat resemble bizarre legged worms with round, shiny eyes. Their legs are proportionately very small, and their feet are greatly reduced. Despite appearances, they are very capable walkers. All slender salamanders have four stumpy toes on both their front and hind feet.

Their small and thin design allows them to enter very small and tight spaces, including the burrows of earthworms and the tunnels of subterranean insects such as termites or ants. This type of microniche habitat is not accessible to most other salamander species.

There are currently understood to be twenty-two species of slender salamanders, two of which are found in coastal Oregon and the rest are limited to California. Because of their subtle morphological differences, identification

The erratic, zigzagging trail of a California slender salamander moving rapidly. Notice few tracks are visible, as most were erased by the lashing tail.

is difficult. This is especially true in the field, and some species are indeed so similar that they are best separated by their exact range.

Some species of slender salamanders, such as the California slender salamander (*Batrachoseps attenuatus*) and garden slender salamander (*Batrachoseps major major*), have demonstrated a considerable resilience in the face of urbanization. Some have even been found in the marginal habitats of freeway right-of-ways (Stebbins 2012). Both species are commonly found in moist spots under rocks, large pieces of bark, and other moisture-holding items in urban and suburban gardens. Here they are likely to be encountered by people, especially during times of wet weather when they are more surface active.

At drier times of year, slender salamanders may disappear into deep underground retreats with constant soil moisture, where they wait out the dangerously dry conditions above.

Surprisingly, one subspecies—the desert slender salamander (*Batrachoseps major aridus*)—is found in desert canyons. This endangered subspecies has a very tenuous existence, as it depends on moist habitats in the depths of canyons in the extremely arid Colorado Desert near Palm Springs, California. The moist habitats it depends on are, of course, already a finite commodity in such an arid landscape. And this habitat is made rarer still by continued suburban growth.

Despite their fragile appearance and seeming defenselessness, these little salamanders have some interesting ways of avoiding predation. One is that their long tails break off relatively easily, meaning that if a predator such as a bird grabs the tail, it will likely detach. The tail goes into violent spasms of motion which can last several minutes, distracting the predator as the

The understep walking trail of a California slender salamander includes the tiny toes. This salamander was moving from left to right across the frame.

salamander slinks off toward its nearest hiding place. This type of defensive strategy is more familiar in lizards, but a few other salamander species, such as ensatinas (*Ensatina eschscholtzii* and *Ensatina klauberi*), also employ it.

Additionally, slender salamanders may also rapidly coil and uncoil or flail their whole bodies back and forth when exposed or prodded by a potential predator. This rapid motion might be startling, and can also propel the animal in unexpected directions. Many, perhaps all, slender salamanders exude sticky mucus from their skin, which fouls up a small predator's mouth and allows the salamander to escape.

Despite their diminutive forms and tiny feet, slender salamanders do leave recognizable sign. The most likely sign left by the passing of these salamanders are their walking trails in fine substrates and, for a few species, their egg masses under cover objects.

California Slender Salamander (*Batrachoseps attenuatus*)

Found around the San Francisco Bay Area and north to the extreme southwestern coast of Oregon, this species is one of the most commonly encountered slender salamanders. Its range includes some intensely developed areas and it is often found in suburban backyards. Its preferred habitats include forests, woodlands, grasslands, chaparral, and edges of marshes, and in these areas they are typically found under moisture-holding cover objects such as boards, bark, leaf litter, and logs.

Adults measure between 3 and 5.5 inches in total length. They are typically dark colored, even black with a brown, red, or yellowish stripe down the length of the

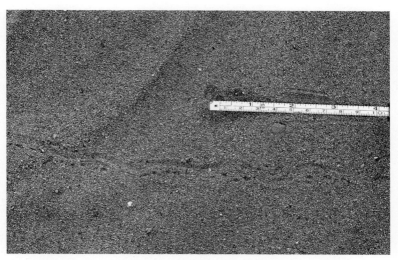

The typical understep walking trail of a California slender salamander showing a gentle undulation, wide tail drag, and the tiny dots left by the feet.

dorsal surface. The legs and feet of this species are small, and the body and tail show obvious grooves, called "costal" and "caudal" grooves.

Watch for their trails and tracks in spots that have extremely fine substrates such as fine dust in protected areas under bridges or in puddles of extremely fine silt in trails or along the edges of waterways. Though not aquatic, slender salamanders are often found near water, especially during the drier parts of the year.

The trails of this species include a wide tail drag mark and very small tracks, using a short stride on either side of the tail drag. The tail drag somewhat resembles the crawling trails of large earthworms.

Track: Front: 0.18–0.25 in (0.46–0.63 cm) x 0.128–0.144 in (0.32–0.36 cm)
These tracks are tiny. Front tracks have four toes. All toes are very stumpy. Toe 3 is slightly longer than toe 2. Toe 1 is a mere bump. Tips of all toes are rounded. The front tracks tend to orient directly forward relative to the center line of the trail.

Hind: 0.21–0.36 in (0.53–0.91 cm) x 0.18–0.31 in (0.46–0.79 cm)
These tracks are tiny. Hind tracks have four toes. All toes are very stumpy. Toe 1 is a mere bump, toe 3 is the longest. Toes 2 and 4 are a similar length. Tips of all toes are rounded. The hind tracks tend to orient directly forward relative to the center line of the trail.

Trail: Understep Walk   Trail Width: 0.433–0.72 in (1.1–1.84 cm)
Stride: 0.2–0.3 in (0.5–0.76 cm)
Tail Drag: 0.08–0.41 in (0.22–1.046 cm)

Slender salamander understep walk trail. SEB BARNETT

The tail drag of this species is very wide relative to the overall trail. At slower speeds, it shows gentle undulations. At higher speeds, the tail drag will show an erratic zigzagging pattern and only a few tracks might be visible.

Notes:

- You'll often encounter the trails of slender salamanders crossing drying muddy pools in trails and dirt roads. These salamanders are not aquatic and do not often enter water, though they are capable of swimming if forced to do so.
- If you expose these salamanders under cover objects, please make sure to replace the objects in the same position to help preserve that microhabitat.

Spotted Salamander (*Ambystoma maculatum*)

A moderately large and distinctly marked salamander of the central and eastern United States, this species typically has a charcoal-gray dorsal surface and paler gray sides and ventral surface. The body is relatively thick, with obvious costal grooves. Scattered across the dorsal surface—from the head to the tail tip—are round, yellow or yellowish-orange dots that appear in pairs on either side. Occasionally, these dots coalesce into several small blotches. The tail is long and round at the base, but subtly becomes more laterally compressed toward the tip.

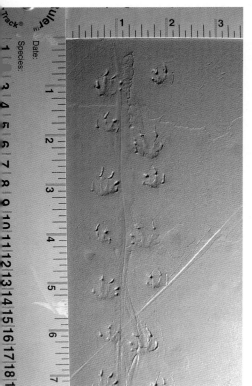

The feet are similar to those of other eastern Ambystoma salamanders, with well-developed, moderately long toes. The hind feet are larger than the front feet, and the fourth toe on the hind foot is the longest.

This species is largely found in eastern deciduous forests. Some populations are also found in coniferous forests and mixed forest types. They primarily breed in woodland vernal pools, and also use ponds on occasion that are free of predatory fish. Though, they may use ponds with predatory fish if there is sufficiently thick aquatic vegetation along the

The rapid overstep walk of an adult spotted salamander under controlled conditions. JOE LETSCHE

The overstep walk of an adult male showing both a tail drag and a drag mark from a swollen vent. Photo made under controlled conditions. JOE LETSCHE

edges, which will help protect the egg masses and young larvae.

Breeding females are, on average, larger than breeding males. Breeding males have significantly larger, more swollen vents. Males deposit spermatophores which are picked up by females to fertilize the eggs. They generally deposit the eggs on sticks, branches, or leaves submerged in the pools. Multiple females often lay egg masses close together and may arrange the oblong masses lengthwise along the same stick. The egg masses of this species develop along with a green, symbiotic algae (*Oophila amblystomatis*) that covers them. The egg masses of this species often contain a protein that gives the mass a milky-white appearance.

Watch for the tracks of this species entering or exiting wetland pools during the breeding season in early spring.

Track: Front: 0.3–0.4 in (0.76–1 cm) x 0.3–0.5 in (0.76–1.27 cm)
The front feet have four toes, all of which tend to register in the tracks. The toes are moderate in length. Toe 3 is the longest, and only slightly longer than toe 2. Toe 2 and toe 4 are similar in length. Toe 1 is the shortest. Overall, the front foot tends to point directly forward relative to the center of the trail or may pitch slightly toward the center of the trail.
Hind: 0.4–0.6 in (1–1.52 cm) x 0.4–0.6 in (1–1.52 cm)
The hind feet have five toes, all of which tend to register in the tracks. The toes are proportionally long. Toe 1 is the shortest. From there, toes 2 through 4 are increasingly longer. Toe 4 is distinctly the longest. Toes 2 and 5 are about the same length. Hind tracks are as wide as they are long, or are slightly wider than long, and tend to face directly forward. The metatarsal surface area may register more than related species with similar hind foot structure.
Trail: Understep Walk   Trail Width: 1.3–2 in (3.3–5.08 cm)
Stride: 1.7–2.3 in (4.32–5.84 cm)
Tail Drag: 0.1–0.2 in (0.25–0.5 cm)
Tail drag is generally narrow, V-shaped in cross section, and is made mostly by the area of the tail near the tip. Occasionally, belly drag marks might also be present, especially in a heavily gravid female or a male with a swollen vent. These can measure between 0.4 and 0.6 inches wide.

Notes:

• A distinct and charismatic salamander of eastern deciduous woodlands. Familiar to many and even sold in some pet stores. It can survive well in captivity, but is best observed thriving in its native habitats.

• Like other salamanders, spotted salamanders have noxious defensive skin secretions. They will head-butt and/or tail-lash an approaching predator, and along with the Jefferson salamander, will commonly play dead when handled. This species may also vocalize.

• The dependency on vernal pools and deciduous woodlands makes this species vulnerable to habitat alteration and habitat loss. If you live in the range of this species, consider creating vernal pools in your area to help this and other vernal-pool-breeding amphibians. If you are lucky enough to already have such biological gems where you live, make sure to protect these vital pools and the woodlands that surround them.

Tiger Salamander (*Ambystoma tigrinum*)

This large terrestrial salamander is the most wide-ranging salamander species in North America. The largest specimens rival the coastal giant salamander (*Dicamptodon tenebrosus*) as some of the largest terrestrial salamanders in the world. Tiger salamanders have large heads with small eyes and wide mouths. Their bodies are thick, and are supported on short, dense legs. Their tails are also thick and rounded at the base, progressively becoming more laterally compressed toward the tip.

This species has six subspecies found over its vast range: gray tiger salamanders (*Ambystoma tigrinum diaboli*), barred tiger salamanders (*A. t. mavortium*), blotched tiger salamanders (*A. t. melanosticum*), Arizona tiger salamanders (*A. t. nebulosum*), Sonoran tiger salamanders (*A. t. stebbensi*), and eastern tiger salamanders (*A. t. tigrinum*). Recently, the barred tiger salamander has been separated into its own species (*Ambystoma mavortium*).

Throughout the tiger salamander's wide range, its preferred habitats include coniferous forests, woodlands, open fields and brushy areas, alpine and subalpine meadows, grasslands, and even deserts. Adults seek out breeding sites in vernal pools, permanent ponds, cattle tanks, and subalpine lakes.

Adult tiger salamanders are capable of digging their own burrows, or they may seek out the burrows of other animals as shelter. True to their family, this species of "mole salamander" spends much of its adult terrestrial life in burrows underground.

Watch for the tracks and trails of this species entering shallow wetland pools during the breeding season, or exiting them as the pools dry up.

Track: Front: 0.36–0.6 in (0.91–1.52 cm) x 0.43–0.7 in (1.09–1.78 cm)
Tracks are medium in size. The track features are proportionally robust overall. Toes are wide and thick. There are four toes on the front foot. Toe 3 is slightly longer than toe 2. Toes 1 and 4 are similar in length. Front tracks align directly forward or pitch at an angle toward the middle of the trail.

Left: The understep walking trail of an adult tiger salamander leaving a drying wetland in Arizona. Look closely for the tubercle at the base of toe 1 on the hind track. JONAH EVANS *Right:* The understep walking trail of an adult tiger salamander. Only the toe tips and subtle tail drags are visible moving toward the top of the photo. JONAH EVANS

Hind: 0.46–0.75 in (1.16–1.9 cm) x 0.59–0.8 in (1.5–2 cm)

Tracks are medium in size. The track features are proportionally robust overall. Toes are wide and thick. There are two circular tubercles on the hind foot. The larger is at the base of toe 1 and the smaller one is at the base of toe 5. Toe 3 is the longest toe. Toes 2 and 4 are similar in length. Toes 1 and 5 are also similar in length. Hind tracks align directly forward or pitch at an angle toward the middle of the trail.

Trail: Understep Walk   Trail Width: 1.6–2.8 in (4.06–7.11 cm)
                         Stride: 2.3–3.8 in (5.84–9.65 cm)

Tail Drag: 0.1–0.2 in (0.25–0.5 cm)

The tail of this species is laterally compressed, with the most acutely compressed section at or near the tip and becoming increasingly more rounded in cross section as you move up toward the tail base.

Notes:

- A closely related species, the California tiger salamander (*Ambystoma cali-forniense*) is restricted in its range to the coastal and interior parts of central California. This species is dependent on both vernal pools for reproduction and the presence of pocket gopher (*Thomomys* sp.) and California ground squirrel (*Otospermophilus beecheyi*) burrows for refuge.

- When housed in an appropriate aquarium or terrarium, tiger salamanders have been known to live for a long time. A gilled adult was reported to have lived as long as twenty-five years (Petranka 1998).

- This species has been widely used as a source of live bait for fishing, which has spread this species into many large bodies of water well outside its native range. This practice is cruel, and generally fatal to the gilled larvae. In addition, those that manage to escape and survive may introduce new diseases to the native amphibian species.

Frogs and Toads 3

Despite their dependency on water, frogs and toads can be found in a wide variety of habitats, including ponds, streams, rivers, lakes, marshes, and swamps, as well as in drier areas, such as deciduous and coniferous forests, oak savannas, grasslands, and even deserts. Finding a toad hopping around on a rainy night while driving through the desert can seem incongruous, yet these animals have adapted to survive in such incredibly harsh places and take advantage of the brief bursts of moisture when they occur.

The majority of frogs and toads are, of course, found in greater abundance and diversity in wetter places. In all of the places where frogs and toads dwell, you are most likely to find their tracks and sign within close proximity to water. Toads, including spadefoots, generally seek out water for more extended periods during breeding time. This is also true of some of the terrestrial frogs, such as wood frogs and red-legged frogs. Some frogs are much more aquatic and spend a majority of their lives in, or very near to, water. Some examples of more water-dependent species are pig frogs, river frogs, and bullfrogs.

Locomotion

Generally speaking, North American frogs and toads are capable of only two types of locomotion: walking and hopping. Frogs walk one leg at a time, alternating a front leg on one side with the hind leg on the opposite side. Toads, especially most species in the genus *Anaxyrus*, walk frequently. Walking is much less common in frogs, which generally move by hopping. Frogs might be seen walking during mating, when the male grasps the female just under the arms and rides on her back, an activity called "amplexus."

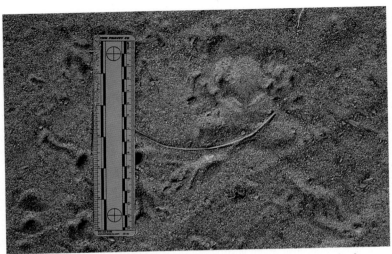

An example of the tracks of a pair of boreal toads in amplexus. Notice the four bunched tracks of the female and the two dragging legs impressed in the sand behind her.

Particularly large and heavy frogs such as bullfrogs can sometimes be observed walking short distances. Tree frogs use these two methods of locomotion to climb, even up sheer vertical surfaces.

Hopping frogs and toads propel themselves forward using their larger rear legs by pushing suddenly downward. When they land, the front feet hit the ground first, and the hind feet register just behind them. The belly region of the frog or toad may also register with each hop. These hops might be short, such as those made by most toad species, while the hops of some of the larger frogs can be impressively long. The power of the leaps is generally proportionate to the size of the legs, and the muscles that move them. Some of the most powerful jumpers are members of the genus *Rana* and *Lithobates*, though some of the larger tree frog species can also make impressive leaps. Cricket frogs (*Acris* ssp.), although very small, are incredible jumpers and are able to propel themselves a distance equal to many times their own body length.

Foot Structure

The front feet of all frog and toad species possess four toes, while the hind feet possess five toes. Most species have some degree of webbing on their hind feet. The length of toes on the front feet relative to each other varies from species to species. In most species the hind feet are larger, as they are the main form of propulsion both in swimming and in hopping.

The toes are numbered from the inside to the outside to allow for easy reference. The toes of the front foot are labeled toe 1–4, with toe 1 being closest to the body. The toes on the hind foot are labeled toe 1–5, with toe 1 being closest to the body, and generally also the shortest. Tree frogs have enlarged toe pads on all of their toe tips.

The front feet of frogs and toads possess several tubercles, which are pads that help them with traction. Those found on the toes are known as "subarticular tubercles." These help the toes grip uneven surfaces. On the palm (more accurately known as the "metacarpal region"), there are several more structures. One or multiple pads located near the lower center of the metacarpal region are called the "palmar tubercles." Located near the base of toe 1 is the "thenar tubercle." On toe 1, male frogs and toads also develop a nuptial pad used during breeding activities to grasp the female. This pad may be subtle most of the year, but becomes much more pronounced during the breeding season.

The hind feet of frogs and toads possess several types of tubercles. Just like on the toes of the front feet, there are subarticular tubercles on the toes of the hind feet. On the heel (more accurately known as the "metatarsal region"), there are two types of tubercles. The larger tubercle, which is located near the inner edge of the foot, is known as the "inner metatarsal tubercle." On the other side of the foot, near the outer edge of the foot, is a somewhat smaller tubercle known as the "outer metatarsal tubercle." This outer metatarsal tubercle is reduced or absent in most frogs, and often more prominent in toad species. The hind feet also tend to possess webbing between the toes, which may be extensive and reach to the very tip of the toes in some highly aquatic species, while some species that are largely terrestrial have greatly reduced webbing.

Frog Track Morphology

Generally speaking, the track groups of frogs and toads appear rather close together, looking roughly square or rectangular in overall shape. The front tracks tend to land facing each other, not unlike the tracks of woodpeckers. The hind tracks land to the outside of the track group, facing generally forward, and nearly always with the toes angled somewhat away from the center of the trail. Although the exact positions of these tracks can vary depending on the way the animal is moving, species, and groups of species, tend to show certain patterns consistently. Along with the presence, size, and arrangement of tubercles on the feet, these track details can help distinguish one species from another, or at least help narrow down the possibilities.

The texture of the skin on the underside of the feet of toads—especially those belonging to the genera *Anaxyrus*, *Incilius*, and *Rhinella*—is distinct. Species of these groups have many abundant, small bumps that cover the undersides of their feet. They have one or several large tubercles on the hind feet, with the largest located just under toe 1. All of these toad species also tend to have reduced webbing, especially on toe 4, which is more than half free of webbing.

In contrast, the spadefoot toads (*Scapiophus* and *Spea*) have almost completely smooth front and hind feet. They possess a large, well-developed, and surprisingly hard tubercle just below the base of toe 1. This hardened tubercle is the "spade" for which this group is named, and is indeed a powerful digging tool. Spadefoot toads also have extensive webbing that reaches to the tip of each toe on the hind foot.

Tree frogs (*Hyla* and *Osteopilus*) and chorus frogs (*Pseudacris*) have toes on both front and hind feet that end in enlarged, sticky pads that help them climb. These climbing frogs also tend to have front and hind feet that are similar in size.

The true frogs—both *Rana* and *Lithobates*—have foot structures that include very long and skinny toes on their hind feet and long, tapered toes on their front feet. The undersides of their feet tend to be fairly smooth, and the small tubercles that are present are limited in most species to only small areas on their toes, and very few, if any, are present on the rest of the foot. Those that are present generally do not show up in their tracks, with the exception of the small tubercle at the base of toe 1, which is present in nearly all frogs and toads.

The males of most frogs and toads also develop some type of nuptial pad. This pad appears in most species on toe 1, at the base, or it can develop to cover more than half of the underside of the toe. These pads aid the male in gaining a good grip on a female during amplexus. In some substrates, these nuptial pads may register clearly enough to allow a tracker to determine the sex of the frog. It is important to distinguish here between the marks made by this enlarged structure and by the tendency for the toe tip—especially the tip of toe 1—to register more deeply. The latter is due to a combination of fine substrates and strong impact upon landing. Toe 1 is usually the first toe to hit the ground.

Some frogs also tend to leave toe drags in loose substrates from the explosive action of their hops. These tend to angle backward—often at roughly 45 degrees—away from the stationary positions of the hind tracks. In situations where the frog is trying to rapidly escape, these toe drags may be so explosive that they make the entire track group obscure. This is especially likely in loose, powdery substrates like dry sand.

Sometimes during the breeding season, the tracks of what appear to be six-legged frogs can be seen, especially near breeding sites. These are the tracks of frogs in amplexus. Look close to see the four closely spaced tracks of the female and the two trailing legs of the hitchhiking male. The male may have his hind legs folded close into the female's

The tracks of an adult bullfrog show the marks made by extra toes on its right hind foot.

hind legs and you may only see the male's hind foot tracks and part of the leg on either side just above the foot. Conversely, the male may have his legs stretched out and dragging to nearly their full length behind the female. In this position, the legs of the male may register with the feet upside down.

Due to a parasite called *Ribeiroia*, some populations of frogs (as well as some toads and salamanders) develop limb deformities. Because this parasite burrows into the flesh of the developing legs in tadpoles, it can cause a variety of limb abnormalities including missing limbs, partially developed limbs, or extra limbs. This strange limb development is believed to be a deliberate ploy on the part of the parasite to make the frogs an easier target for herons, which are the parasite's main hosts. This type of infection is more likely to occur in wetland areas that are more disturbed, less diverse, and impacted by pollutants such as pesticides. Frogs affected by this parasite may leave tracks that are very unusual and more difficult to decipher.

Family Hylidae

This well-known group of frogs features expanded toe pads that allow them to climb almost any surface. This family includes chorus frogs (*Pseudacris*), cricket frogs (*Acris*), tree frogs (*Hyla*), and Cuban tree frogs (*Osteopilus septentionalis*).

Tree frogs are often able to jump far relative to their size, and some species are partially or largely arboreal. Ten commonly recognized species belonging to the genus *Hyla* are found in the United States. These tend to have larger toe pads than those of the chorus frogs.

Canyon Tree Frog (*Hyla arenicolor*)

The canyon tree frog inhabits the flowing water courses through desert canyons. Though it can be common in its rocky habitat, it is also easily overlooked. This species tends to match the rock it lives on, with gray, light brown, or even pinkish coloration. The color may be uniform or show subtle dark gray, brown, or green spots.

Left: The front foot of a common gray tree frog. JASON KNIGHT *Right:* The hind foot of a common gray tree frog. JASON KNIGHT

 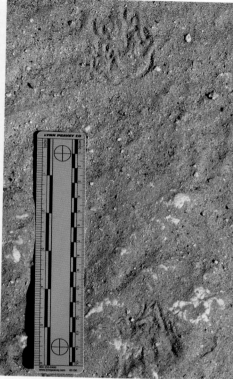

Left: The track group from an adult canyon tree frog. Notice the drag lines from the toes as the frog kicked explosively when hopping. *Right:* The hopping trail of an adult canyon tree frog in sand near the edge of a stream.

During the day, it often rests in rock crevices with its feet and legs tucked under its body and its eyes closed. This position helps minimize water loss to evaporation. It is not uncommon to find several canyon tree frogs perched in the same crevice or on the same rock face. These frogs often reuse the same shelter spot over and over again. Such crevices are good places to look for their scat, when the frogs themselves are not visible. They may also congregate on large boulders directly at the streamside or even in the stream. A similar frog, the California tree frog (*Pseudacris cadaverina*), fills a comparable habitat niche along the rocky streams of northern Baja California and the canyons of southern California.

These frogs are capable of climbing shear rock faces with relative ease due to their sticky toe tip pads.

Track: Front: 0.5–0.9 in (1.3–2.29 cm) x 0.27–0.5 in (0.7–1.27 cm)
Tracks are small. There are four toes, and the feet are K- or X-shaped overall. All toes tend to register in the tracks, and all possess well-developed circular pads on toe tips, with the largest appearing on toes

3 and 4. There is no webbing present on the front foot. Toe 1 is the shortest. Toe 3 is the longest. Toe 4 is slightly longer than toe 2. There are two small, paired palmar tubercles. The thenar tubercle is oval in shape, moderately large, and located at the base of toe 1. Front tracks often register directly ahead of the hind tracks. Toe 4 often points directly opposite of toe 1, creating a nearly straight line along the outer edge of the front track. The angle of the overall track has a pitch of between 45 and 90 degrees.

Hind: 0.5–1.1 in (1.3–2.79 cm) x 0.31–0.7 in (0.8–1.78 cm)

Tracks are small. There are five toes, and all toes tend to register in the tracks. Webbing is present, and toe 4 extends beyond the webbing by two phalanges. Toe 4 is the longest. Toe 1 is the shortest. Enlarged toe tip pads are largest on toes 3, 4, and 5. Subarticular tubercles are small and round, with three present on toe 4. The inner metatarsal tubercle is very small, much smaller than the size of the smallest toe tip pad, located below the base of where toes 4 and 5 meet. The outer metatarsal tubercle is oval in shape, about the size of the toe tip pad on toe 5 and on the same plane as the inner metatarsal tubercle. The outer metatarsal tubercle is located at the base of toe 1. Skin texture is coarsely granular on the hind track, especially on the metatarsal region. Hind tracks register with the toes pitching outward away from the center line of the trail up to about 45 degrees.

Track Group Size: 1.22–1.9 in (3.1–4.83 cm) x 1.25–1.9 in (3.2–4.83 cm)

Trail: Hop   Stride: 5.5–14 in (13.97–35.56 cm)

Notes:

• This tree frog often goes unnoticed along rock creeks throughout its range. You may find several frogs sharing the same rock face crevice, and you can identify such sites even when the frogs are not there by the presence of scat pellets.

Common Gray Tree Frog (*Hyla versicolor*) and Cope's Gray Tree Frog (*Hyla chrysoscelis*)

These two closely related tree frogs are both highly cryptic species, matching surfaces such as tree bark perfectly. They are usually gray, light tan, or greenish, and may be uniform in color or have a series of uneven lichen-like blotches on their head, back, and limbs. These blotchy patches are generally grayish or occasionally greenish.

Telling these two species apart by sight alone is extremely difficult; you should also consider their voices and genetics. Their ranges overlap around the Great Lakes and along parts of the Mississippi watershed. In the Upper Midwest where these two species are found together, you'll find the common gray tree frog (*Hyla versicolor*) mostly in forests, and Cope's gray tree frog (*Hyla chrysoscelis*) in oak savannas, prairies, and grasslands (Knutson et al. 2000).

Left: Common gray tree frog range. *Right:* Cope's gray tree frog range.

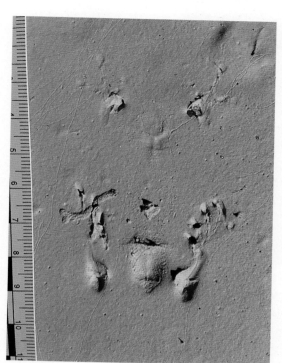

The beautiful tracks in clay of a Cope's gray tree frog.
STEVE FORTIN

Track: Front: 0.7–1 in (1.78–2.54 cm) x 0.5–0.7 in
(1.27–1.78 cm)

Tracks are small. There are four toes, and the
feet are K- or X-shaped overall. All toes tend
to register in the tracks. All toes possess well-
developed circular pads on toe tips, with the
largest on toes 2 and 3. There is no webbing
present on the front foot. Toe 1 is the shortest.
Toe 3 is the longest. Toe 4 is slightly longer than
toe 2. Slight webbing occurs between all toes.
There are two small, paired palmar tubercles.
The thenar tubercle is oval in shape, moderately
large, and located at the base of toe 1. Front
tracks often register directly ahead of the hind

**Common gray tree
frog hopping trail.**

tracks. Toe 4 often points directly opposite of toe 1, creating a nearly
straight line along the outer edge of the front track. The angle of the
overall track has a pitch of between 45 and 90 degrees.

Hind: 0.8–0.9 in (2–2.29 cm) x 0.5–0.7 in (1.27–1.78 cm)

Tracks are small. There are five toes, and all toes tend to register in the
tracks. Webbing is present, and toe 4 extends beyond the webbing by two
phalanges. Toe 4 is the longest. Toe 1 is the shortest. Enlarged toe tip pads
are largest on toes 3, 4, and 5. Subarticular tubercles are small and round,

The track group from an adult common gray tree frog. JASON KNIGHT

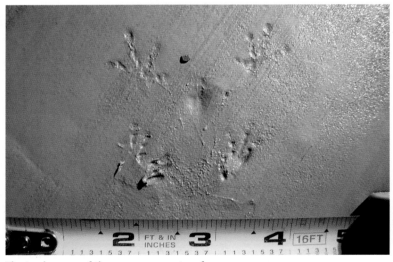

The track group of the common gray tree frog. JOE LETSCHE

with three present on toe 4. The inner metatarsal tubercle is very small, much smaller than the smallest toe tip pad, and is located below the base of where toes 4 and 5 meet. The outer metatarsal tubercle is oval in shape, about the size of the toe tip pad on toe 5 and on the same plane as the inner metatarsal tubercle. An outer metatarsal tubercle is located at the base of toe 1. Skin texture is coarsely granular on the hind track, especially on the metatarsal region. Hind tracks register with the toes pitching outward away from the center line of the trail up to about 45 degrees.

Track Group Size: 2–2.3 in (5.08–5.84 cm) x 2.3–2.5 in (5.84–6.35 cm)
Trail: Hop   Stride: 5–14 in (12.7–35.56 cm)
Notes:

- You'll encounter the tracks of tree frogs most commonly as they seek shallow wetlands for breeding sites.

Cuban Tree Frog (*Osteopilus septentrionalis*)

This giant tree frog species is a tropical invader from the island of Cuba. This species shows sexual dimorphism, with females being about one-third larger than males. Cuban tree frogs are generally gray or some shade of light brown. Some frogs are uniform in color, while others can have mottling of green, gray, or dark brown.

This skin of this tree frog is granular, and the skin on the head is actually fused to the skull. This allows these frogs to back into tight places and use their heads to block the entrance (Dodd Jr. 2013).

Given the frog's great size, it eats a wider variety of prey items than native tree frogs in its range in Florida. Indeed, this species does sometimes consume native frogs and even small lizards. It also feeds on various insects such as beetles, roaches, moths, and spiders. It is not uncommon to find one of these large frogs lurking near a porch light or other artificial light as it waits to ambush the arriving insects. Snakes, turtles, and some birds prey on this species.

This frog leaves impressively large tracks (can be comparable in track size to a large adult *Lithobates* frog), and shows long, flexible toes with enormous toe tip pads.

The impressively large track group of an adult Cuban tree frog.

Track: Front: 1.6–2.3 in (4.06–5.84 cm) x 1–1.3 in (2.54–3.3 cm)

Tracks are small to medium. There are four toes, and the feet are K- or X-shaped overall. All toes tend to register in the tracks. All toes possess very large, well-developed circular pads on toe tips. The largest toe tip pads are on toes 2 and 3. Slight webbing appears between toes 2 and 3, and between toes 3 and 4. Toe 1 is the shortest. Toe 3 is the longest. Toe 4 is slightly longer than toe 2. There is a very small palmar tubercle where the metacarpal bones below toes 3 and 4 come together. The thenar tubercle is oval in shape, moderately large, and located at the base of toe 1. Front tracks often register directly ahead of the hind tracks. Toe 4 often points directly opposite of toe 1, creating nearly a straight line along the outer edge of the front track. The angle of the overall track has a pitch of between 45 and 90 degrees from the center line of the trail. In the track, toes 3 and 4 often point in the same direction and appear relatively parallel. This can vary sometimes, however, as the long, flexible toes of this species may bend considerably.

Hind: 1.8–2.1 in (4.57–5.33 cm) x 1.3–1.6 in (3.3–4.06 cm)

Tracks are small to medium. There are five toes, and all tend to register in the tracks. Webbing is moderate, and toe 4 extends beyond the webbing by two phalanges. Toe 4 is the longest. Toe 1 is the shortest. Enlarged toe tip pads are largest on toes 3, 4, and 5. Subarticular tubercles are small and round, with three present on toe 4. The inner metatarsal tubercle is very small, and likely will not register in the track. Outer metatarsal tubercle is oval in shape, about the size of the toe tip pad on toe 5 and on the same plane as the inner metatarsal tubercle. The outer metatarsal tubercle is located at the base of toe 1. Skin texture is coarsely granular on the hind track, especially on the metatarsal region. Hind tracks register with the toes pitching outward away from the center line of the trail up to about 45 degrees.

Track Group Size: 3–3.5 in (7.62–8.89 cm) x 4.5–5 in (11.43–12.7 cm)

Trail: Hop   Stride: 4–48+ in (10.16–121.92 cm)

Notes:

- This nonnative tree frog has similar overall foot structure to our native tree frogs, but with proportionally larger toe tip pads. These toe tip pads are largest on the front feet, and make sizable indentations in the tracks.

- Cuban tree frogs can leap several feet horizontally with each hop when escaping potential danger.

- This species produces a noxious skin secretion when handled that causes mild allergic reactions such as sneezing and a runny nose. I personally experienced a burning sensation in my eyes when I rubbed them after handling one of these frogs. This lasted approximately ten minutes.

Cuban tree frog trail.

Chorus Frogs (Genus *Pseudacris*)

Chorus frogs belong to the tree frog family, Hylidae. Frogs in this genus typically have expanded toe tips that aid in climbing. They possess intercalary cartilage that provides additional flexibility to their digits. Chorus frogs tend also to have reduced webbing on their hind feet. The exact number of species in this group is under some debate. Sixteen species are commonly recognized in this group.

Many chorus frogs are terrestrial, and those that climb often remain relatively close to the ground. They are small frogs, with the adults of most species growing between 1 and 2 inches in total length. The smallest frog in the United States, the little grass frog (*Pseudacris ocularis*) is a chorus frog. This species has an adult length of only 0.3 inches to a maximum of 0.7 inches.

Chorus frogs create track groups that make a square shape when outlined. The front tracks often land directly ahead of the hind tracks.

Northern Pacific Chorus Frog (*Pseudacris regilla*)

This species is well known to many, even those far outside of its native range, because many Hollywood movies use the sounds of its chorusing in the background. This small frog comes in a variety of colors, including different shades of green, brown, gray, and red. Many also have various blotches of dark brown on the body. All possess an obvious dark-colored stripe that passes through the eye and extends at least to the shoulder. Often there is a dark V- or Y-shaped marking between the eyes. This species and its two close relatives are polyphenic,

Left: The track group of a pair of northern Pacific chorus frogs in amplexus. ROBERT MELLINGER *Right:* The hopping trail of a northern Pacific chorus frog. SEB BARNETT

meaning they can change color and pattern to match substrate, as well as in response to temperature and light conditions (Marimon 1923). Therefore, identifying this species by one distinctive pattern or color is futile.

This once single species has been divided into three distinct species according to mitochondrial genetics: the northern Pacific chorus frog (*Pseudacris regilla*), Sierran chorus frog (*Pseudacris sierra*), and Baja California chorus frog (*Pseudacris hypochondriaca*) (Dodd 2013).

Northern Pacific chorus frogs are found in a broad spectrum of wetland habitats, including open forests and woodlands, chapparal, sagebrush steppe, grasslands, agricultural areas, and even suburban streams and moist yards. You'll most likely notice them in the spring, when many males participate in choruses that can last for hours.

These frogs feed on a variety of small invertebrates, such as flies, spiders, beetles, termites, and moths. They will hunt for prey both on the ground and

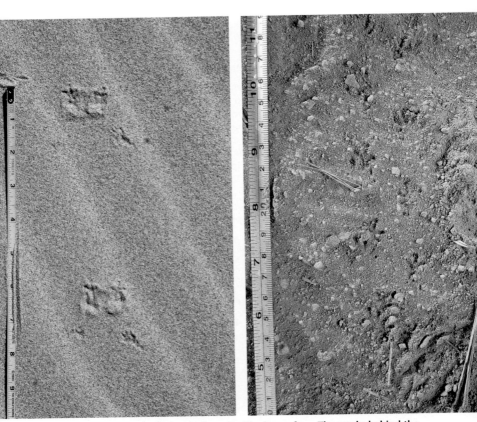

Left: The hopping trail of a northern Pacific chorus frog. The marks behind the main track group were made by the toes of the rear feet as the frog jumped.
Right: The short walking trail of a northern Pacific chorus frog.

The front foot of a northern Pacific chorus frog.

among the branches and leaves of low-growing vegetation. This species may sleep in crevices low down on trees or rock faces, under cover objects such as bark or boards, or even on the surface of leaves in thickets. Look closely on large leaves along wetlands where this frog regularly rests for their small, brown scats.

This frog species and its recently divided close relatives are some of the most commonly encountered amphibians in their preferred habitats. You'll see their hopping tracks in fine silt or mud at the edges of creeks, ponds, and shallow wetlands.

The hind foot of the northern Pacific chorus frog.

Track: Front: 0.5–0.6 in (1.27–1.52 cm) x 0.3–0.6 in (0.76–1.52 cm)

Tracks are small. There are four toes, and the feet are K-shaped overall. All toes tend to register in the tracks. All toes possess well-developed circular pads on toe tips, largest on toes 3 and 4. There is no webbing present on the front foot. Toe 1 is the shortest. Toe 3 is the longest. Toe 4 is slightly longer than toe 2. Multiple small palmar tubercles are present. The thenar tubercle is oval in shape, relatively small, and located at the base of toe 1. Front tracks often register directly ahead of the hind tracks. Toe 4 arches along the same general angle as that of the overall hind track—roughly 45 degrees.

Hind: 0.6–0.8 in (1.52–2 cm) x 0.4–0.7 in (1–1.78 cm)

Tracks are small. There are five toes, and all toes tend to register in the tracks. Webbing is present, and toe 4 extends beyond the webbing by two phalanges. Toe 4 is the longest. Toe 1 is the shortest. Enlarged toe tip pads are progressively larger from toe 1 to toes 4 and 5. Subarticular tubercles are small and round, with three present on toe 4. The inner metatarsal tubercle is relatively small, roughly the size of the smallest toe tip pad, and located at the base of toe 5 and oriented in line with the plane of toe 1. The outer metatarsal tubercle, oval in shape, is about the size of the largest toe tip pad and slightly higher up than the inner metatarsal tubercle. Hind tracks register with the toes pitching outward away from the center line of the trail up to about 45 degrees.

Track Group Size: 1.5–1.8 in (3.81–4.57 cm) x 1.3–1.9 in (3.3–4.83 cm)

Like other *Pseudacris* and *Hyla* species, Pacific chorus frogs have a grouping of tracks that appear roughly in the shape of a square. This is because the front tracks land almost perfectly ahead of the hind tracks and are spaced about the same width apart.

Trail: Hopping   Stride: 3.5–10 in (8.9–25.4 cm)
         Walking   Stride: 1.2–1.4 in (3–3.55 cm)

Notes:

• Being small and lightweight, this frog's tracks generally only register in very fine substrates.

• This species may travel overland for considerable distances when moving from one wetland to another. I have observed the trail of a single Pacific chorus frog traveling for over a mile across open sand dunes along the Oregon coast.

• You can find this species in some very isolated desert oasis sites in California. These moisture-loving amphibians are surprising to encounter in these remote desert water holes.

• Northern Pacific chorus frogs rest during the day in partial shade on a relatively exposed surface, with their legs tucked under their bodies and their head and body pressed down onto the surface of the object. This helps reduce water loss and disguises their shape.

• Until recently, this frog was referred to as the Pacific tree frog (*Hyla regilla*), but is now more accurately considered a chorus frog.

Cricket Frogs

Cricket frogs belong to the family Hylidae along with chorus frogs and tree frogs. They have streamlined bodies and long hind legs, and their pointed heads end in a blunt snout. Although related to tree frogs, they have only a slight suggestion of enlarged toe tip pads.

Regarding track morphology, cricket frogs share with chorus frogs and tree frogs the tendency to create a square track group. Also, the front feet do not face each other directly across the center line of the trail; rather, they

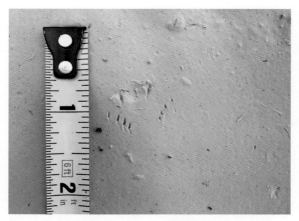

The track group of a Blanchard's cricket frog. CHRIS HYDE

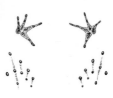

pitch outward between 45 and 90 degrees away from the center line. This is unlike the true frogs (*Lithobates*, etc.), true toads (*Anaxyrus*, etc.), and spadefoots (*Spea*, *Scaphiopus*), all of which have front feet that face each other across the central line of the trail.

There are three species of cricket frogs in the United States: the northern cricket frog (*Acris crepitans*), southern cricket frog (*Acris gryllus*), and Blanchard's cricket frog (*Acris blanchardi*). Until relatively recently, it was considered that there was only a single species with three subspecies.

Cricket frogs have the largest tadpoles in proportion to the size of adults. Large tadpoles can actually grow longer than fully mature adult cricket frogs.

As the name implies, observers often confuse the calls of cricket frogs for those of crickets. This similiarity between the cricket frogs and crickets has likely led to some observational bias when surveying for this species using calls alone.

Blanchard's Cricket Frog (*Acris blanchardi*)

This species of cricket frog is found farthest west, with a range reaching into west Texas, southeastern New Mexico, and a small area of Colorado and southeastern South Dakota. It is also found across the southern Great Lakes region and south to Louisiana and western Mississippi. It ranges as far east as central Ohio and western West Virginia.

Blanchard's cricket frogs prefer relatively open habitats, especially around permanent water sources such

Cricket frog hopping trail. SEB BARNETT

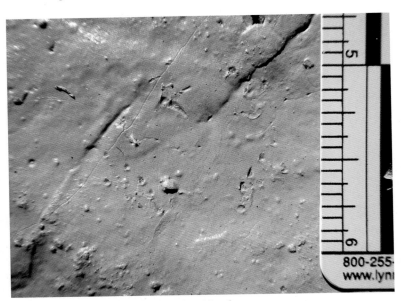

The track group of another Blanchard's cricket frog. JASON KNIGHT

as small streams, marshes, ponds, and lakes. They may be found in temporary wetlands, but generally only when they are directly next to permanent water sources.

This species prefers moist and often muddy substrates near sheltered spots such as roots or rocks. Seek out their tracks in these spots. They are very lightweight, so they leave clear tracks only on very fine mud with the right moisture content.

Track: Front: 0.2–0.3 in (0.5–0.76 cm) x 0.2–0.3 in (0.5–0.76 cm)

Tracks are very small. There are four toes, and feet are K-shaped overall. All toes tend to register in the tracks. In firmer substrates only the major pads on the metacarpal region and the toe tips may register. Toe 3 is the longest. Toes 2 and 4 are similar length. Toe 1 is the shortest. The palmar tubercle is relatively large, oval shaped, and is oriented pointing lengthwise toward the base of toe 4. The subarticular tubercles are relatively small and rounded. The oval thenar tubercle is nearly as large as the palmar tubercle, and is located at the base of toe 1, oriented pointing lengthwise into the toe. The toe tips are rounded and flexed down into the substrate when the track is made. The front tracks land directly ahead of the hinds, equally far out from the center line of the trail or somewhat closer to the center. Toe 4 often points directly opposite of toe 1, creating nearly a straight line along the outer edge of the front track. The angle of the overall track has a pitch of between 45 and 90 degrees.

Hind: 0.4–0.5 in (1–1.27 cm) x 0.2–0.4 in (0.5–1 cm)

Tracks are small. There are five toes, and all toes tend to register in the tracks. Webbing is present, and toe 4 extends beyond the webbing by two phalanges. Toe 4 is the longest. Toe 1 is the shortest. Toes 3 and 5 are similar in length, though toe 5 reaches slightly further. Toe 2 is shorter

The ventral surface of a Blanchard's cricket frog, showing the details of the feet. JASON KNIGHT

than toe 3. Subarticular tubercles are small and rounded. The outer metatarsal tubercle is small, about the same size as a toe tip, and is located at the base of toe 1. The inner metatarsal tubercle is slightly smaller than the outer metatarsal tubercle and is located below the base of toes 4 and 5. The inner metatarsal tubercle is on the same plane as the outer metatarsal tubercle. Skin texture is relatively smooth, with a few tiny, dispersed tubercles. The hind tracks are oriented outward at about 45 degrees, and often register with the feet relatively close together, giving a narrow impression overall.

Track Group Size: 0.7–0.9 in (1.78–2.29 cm) x 0.7–1 in (1.78–2.54 cm)

Trail: Hop   Stride: 6–48 in (15.24–121.92 cm)

Notes:

• These frogs are incredible jumpers, and will often jump a distance of several feet with each leap when startled. This is a pretty remarkable feat given that their entire body is generally only around 1 to 1.5 inches long.

• Not only are they able to jump far, but they can also turn with incredible agility. The frogs often switch directions sharply several times in the span of just a few hops.

• This species has declined in many parts of its range, likely due to a combination of habitat destruction, increasing pesticide and herbicide use, urbanization, and diseases. Blanchard's cricket frogs are more likely to use restored or created wetland habitats than the other two cricket frog species.

Spadefoots (*Scaphiopodidae*)

Spadefoots, or spadefoot "toads," are plump, short-legged frogs with relatively smooth skin (with only small bumps) and vertical pupils. They are so named because of a sharp-edged, cornified tubercle found on the underside of each hind foot. These hardened projections do indeed make very effective spades, allowing these amphibians to dig effectively even through surprisingly hard-packed desert soils. Most species prefer to make burrows in sandier soils whenever possible, however.

The North American spadefoots belong to the two genera: *Spea* and *Scaphiopus*. They are not particularly powerful jumpers, and tend to go only a few inches with each hop.

The following species have wedge-shaped spades that lay nearly flat (parallel) to the surface of the foot: *Spea bombifrons*, *Spea intermontana*, *Spea hammondii*, and *Spea multiplicata*. The following species have crescent-shaped

The front feet of a Great Basin spadefoot.

The hind foot of a Great Basin spadefoot.

spades, with edges angled perpendicular to foot surface: *Scaphiopus hurterii*, *Scaphiopus holbrookii*, and *Scaphiopus couchii*.

In the hind tracks of *Spea* members, the spade aligns so that its longest side is in line with the length of the foot. Meanwhile, in the hind tracks of *Scaphiopus* members, the spade aligns so that its longest side points across the width of the foot. Always look for the clearest tracks to observe this feature accurately.

The smooth-skinned appearance of the feet and the extensive webbing makes distinguishing spadefoots from true toads via tracks relatively simple, as long as you can find clear tracks. True toads tend to have very bumpy feet, and you can see this in their tracks in fine substrates. They also all have a toe 4 that reaches out well beyond the other toes and is only webbed about midway up the toe.

Desert-dwelling spadefoot species have been shown to emerge and become surface active in response to low-frequency sounds that accompany rainfall, and not necessarily in response to the change in soil moisture (Dodd 2013).

Couch's Spadefoot (*Scaphiopus couchii*)

This is potentially the largest spadefoot species in the United States, with adults growing between 2.25 and 3.5 inches long. Couch's spadefoot inhabits some of the most arid landscapes, including the extreme eastern edge of the Colorado desert, throughout the Sonoran desert, and east into the Chihuahuan  desert. This species is found in habitats including dry washes, shortgrass prairies, creosote flats, open mesquite woodlands, and along the edges of sand dunes.

This species varies somewhat in appearance, but often has dorsal spots or mottling in alternating black and green or yellowish. In some specimens, blotches form into reticulations. Ventral surface is whitish cream in color. Eyes are large, with vertical pupils. No bump (called a "boss") appears between the eyes.

A variety of small invertebrates, such as beetles, mantids, crickets, spiders, grasshoppers, termites, and moths make up the primary diet. These frogs

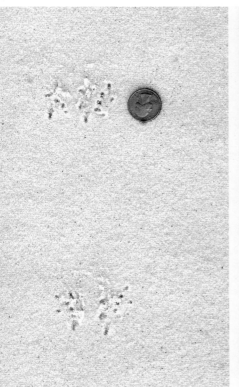 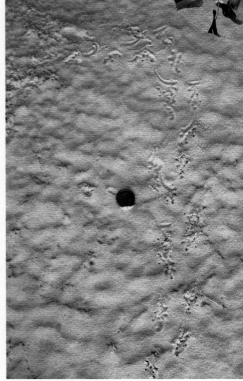

Left: The hopping tracks of an adult Couch's spadefoot. *Right:* The sloppy walking trail of an adult Couch's spadefoot.

typically sit and wait for their prey to come. You'll most likely encounter them during and shortly after desert thunderstorms, which trigger breeding.

Watch for tracks in open, sandy areas during or just after rains. Numerous trails converging will likely lead to a breeding pool. Although this species usually hops, it may also sometimes walk.

Track: Front: 0.8–1.2 in (2–3 cm) x 0.4–0.6 in (1–1.52 cm)

Tracks are small. There are four toes, and the feet are K-shaped overall. All toes tend to register in the tracks. No webbing is present on the front feet. Toes 2 and 4 are similar in length, both shorter than the other toes. Toe 1 can appear thicker, especially in breeding males. During breeding males develop a dry, swollen nuptial pad across much of toe 1 and a small area of toe 2. The palmar tubercle is present, and is relatively flat and wide. No obvious subarticular tubercles are present. The thenar tubercle is present near the base of toe 1, and is relatively flattened. Toe tips are rounded. Front tracks land partially or fully in the space between the hind tracks. Front tracks have the toes pointed toward each other across the central line of the trail.

Hind: 0.9–1.2 in (2.28–3 cm) x 0.4–0.6 in (1–1.52 cm)

Tracks are small. There are five toes, and all tend to register in the track. Toe 4 is the longest. Toe 1 is shortest. There is obvious webbing present and it extends to the tip of each toe. There are no obvious subarticular tubercles present on the toes. There is no visible outer metatarsal tubercle. The inner metatarsal tubercle is a cornified, thick, crescent-shaped spade, its edge angled perpendicular to the foot surface. The spade will align with the longest side pointing across the width of the foot. Hind feet orient directly forward when hopping, but may pitch slightly to the outside when walking.

Track Group Size: 1–1.8 in (2.54–4.57 cm) x 1.7–2.7 in (4.3–6.85 cm)

Trail: Hop   Stride: 2–7.5 in (5–19 cm)

Walk   Stride: 1.5–3 in (3.81–7.62 cm)

Notes:

- This species is often associated with the edge of sand dune systems and in dry washes in California and New Mexico.

- It is an explosive breeder and may be done with reproduction in just a few short nights under the right rainy conditions. This adaptation is vital to making use of the ephemeral pools of water in the arid lands they inhabit.

Eastern Spadefoot (*Scaphiopus holbrookii*)

This familiar spadefoot lives in the eastern United States, but compared to other frogs that share its range, it often passes unobserved because it is nocturnal and seldom seen outside the breeding season. Adults are between 1.5 and 3 inches in length. The eastern spadefoot has a short, blunt snout and large eyes with vertical pupils. The dorsal color is brown with yellow to greenish blotches and two yellowish stripes running parallel down the middle of the back.

It favors habitats with loose, sandy soils, including sandhills, upland hardwoods, pinewood flats, and some disturbed habitats like pastures, pine plantations, and agricultural fields. This species is largely nocturnal and commonly found only during the breeding season between March and July.

The clear track group of an eastern spadefoot. Notice the webbing all the way to the tip of each toe on the hind foot. JOE LETSCHE

Track: Front: 0.4–0.8 in (1–2 cm) x 0.3–0.6 in (0.76–1.52 cm)

Tracks are small. There are four toes, and the feet are K-shaped overall. All toes tend to register in the tracks. No webbing is present on the front feet. Toes 2 and 4 are similar in length, both shorter than the other toes. Toe 3 is noticeably longer than the other toes. Toe 1 can appear thicker, especially in breeding males. During breeding, males develop a dry, swollen nuptial pad across much of toe 1 and a small area of toe 2. Two palmar tubercles are present, one relatively flat and wide, and one small and round, closer to the base of toes 2 and 3. No obvious subarticular tubercles are present. The thenar tubercle is present near the base of toe 1, and is relatively flattened. Toe tips are rounded. Front tracks land partially or fully in the space between the hind tracks. Front tracks have the toes pointed toward each other across the central line of the trail.

Hind: 0.4–0.8 in (1–2 cm) x 0.3–0.5 in (0.76–1.27 cm)

Tracks are small. There are five toes, and all tend to register in the track. Toe 4 is the longest. Toe 1 is shortest. There is obvious webbing present and it extends to the tip of each toe. Subarticular tubercles are present on the toes, but they are not obvious. There is no visible outer metatarsal tubercle. The inner metatarsal tubercle is a cornified, thin, wedge-shaped spade that lays nearly flat to the surface of the foot. This tends to orient vertically on the foot in line with toe 2, similar in appearance to the large inner metatarsal tubercles of *Anaxyrus* species. Hind feet tend to be oriented directly forward when hopping, but may pitch slightly to the outside when walking.

Track Group Size: 1–1.5 in (2.54–3.81 cm) x 1.5–2 in (3.81–5 cm)

Trail: Hop   Stride: 2–5 in (5–12.7 cm)

Notes:

• Look for the extensive proximal webbing to the tips of all the toes on the hind feet and the lack of subarticular tubercles on both front and hind feet to help differentiate this species from *Anaxyrus* toads that share its range.

- Watch for the trails of this species around vernal pools or other temporary wetlands in areas of sandy soil, even in mud puddles in agricultural fields.
- During the breeding season, multiple burrows might be found in sandy areas adjacent to shallow wetland pools.

Great Basin Spadefoot (*Spea intermontana*)

This is the spadefoot species of the dry, intermountain regions of the western United States. It prefers habitats such as shrub-steppe, arid sagebrush plains, open coniferous forests, and shortgrass prairies. It is also found in some agricultural areas. It requires relatively loose, sandy soils to dig into when not surface active.

This species is largely nocturnal and generally only surface active during the breeding season, most often between March and July. Adults feed on ants, beetles, crickets, and grasshoppers. A single adult spadefoot can eat up to 18 percent of its body weight in a single feeding.

Watch for tracks in sandy areas or mud near shallow, rain-filled ponds. Following a trail away from a breeding pool may lead to the dig-in site.

Track: Front: 0.6–1.1 in (1.52–2.79 cm) x 0.4–0.7 in (1–1.78 cm)

Tracks are small. There are four toes, and the feet are K-shaped overall. All toes tend to register in the tracks. No webbing is present on the front feet. Toes 2 and 4 are similar in length, both shorter than the other toes. Toe 3 is noticeably longer than the other toes. Toe 1 can appear thicker, especially in breeding males. During breeding, males develop a dry, swollen nuptial pad across much of toe 1 and a small area of toe 2. Two palmar tubercles are present, one relatively flat and wide, and one small and round, closer to the base of toes 2 and 3. No obvious subarticular tubercles are present. The thenar tubercle is present near the base of toe 1, and is relatively flattened. Toe tips are rounded. Front tracks land partially or fully in the space

The hopping trail of the Great Basin spadefoot.

The hopping trail of a Great Basin spadefoot ending at a dig-in spot.

between the hind tracks. Toes point toward each other across the central line of the trail.

Hind: 0.7–1 in (1.78–2.54 cm) x 0.3–0.5 in (0.76–1.27 cm)

Tracks are small. There are five toes, and all tend to register in the track. Toe 4 is the longest. Toe 1 is shortest. There is obvious webbing present and it extends to the tip of each toe. There are no obvious subarticular tubercles present on the toes. There is no visible outer metatarsal tubercle. The inner metatarsal tubercle is a cornified, thin, wedge-shaped spade that lays nearly flat to the surface of the foot. This orients vertically on the foot in line with toe 2, similar in appearance to the large inner metatarsal tubercles of *Anaxyrus* species. Hind feet tend to be oriented directly forward when hopping, but may pitch slightly to the outside when walking.

Track Group Size: 1.2–2.3 in (3–5.84 cm) x 1.4–2.2 in (3.56–5.59 cm)

Trail: Hop   Stride: 2.3–7.5 in (5.84–19 cm)

Notes:

• You may encounter this species on roads at night during periods of rainfall. I encountered between thirty and forty of them slowly crossing one small quarter mile of roadway in central Washington State. All of them were traveling toward the edge of an irrigated orchard that was being sprayed by a large sprinkler.

• This little spadefoot species tends to hop with short, rapid hops. Its nocturnal trails often lead to a dig-in site or to a daytime shelter under moisture-holding debris.

True Frogs: Family Ranidae

Members of the true frogs found in North America are now described as belonging to the genus *Lithobates*, though many older texts refer to them as all belonging to *Rana*. Twenty-one commonly recognized species of *Lithobates* live in the United States, as well as eight species of *Rana*.

Lithobates frogs tend to hop most of the time, and rarely walk. They are also great swimmers, and many spend a great deal of time in or near water. You're likely familiar with this typical frog's appearance: long, muscular rear legs, pointed heads, sleek bodies, and large eyes. They have long toes on both their front and hind feet, and have especially powerful hind legs. Their hind feet tend to have extensive webbing, though it may be reduced in more terrestrial species.

The tracks of a bullfrog hopping toward the camera, then walking away from the camera.

The large track group from an adult bullfrog in dried mud.

For many years the northern leopard frog (*Lithobates pipiens*) served as a staple animal for dissection and for teaching anatomy. Though still a common frog in much of the northern United States, this species has experienced local extinctions in the westernmost parts of its native range.

This group of frogs acts as a vital link in maintaining healthy ecosystems by consuming many small invertebrates, and being food for many other animals.

The tracks of true frogs typically register with the front tracks partially in between the hind tracks. When outlined, this makes for a roughly rectangular grouping.

Bullfrog (*Lithobates catesbeianus*)

This is the largest native frog in North America, and has been introduced widely outside of its native range to places as far west as California, Oregon, and Washington. It has also been introduced in several of the Hawaiian islands and even as far west as Japan. This species was specifically introduced in many areas as a source of frog legs for human consumption, but has escaped and spread into many new areas. It has, unfortunately, helped speed the decline of other amphibians through direct predation, indirect competition, and sometimes by spreading disease.

Adults are dark green, olive, or brownish and lack dorsolateral folds. Skin may appear mottled in brown or uniform in color. Males have yellowish throats and tympanums larger than their eyes. There is a distinct fold of skin behind and around the tympanum. The loud snoring or bellowing call of large males likely earned this species its name, as the call can resemble the bellowing of a faraway bull.

Adults range in size between 3.5 and 8 inches in total length. This frog is a dominant amphibian predator, and will eat anything small enough to swallow, including other frogs, salamanders, small snakes, large insects, centipedes, spiders, and even sometimes small birds and rodents.

The front foot of a subadult bullfrog.

Left: **The hind foot of a subadult bullfrog, showing the webbing spread.** *Right:* **The track group of a large, adult bullfrog.**

You can often find tracks near the edges of slow-moving streams, ponds, or low-elevation lakes where bullfrogs bask. Because of their size, finding the tracks of a large adult bullfrog can be awe inspiring.

Track: Front: 1.96–2.7 in (4.97–6.86 cm) x 0.86–1.21 in (2.18–3.07 cm)

Small to medium tracks. There are four toes and tracks are K-shaped. Toes are long, tapered, and may show some curving near their tips. The subarticular tubercles are very small and generally don't register in the track. There are two palmar tubercles; the one positioned centrally on the back of the palm area is larger and more oval. The smaller, more oblong palmar tubercle is found closer to the base of toe 4. The front tracks face each other, and are found between or slightly ahead, but to the inside, of the hind tracks.

Hind: 2.25–3.9 in (5.71–9.9 cm) x 1.02–2.5 in (2.59–6.35 cm)

Medium tracks. There are five toes; toe 4 is the longest. Tracks generally angle in the direction of travel, or may angle outward at about 45 degrees. The subarticular tubercles are small and typically do not register in the track except in the finest substrates. The inner metatarsal tubercle is found at the base of toe 1, and is similar in size to the impression made by the tip

Close-up of a walking trail of a bullfrog.

Bullfrog hop trail. SEB BARNETT

of toe 1. Toes are long and tapered and ends are rounded. Webbing is extensive, and toe 4 extends beyond the webbing by about one phalange (Powell, Collins, Hooper Jr. 2012). Hind tracks register with the toes pitching outward away from the center line of the trail up to about 45 degrees.

Track Group Size: 2.36–5 in (5.99–12.7 cm) x 3.42–8.2 in (8.68–20.83 cm)

Trail: Hop   Stride: 3.18–36+ in (8.07–91.44 cm)
        Understep Walk   Stride: 6–9.4 in (15.24–23.88 cm)

Notes:

- Bullfrogs spend significant amounts of time basking in shallow water, on a log, or on the bank. These frogs generally only bask about one jump length from the water's edge. When disturbed, bullfrogs give a characteristic "eerp!" call and dive headfirst into the nearest water.

- In some substrates, only the hind feet will register clearly; sometimes you'll only see the marks made by the toe tips during explosive jumps.

Foothills Yellow-legged Frog (*Rana boylii*)

This small to medium-sized frog ranges between 1.5 and 3 inches in total length as an adult. It's dependent on rocky streams and rivers, and you can find it along coastal California and southern coastal Oregon. It breeds and lays its eggs in these same creeks. The appearance of this frog reflects where it lives—its slightly bumpy skin can come in shades of gray, brown, and red depending on the colors of the rocks in the streams of its habitat. The voice of the male sounds like a brief, guttural growl.

This frog depends on water (or its edge) for food, cover, and practically all of its needs. It feeds on a variety of small invertebrates found in this habitat. Look for tracks in the fine silt along quiet pools and back eddies of rivers and creeks. Foothills yellow-legged frogs generally do not travel very far from the water's edge.

Track: Front: 0.6–0.86 in (1.52–2.18 cm) x 0.2–0.4 in (0.5–1.01 cm)
   Tracks are small. Front feet have four toes, all four of which tend to register in the track. Toe 1 is the longest. Toes 2 and 3 are nearly equal length. Toe 4 is the shortest. All the toes end in bluntly rounded tips. Males have a

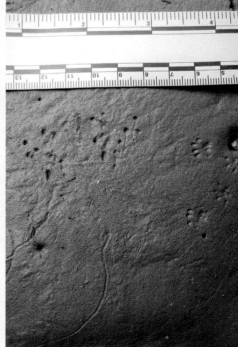

Left: The front foot of an adult foothills yellow-legged frog. *Right:* The tracks of a foothills yellow-legged frog and adjacent deer mouse tracks. This frog was likely a male, as indicated by the swollen pads on toe 1 of the front tracks.

The hind feet of an adult foothills yellow-legged frog.

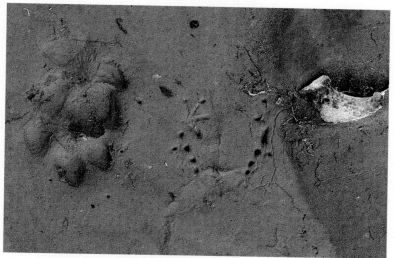

The tracks of a foothills yellow-legged frog next to the direct register tracks of a gray fox.

swollen pad that extends from the base halfway up toe 1. Toe 1 also possesses a moderate subarticular tubercle near the tip, which is noticeably larger than those found on other toes. The other subarticular tubercles are small in size and may or may not register in the track. There is a small, oval-shaped palmar tubercle. The front tracks face each other and generally register partially between the hind tracks.

Hind: 0.7–0.9 in (1.78–2.29 cm) x 0.3–0.5 in (0.76–1.27 cm)

Tracks small. Hind feet have five well-developed toes, and all five tend to register in the tracks. The hind tracks generally angle in the direction of travel, or may angle outward usually at less than 45 degrees. The inner metatarsal tubercle is small, registers about the same size as a toe tip, and is found at the base of toe 1. The fourth toe is the longest. Hind tracks register with the toes pitching outward away from the center line of the trail up to about 45 degrees.

Track Group Size: 0.9–1.5 in (2.28–3.81 cm) x 1.4–1.8 in (3.5–4.57 cm)

Trail: Hop   Stride: 6-12+ in (15.24–30.48 cm)

           Understep Walk   Stride: 2–2.6 in (5.08–6.6 cm)

Notes:

- This species has proportionately thick toes, with relatively blunt tips compared to other *Rana* or *Lithobates* frogs that share its range. The thick proportions are more likely to be confused with *Anaxyrus* toads. Note however, that this species has a small tubercle at the base of toe 1 (which sometimes does not register), more extensive webbing on toe 4, and an otherwise smooth texture to the underside of the feet.

- This stream-dependent species must precisely time its breeding and egg-laying with the diminishment of the spring stream flow. This is problematic on drainages with dams, which alter flow artificially, as poorly timed high flow caused by extensive water release will sweep eggs and tadpoles out of key breeding areas.

- A related species, the Sierra Nevada yellow-legged frog (*Rana sierrae*) is a dweller of high elevations. Both it and the southern mountain yellow-legged frog (*Rana muscosa*) share the foothills yellow-legged frog's foot morphology, including relatively thick toes and round, blunted toe tips. Further scientific research is needed to see how their tracks vary.

Green Frog (*Lithobates clamitans*)

This small to medium-sized frog is common through the eastern half of the United States. Adults measure 2–4.25 inches in total length. Green frogs have two dorsolateral folds that stretch from behind the eye and halfway down the body. There are two color phases or two subspecies: the northern green frog (*Lithobates clamitans melanota*) and the bronze frog (*Lithobates clamitans clamitans*). The northern subspecies is green to olive brown with a somewhat bumpy skin surface. It often has spots between the dorsolateral folds. The head is usually bright green, and it lacks a pale line on the upper lip; the middle of the tympanum is pale yellow or whitish. The male's throat is yellow or light grayish white. The southern subspecies, the bronze frog, is typically bronze or reddish brown. It does not have spots between the dorsolateral folds. The upper jaw is often bright green or sometimes bright bronze.

You can find this frog in a variety of habitats, including small to large ponds, large lakes, woodland pools, streams, swamps, rivers, and streamside riparian areas. This species tends to prefer habitats not strongly affected by agriculture or urbanization. The green frog will stay near the water's edge as an adult, spending time along the banks of various wetland environments where their tracks are most likely to be encountered.

Males call with a sound similar to that made by plucking a loose banjo string. This species will feed on almost any vertebrate or invertebrate they can get in their mouth, and some have been observed targeting the largest prey they can ingest.

Track: Front: 0.78–1.3 in (2–3.3 cm) x 0.3–0.7 in (0.76–1.78 cm)

         Tracks are small. Front feet have four toes, all four of which tend to register in the track. Toe 1 is the longest, and about half the toe tapers

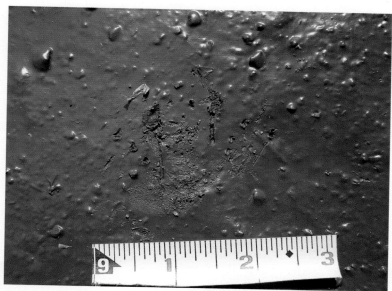

The track group of an adult green frog. GEORGE LEONIAK

significantly, while the base is noticeably thicker, especially in breeding males. Toe 3 is the second longest. Toe 2 is shorter than toe 3. Toe 4 is the shortest. All toes end in bluntly rounded tips. Males have a swollen pad that extends from the base halfway up toe 1. The subarticular tubercles are medium sized and round to pill-shaped, and frequently register in the track. The largest and most deeply registering subarticular tubercle is found on toe 1. The palmar tubercles are very small and may not register in the track. The front tracks face each other and generally register partially between the hind tracks.

Hind: 0.8–1.5 in (2–3.81 cm) x 0.3–0.7 in (0.76–1.78 cm)

The hind feet have five well-developed toes, and all five tend to register in the tracks. The hind tracks generally angle in the direction of travel, or may angle outward at about 45 degrees. The inner metatarsal tubercle is narrow, elongated, and found at the base of toe 1. The fourth toe is the longest. Webbing on toe 4 ends before the last two joints. Hind tracks register with the toes pitching outward away from the center line of the trail up to about 45 degrees.

Track Group Size: 1.3–2.4 in (4.19–6 cm) x 1.9–3 in (4.83–7.62 cm)

Trail: Hop   Stride: 6–24+ in (15.24–60.96 cm)

Notes:

• The front feet of green frogs can be distinguished from other frogs in their range by the especially long toe 1 on the front foot.

• This frog species will often freeze and crouch at the close approach of a predator, or will jump into water with an "arrp!" call. A series of such alarm sounds might indicate the presence of a hunting mink, otter, or raccoon nearby.

Northern Red-legged Frog (*Rana aurora*)

This small to medium-sized frog is the largest native frog in the Pacific Northwest region of the United States. Its range includes southwestern British Columbia, western Washington, western Oregon, and northwestern California, preferring cool, moist coniferous forests.

Adults range in size between 2 and 5 inches in total length. They are typically tan to dark brown, with a pale upper lip line that reaches beyond the tympanum. Black and gray mottling appears in the crouch, with red coloring on the undersides of forearms, rear legs, and abdomen. The male calls with a short series of stuttering knocking sounds while fully submerged, and is difficult to hear except at close range.

You'll be able to observe the northern red-legged frog more easily when it gathers to breed in ponds during the late winter and early spring. Outside of the breeding season, you may see these frogs sitting quietly in the forest understory or rapidly escaping by taking long leaps under the cover of ferns on the forest floor. These frogs are somewhat terrestrial, and you may find them hundreds of meters from the nearest water, especially during periods of rain.

Left: The front foot of a breeding male northern red-legged frog, showing the swollen nuptial pad on toe 1. *Right:* The hind foot of an adult northern red-legged frog. Look close for the subarticular tubercles and the inner metatarsal tubercle at the base of toe 1. Also notice how far the webbing reaches up the length of toe 4.

The track group of an adult northern red-legged frog. The low webbing on toe 4 of the hind feet and the subarticular tubercles are visible.

Track: Front: 0.75–1 in (1.9–2.54 cm) x 0.4–0.6 in (1.01–1.53 cm)

The front feet have four well-developed toes, and all four tend to register in the tracks. The front tracks tend to land directly between the hind tracks or just ahead of them. Front tracks land closer to the center line of the trail than the hind tracks. The subarticular tubercles are small, with the largest one found near the tip of toe 1. Males of this species develop two swollen, aligned pads at the base of toe 1 that reach midway up the toe along the inside. This species does not have an obvious palmar tubercle. There is a single, small round tubercle just above the base of each toe. The toe tips have small, rounded ends.

Hind: 1–1.3 in (2.54–3.3 cm) x 0.5–1 in (1.27–2.54 cm)

The hind feet have five well-developed toes, and all five tend to register in the tracks. The hind tracks generally angle in the direction of travel, or may angle outward at about 45 degrees. There is also a small, round tubercle on the hind foot just at the base of toe 1. Toe 4 is the longest. Webbing on toe 4 reaches only slightly above the third phalange. Hind tracks register with the toes pitching outward away from the center line of the trail up to about 45 degrees.

Track Group Size: 1–2.4 in (2.54–6.1 cm) x 2–3.7 in (5.08–9.4 cm)

Trail: Hop   Stride: 3–16+ in (7.62–40.64 cm)

This species tends to hop most of the time, though it is capable of walking. I've only observed this species walking when a male and female were in amplexus.

Notes:

• This is the classic "forest frog" on the rainy, coniferous Pacific coast. It is usually very alert, and often responds quickly to approach by diving into thick cover or into water.

• This species makes a soft, rhythmic, puttering sound. The males call while fully submerged and resting in the shallows of a pond. A group of males chorusing may be difficult to hear. You'll be able to hear this sound best while sitting still at the edge of a breeding pond.

- You can distinguish the tracks of this frog species from tracks of bullfrogs of a similar size by observing the difference in the webbing on toe 4 on the hind foot. Northern red-legged frogs have more reduced webbing, which may aid in their more terrestrial lifestyle.

Pickerel Frog (*Lithobates palustris*)

This is a small to medium-sized frog found in the eastern half of the United States. Adults grow to a length of 1.5 to 3.5 inches. At first glance, it's easy to mistake the pickerel frog for a leopard frog, which it resembles in general appearance. The pickerel frog's dorsal spots, however, are squared rather than rounded, and tend to appear in two parallel rows down the middorsal region. This species also has orange or yellow coloration on the ventral region, unlike the immaculate white or cream of leopard frogs. There is no pale spot in the center of the tympanum. The male vocalizes with a vibrating growl.

The habitats that this species prefers include lakes and ponds within woodland or forested landscapes, meadows and marshes, and occasionally agricultural fields. These frogs have a strong preference for colder water and spring-fed streams, and because of their sensitivity to various pollutants, also act as good indicators of high-quality wetland habitats (Dodd 2013).

The track group of a pickerel frog. STEVE FORTIN

Hopping trail
of a pickerel frog.

Adult pickerel frogs possess a strongly distasteful skin secretion that repels some predators. The secretions elicit a strong olfactory reaction, and are sometimes described as smelling like a latex glove.

Adult frogs feed largely on a variety of invertebrates, and are themselves preyed upon by mink, a few large fish species, and bullfrogs. Some predators will reject them after grabbing them and finding their taste repugnant.

Track: Front: 0.7–1.3 in (1.78–3.3 cm) x 0.4–0.6 in (1–1.52 cm)

Tracks are small. Front feet have four toes, all four of which tend to register in the track. Toe 1 is the longest. Toes 2 and 3 are nearly equal length. Toe 4 is the shortest. All toes end in bluntly rounded tips. Males have a swollen pad that extends from the base halfway up toe 1. The subarticular tubercles are moderately large and rounded, and frequently register in the track. The palmar tubercles are very small and may not register in the track. The front tracks face each other and generally register partially between the hind tracks.

Hind: 0.8–1.5 in (2–3.81 cm) x 0.5–0.9 in (1.27–2.3 cm)

Tracks are small. The hind feet have five well-developed toes, and all five tend to register in the tracks. The hind tracks generally angle in the direction of travel, or may angle outward, usually at less than 45 degrees. The inner metatarsal tubercle is small and found at the base of toe 1. Toe 4 is the longest. It has webbing that reaches only about halfway up the toe. The subarticular tubercles are medium in size and may or may not register in the track. Hind tracks register with the toes pitching outward away from the center line of the trail up to about 45 degrees. Often, toe 5 is angled outward away from toe 4 at less than 45 degrees.

Track Group Size: 1–2 in (2.54–5.08 cm) x 1.5–2.6 in (3.81–6.6 cm)

Trail: Hop   Stride: 6–24+ in (15.24–60.96 cm)

Notes:

• Use caution if this frog is placed in a container with other amphibians, as its toxic skin secretions can lead to its death as well as that of the other amphibians in the container.

• Like its relative the gopher frog, this species is semiterrestrial. You'll most often encounter its tracks along the edges of cold, spring-fed streams, but you may find it farther from such permanent water sources.

Pig Frog (*Lithobates grylio*)

This is the second-largest true frog in the United States. Adults grow to a length of between 3.25 and 6.5 inches long. Females tend to be larger than males, and males tend to have a bright yellow throat and tympanum larger than the eye. The snout of this species is narrower and more pointed than the bullfrog's (*Lithobates catesbeianus*). The overall colors are similar to the bullfrog, olive or dark green with a few dark spots. Pig frogs are so named because males call with a sound somewhat like the multiple grunts of a pig.

This species is largely aquatic, and in the rare instances when it ventures onto land, it does not wander far from the water's edge. Its highly aquatic nature is reflected in the webbing on its hind feet, which is even more extensive than that of the bullfrog. You'll typically find this species in ponds, lakes, freshwater marshes, wet prairies, sloughs, canals, and sawgrass marshes. It appears to prefer well-vegetated habitats with lots of floating and emergent

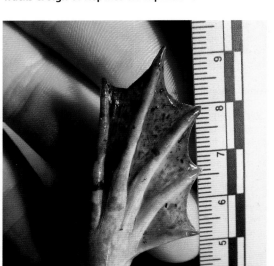

The hind foot of an adult pig frog, showing the extensive webbing reaching to the tip of toe 4.

plants. It can survive in areas where uplands have been altered or disturbed by human activity, as long as there is no pollution runoff from these areas.

Adults feed on many different types of invertebrates, including true bugs, adult dragonflies, and crayfish. In the Everglades, crayfish make up 75 percent of the pig frog's diet (Dodd 2013). Pig frog adults also eat some vertebrates such as smaller frogs, snakes, salamanders, fish, and lizards.

You'll most likely encounter the tracks of this species during low-water periods at the edge of large wetlands.

Hopping trail of a pig frog.

Track: Front: 0.8–2.3 in (2–5.84 cm) x 0.7–1.5 in (1.78–3.81 cm)

Tracks are small to medium in size. There are four toes and tracks are K-shaped. Toes are long, very tapered, and may show some curving near their tips. Toe tips are somewhat more tapered than those of bullfrogs and may appear slightly rounded or pointed. The subarticular tubercles are very tiny and generally don't register in the track. There are typically no obvious palmar pads visible in the track. Front tracks face each other, and are found in between or slightly ahead, but to the inside, of the hind tracks.

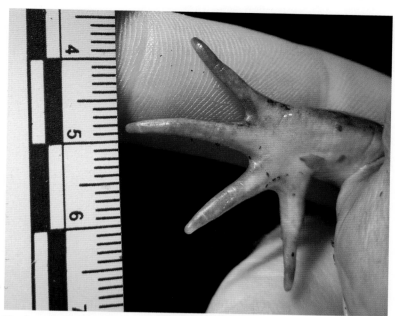

The front foot of an adult pig frog. Notice the very smooth surface relative to other true frogs.

The track group from an adult pig frog in loose sand. Notice the tapered toes of the frog tracks.

Hind: 1.5–2.7 in (3.81–6.86 cm) x 1–2 in (2.54–5.08 cm)

Tracks are small to medium in size. There are five toes, toe 4 being the longest. Tracks generally angle in the direction of travel, or may angle outward at about 45 degrees. The subarticular tubercles are very small and typically do not register in the track except in the finest substrates. The inner metatarsal tubercle is small and found at the base of toe 1. Webbing is extensive, and reaches to the tip of each toe, including the tip of toe 4. Hind tracks register with the toes pitching outward away from the center line of the trail up to about 45 degrees.

Track Group Size: 2–4 in (5.08–10.16 cm) x 2.5–5.2 in (6.35–13.2 cm)

Trail: Hop   Stride: 4–36+ in (10.16–91.44 cm)

Notes:

• This species leaves large tracks, similar to those of bullfrogs with some subtle differences. Look for the more tapered toes, webbing to the tips of all toes on the hind feet, and tiny subarticular tubercles.

• You'll most likely encounter the pig frog's tracks during the dry season or a low-water event. You may confuse the tracks of large adults with those of large bullfrogs.

Southern Leopard Frog (*Lithobates sphenocephalus*)

This familiar southern species of true frog is alert and moves quickly when startled. Typical habitats include shallow wetlands, flooded fields and meadows, marshes, swamps, cattail and sawgrass marshes, beaver ponds, lakeshores, and ditches. This species appears to prefer densely vegetated habitats.

Adults of this species grow between 2 and 4.7 inches long. They have pointed heads and pale dorsolateral folds that run from the back of the eye and end before the vent. The color varies from green to tan, brown, or bicolored. This species tends to have a pale spot in the center of each tympanum. Adults typically have two or three rows of brown spots between the dorsolateral folds with more scattered spots on the sides, and often on the dorsal surface of the front and hind limbs. The ventral surface is creamy white in color. The males call in a series of chuckling sputters.

Southern leopard frogs feed on almost anything small enough that fits in their mouths, including invertebrates such as beetles, flies, and moths, as well as small vertebrates such as frogs, fish, and even salamanders. This species typically hunts in an ambush style, lunging at prey as it comes into range. This frog is both diurnal and nocturnal.

Watch for the tracks of this frog in mud along wetland edges, in puddles in dirt roads, and in silt or dust under bridges.

Hopping trail of
a leopard frog.

The front foot of an adult southern leopard frog showing the tendons that create lines in the tracks.

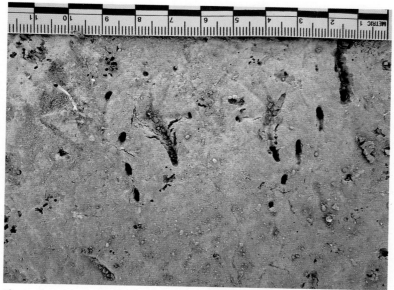

The track group of an adult southern leopard frog. Notice the lines made by the tendons in the front tracks.

Track: Front: 0.78–1.5 in (2–3.81 cm) x 0.43–1 in (1.1–2.54 cm)

Tracks are small. Front feet have four toes, all of which tend to register in the track. Toe 1 is the longest. Toes 2 and 3 are nearly equal length. Toe 4 is the shortest. All the toes end in bluntly rounded tips. Males have a swollen pad that extends from the base halfway up toe 1. The subarticular tubercles are small and may or may not register in the track. The palmar tubercles are very small and may not register in the track. The front tracks face each other and generally register partially between the hind tracks. The thin tendons on the ventral surface of the front feet frequently register for the length of the toes in the tracks.

Hind: 1.3–1.96 in (3.3–5 cm) x 0.6–1.18 in (1.52–3 cm)

Tracks are small to medium. The hind feet have five well-developed toes, and all five tend to register in the tracks. The hind tracks generally angle in the direction of travel, or may angle outward, usually at less than 45 degrees. The inner metatarsal tubercle is small and found at the base of toe 1. The fourth toe is the longest. Toe 4 has webbing that reaches only about halfway up the toe. The subarticular tubercles are small and may or may not register in the track. Hind tracks register with the toes pitching outward away from the center line of the trail up to about 45 degrees. Often, toe 5 angles outward away from toe 4 at about 45 degrees.

Track Group Size: 1.65–2.6 in (4.19–6.8 cm) x 2.44–3.62 in (6.2–9.2 cm)

Trail: Hop Stride: 6–36+ in (15.24–91.44 cm)

Notes:
- This frog may be abundant on rainy nights along roads that run through its prime habitats, such as flooded grasslands or well-vegetated swamps.

- The tracks of this species are some of the most common amphibian tracks you'll encounter in muddy spots throughout its range.
- The northern leopard frog (*Lithobates pipiens*) is a closely related species with a wider range. It has many track and trail features in common with this frog species.

True Toads: Family Bufonidae

You can find a great diversity of species within this family throughout the world. At one time, many toads in North America and Europe were labeled in the genus *Bufo*. The genus found in North America has been renamed *Anaxyrus* from the more well known *Bufo* (Dodd 2013).

Members of *Anaxyrus* tend to have bumpy skin, fairly short limbs, and enlarged parotid glands behind their eyes that contain poison. These toads

The walking trail of an American toad.
KERSEY LAWRENCE

are generally not known to be great jumpers, and most walk or take short hops as they move around. Bumpy tubercles cover the surface of their feet, and the pattern and arrangement of these tubercles can help differentiate similar species.

These toads are more tolerant of drier conditions than many frogs, and they avoid the heat by being largely nocturnal. They will dig daytime retreats in sandy soils or seek cover deep in cool, moist crevices in rock or log piles.

Western Toad (*Anaxyrus boreas*)

The large western toad is arguably one of the most familiar toads. Adults measure between 2 and 5 inches long. It lives in a wide variety of habitats, including coniferous forests, woodlands, moist meadows, grasslands, and along creeks and in riparian areas in drier habitats such as chaparral. You may also find them in suburban parks and large, moist yards.

There are two subspecies of the western toad: the boreal toad (*Anaxyrus boreas boreas*) and the California toad (*Anaxyrus boreas halophilus*). The boreal toad ranges from

Left: The front foot of a large adult female boreal toad. *Right:* The hind foot of a large adult female boreal toad.

The left hind and left front tracks of a walking boreal toad. KIM CABRERA

northern California as far north as Anchorage, along the coast of Alaska. The California subspecies ranges farther south through most of California and is often paler and less blotched.

Adults vary in color from gray to green or brownish, with dark blotches. A thin, pale, middorsal stripe runs from the nostrils to the vent. Adult females grow larger and adult males tend to have fewer blotches and smoother skin. There are no cranial crests in this species. The western toad is both diurnal and nocturnal, becoming mostly nocturnal during warmer and drier periods.

A variety of invertebrates make up this toad's diet, including grasshoppers, flies, beetles, and ants. You can commonly find scats of these toads where they have been feeding, and you can inspect them to glean further details on their local and seasonal diets.

Watch for the tracks of this species on dirt roads, dusty trails, in sandy spots along rivers or creeks, and in mud at shallow wetland sites. This species is more likely to walk than to hop, but it moves regularly both ways.

Track: Front: 1–1.75 in (2.54–4.44 cm) x 0.45–1.23 in (1.14–3.12 cm)

Tracks are medium sized. There are four toes and all tend to register clearly. The front tracks are K-shaped overall. Toe 4 is the shortest. Toe 2 is slightly shorter than toe 3. Toe 1 is the longest. No webbing is present on the front foot. The palmar tubercle is large and ovoid in shape. The thenar tubercle is small, only half the size of the palmar tubercle. Toe tips are bluntly rounded. Subarticular tubercles are moderately large and may be divided, especially on toes 1 and 4. A nuptial pad is present along the inner edge of toe 1 on adult males, and may become prominent enough to register in the track during breeding season. The metacarpal area has granular skin. Front tracks are oriented facing each other directly across

**Boreal toad
walking trail.**
SEB BARNETT

**The track group of a hopping toad,
showing a wide belly drag ahead
of the track group.**

the center line of the trail. The fronts may land to the inside of the hind
tracks during a walk. When hopping rapidly, fronts tend to land ahead
of the hind tracks, but still align somewhat closer to the center line of
the trail.

Hind: 0.6–2.4 in (1.5–6.09 cm) x 0.4–2.05 in (1–5.2 cm)

Tracks are small to medium. There are five toes, and all tend to register
clearly. Hind tracks are asymmetrical. Webbing is present between all toes.
Toe 4 is the longest toe and webbing reaches less than halfway up this toe,
and may sometimes extend to the toe tip as a thin flange on either side of
toe 4. Toe 1 is the shortest toe. Subarticular tubercles are small and widely
spaced. The inner metatarsal tubercle is moderately large, raised, and
ovoid in shape, and is located at the base of toe 5. The outer metatarsal
tubercle is small, raised, and only half the size of the inner metatarsal
tubercle. It is located inset from the outer edge of the foot and in line
with the base of toe 4. Metatarsal area has granular skin.

Track Group Size: 1–3 in (2.54–7.62 cm) x 1.48–4 in (3.759–10.16 cm)

Trail: Hop   Stride: 1.95–8 in (4.95–20.32cm)

      Walk   Stride: 3–5 in (7.62–12.7 cm)

Notes:

- This is one of the most commonly encountered amphibians in the Pacific coast states. Large black schools of its tadpoles are a common sight in many large ponds and lakes.
- This species has declined in parts of its range due to habitat loss, as well as to the introduction of diseases such as chytrid fungus.
- You'll often find the tracks of this species in coastal sand dunes, especially wetland areas at such sites. Watch for the tracks of toads in amplexus around breeding sites in all habitat types.

Red-spotted Toad (*Anaxyrus punctatus*)

This small toad lives in some of the most arid landscapes in the entire United States. You'll find it in oak woodlands, open grasslands, palm oases, and in deserts. Typically, it lives in dry and rocky areas within these habitats.

Adults measure between 1.5 and 3 inches in total length. They are typically tan or grayish with many small, red spots on the dorsal surface of the face, body, and upper limbs. There are no cranial crests in this species.

Adults feed on bees, crickets, ants, beetles, and other small invertebrates.

You'll likely find the tracks of the red-spotted toad in patches of silt along desert creeks or the edges of palm oases. Look for its tracks during periods of rain.

Track: Front: 0.4–0.6 in (1–1.52 cm) x 0.2–0.4 in (0.5–1 cm)

Tracks are small. There are four toes and all tend to register clearly. The front tracks are K-shaped overall. Toe 4 is the shortest. Toe 2 is slightly shorter than toe 3. Toe 1 is the longest. No webbing is present on the front foot. The palmar tubercle is large and ovoid in shape. The thenar

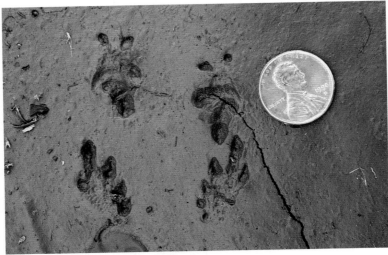

The track group of an adult red-spotted toad. JONAH EVANS

tubercle is small, and less than half the size of the palmar tubercle. Toe tips are bluntly rounded. Subarticular tubercles are small. A nuptial pad is present along the inner edge of toe 1 on adult males, and may become prominent enough to register in the track during the breeding season. The metacarpal area has granular skin. Front tracks are oriented facing each other directly across the center line of the trail. The fronts may land to the inside of the hind tracks during a walk. When hopping rapidly, fronts tend to land ahead of the hind tracks, but still align somewhat closer to the center line of the trail.

Hind: 0.6–0.9 in (1.52–2.29 cm) x 0.3–0.6 in (0.76–1.52 cm)

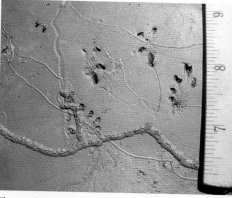

The track group of a red-spotted toad, showing the sink texture on the underside of the feet. CONNOR O'MALLEY

Tracks are small. There are five toes, and all tend to register clearly. Hind tracks are asymmetrical. Webbing is present between all the toes. Toe 4 is the longest toe and webbing reaches less than halfway up this toe. Toe 1 is the shortest toe. Subarticular tubercles are small and widely spaced. The inner metatarsal tubercle is small, raised, and ovoid in shape. It is not much larger than the mark left by the tip of toe 4. The inner metatarsal tubercle is located at the base of toe 5. The outer metatarsal tubercle is small, raised, and about two-thirds the size of the inner metatarsal tubercle. It is located slightly inset from the outer edge of the foot and in line with the base of toe 4. Metatarsal area has granular skin. Hind tracks register with the toes pitching outward away from the center line of the trail up to about 45 degrees.

Track Group Size: 1.5–1.8 in (3.81–4.57 cm) x 1.7–2 in (4.32–5.1 cm)

Trail: Hop Stride: 1.5–4 in (3.81–10.16 cm)

Notes:

• This toad's tracks tend to be stubby, with wide, rounded toe tips and distinctly small inner metatarsal tubercles. This will help you distinguish it from many other toad species.

Southern Toad (*Anaxyrus terrestris*)

The adults of this common toad species grow between 1.75 and 4.5 inches in total length, and have conspicuous cranial crests with a large knob at the posterior end of the parallel interorbital crests. They vary in color from red to grayish or olive green. The parotoid glands can be oval, elongated, or kidney-shaped.

This species lives in a variety of habitats, including seashore scrubs and dunes, pine woods, swamps, wet prairies, and upland forests. You'll also find them in disturbed habitats such as agricultural, suburban, and even urban areas. Adults seek out shallow wetlands, ponds, and even the shallows of large lakes as breeding sites. The southern toad is largely nocturnal.

The track group of a southern toad in loose sand.

It feeds on a variety of invertebrate prey, and tends to hunt using the sit-and-wait ambush style. Snakes, such as hog-nosed snakes (*Heterodon* ssp.), and herons are likely their main predators.

Look for the tracks of this toad in sandy areas of dunes, trails, or sand roads. These toads hop more often than they walk, though they are capable of doing both.

Track: Front: 0.6–1 in (1.52–2.54 cm) x 0.4–0.6 in (1–1.52 cm)
Tracks are small to medium. There are four toes and all tend to register clearly. The front tracks are K-shaped overall. Toe 2 is the shortest, with toe 4 only slightly longer. Toe 1 is slightly shorter than toe 3. Toe 3 is the longest. Toe 1 is the thickest and possesses the largest subarticular tubercles. No webbing is present on the front foot. The palmar tubercle is very large and tear-drop shaped. The thenar tubercle is small, only one-quarter the size of the palmar tubercle. Toe tips are bluntly rounded. The subarticular tubercles are very large. A nuptial pad is present along the inner edge of toe 1 on adult males, and may become prominent enough to register in the track during breeding season. Front tracks are oriented facing each other directly across the center line of the trail. The fronts may land to the inside of the hind tracks during a walk. When hopping rapidly, fronts tend to land ahead of the hind tracks, but still align somewhat closer to the center line of the trail than the hind tracks. The metacarpal area has granular skin.

Southern toad hopping trail.

Hind: 0.8–1.5 in (2–3.81 cm) x 0.4–0.8 in (1–2 cm)

Tracks are medium in size. There are five toes, and all tend to register clearly. Hind tracks are asymmetrical. Webbing is present between all toes. Toe 4 is the longest, and webbing reaches less than halfway up this toe. Toe 1 is the shortest. Subarticular tubercles are very small and widely spaced. The inner metatarsal tubercle is large, raised, and ovoid in shape. The inner metatarsal tubercle is located at the base of toe 5. The outer metatarsal tubercle is small, raised, and only half the size of the inner metatarsal tubercle. It is located inset from the outer edge of the foot and in line with the base of toe 4. The outer metatarsal tubercle is positioned lower on the foot than the inner metatarsal tubercle. The metatarsal area has granular skin. The hind track is generally oriented directly forward or may pitch out up to 45 degrees away from the center line of the trail.

Track Group Size: 1.5–2.3 in (3.81–5.84 cm) x 1.8–2.5 in (4.57–6.35 cm)

Trail: Hop   Stride: 4–6 in (10.16–15.24 cm)

Notes:

• In areas where the substrate is not conducive to leaving tracks, keep an eye out for the large scats of these toads on trails and roadways.

Fowler's Toad (*Anaxyrus fowleri*)

A medium-sized toad of the Midwest and eastern United States, the adult Fowler's toad measures between 1.75 and 3.6 inches in total length. This species is similar to the American toad (*Anaxyrus americanus*), though is smaller on average. Fowler's toads have low parotoid glands, the cranial crest lacks a posterior knob, and the

The track group of an adult Fowler's toad. JOE LETSCHE

Another example of a track group of a Fowler's toad. JOE LETSCHE

postorbital crests directly touch the parotoid glands. They tend to have three to five warts per dark spot. Body color appears tan, reddish brown, or gray.

You can find this species in open habitats that have some sandy or gravelly areas, such as woodlands, lakeshores, sand scrub, beach dunes, and riverbanks. They also inhabit disturbed habitats such as fields and pastures.

Like other toads, this species feeds largely on invertebrates and is beneficial to both gardens and farmlands.

Fowler's toads hop more often than they walk, in relatively short strides.

Track: Front: 0.5–0.8 in (1.27–2 cm) x 0.4–0.6 in (1–1.52 cm)
Tracks are medium in size. There are four toes and all tend to register clearly. The front tracks are K-shaped overall. Toe 4 is the shortest. Toe 2 is slightly shorter than toe 3. Toe 1 is the longest. No webbing is present on the front foot. The palmar tubercle is very large and teardrop shaped. The thenar tubercle is small, only one-quarter the size of the palmar tubercle. Toe tips are bluntly rounded. The subarticular tubercles are very large. A nuptial pad is present along the inner edge of toe 1 on adult males, and may become prominent enough to register in the track during the breeding season. Front tracks are oriented facing each other directly across the center line of the trail. The fronts may land to the inside of the hind tracks during a walk. When hopping rapidly, fronts tend to land ahead of the hind tracks, but still align somewhat closer to the center line of the trail than the hind tracks. The metacarpal area has granular skin.
Hind: 0.4–1 in (1–2.54 cm) x 0.3–0.6 in (0.76–1.52 cm)
Tracks are medium in size. There are five toes, and all tend to register clearly. Hind tracks are asymmetrical. Webbing is present between all toes. Toe 4 is the longest and webbing reaches less than halfway up this toe. Toe 1 is the shortest toe. Subarticular tubercles are very small and widely spaced. The inner metatarsal tubercle is large, raised, and ovoid in shape. The inner metatarsal tubercle is located at the base of toe 5. The outer

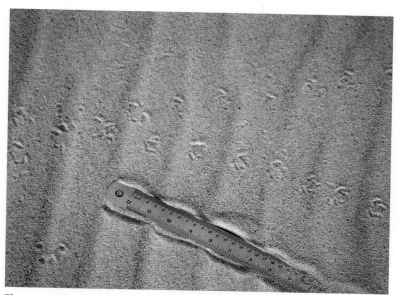

The crossing hopping trails of Fowler's toads. A single set of bounding mouse tracks is in the lower left. JOHN VANEK

metatarsal tubercle is small, raised, and only one-third the size of the inner metatarsal tubercle. It is located inset from the outer edge of the foot and in line with the base of toe 4. Metatarsal area has granular skin. Hind tracks register with the toes pitching outward away from the center line of the trail up to about 45 degrees.

Track Group Size: 0.8–1.5 in (2–3.81 cm) x 1.2–2 in (3–5 cm)

Trail: Hop   Stride: 0.4–4.5 in (1–11.23 cm)

Notes:

• Watch for short hopping trails in sandy areas such as coastal dunes along large lakes and the shorelines of the Gulf of Mexico.

Woodhouse's Toad (*Anaxyrus woodhousii*)

This toad lives in the grasslands of the Great Plains and the landscapes of the semiarid western United States. You'll find it in grasslands, pinyon-juniper woodlands, pine forests, open fields, and even some suburban environments. It tends to be one of the most abundant toads in sandy habitats. Given its preference for hot and dry habitats, it should not surprise you that adults of this species are largely nocturnal; they are, however, active during the day on rainy days. There are two subspecies: the Rocky Mountain toad (*Anaxyrus woodhousii woodhousii*), and southwestern Woodhouse's toad (*Anaxyrus woodhousii australis*).

Adults of this species grow to between 2 and 5 inches in total length. Adults are tan, gray, or olive with a pale dorsal stripe and small, dark-brown blotches. It has a prominent cranial crest, shaped like an L. The cranial

The track group of an adult Rocky Mountain toad.

crests touch the elongated parotoid glands. Males have a dark throat during the breeding season.

The diet of this toad includes a variety of invertebrates, such as centipedes, spiders, true bugs, cicadas, and especially beetles.

Watch for the tracks of this species in sandy areas along rivers, edges of ponds, and in dusty spots under streetlights where adults often gather to feed on insects at night.

Track: Front: 0.9–1.3 in (2.29–3.3 cm) x 0.5–0.7 in (1.27–1.78 cm)
Tracks are medium in size. There are four toes and all tend to register clearly. The front tracks are K-shaped overall. Toe 4 is the shortest. Toe 2 is slightly shorter than toe 3. Toe 1 is the longest. No webbing present on the front foot. The palmar tubercle is very large and teardrop shaped. The thenar tubercle is small, only one-quarter the size of the palmar tubercle. Toe tips are bluntly rounded. The subarticular tubercles are very large. A nuptial pad is present along the inner edge of toe 1 on adult males, and may become prominent enough to register in the track during the breeding season. Front tracks are oriented facing each other directly across the center line of the trail. The fronts may land to the inside of the hind tracks during a walk. When hopping rapidly, fronts tend to land ahead of the hind tracks, but still align somewhat closer to the center line of the trail than the hind tracks. The metacarpal area has granular skin.

Hind: 1–1.5 in (2.54–3.81 cm) x 0.5–0.7 in (1.27–1.78 cm)
Tracks are medium in size. There are five toes, and all tend to register clearly. Hind tracks are asymmetrical. Webbing is present between all the toes. Toe 4 is the longest and webbing reaches less than halfway up this toe. Toe 1 is the shortest. Subarticular tubercles are very small and widely spaced. The inner metatarsal tubercle is large, raised, and ovoid in shape. The inner metatarsal tubercle is located at the base of toe 5. The outer metatarsal tubercle is small, raised, and only one-third the size of the inner metatarsal tubercle. It is located inset from the outer edge of the foot and

in line with the base of toe 4. The metatarsal area has granular skin. Hind tracks register with the toes pitching outward away from the center line of the trail up to about 45 degrees.

Track Group Size: 1.25–3 in (3–7.62 cm) x 0.66–3.42 in (1.68–8.7 cm)

Trail: Hop   Stride: 4.5–9.2 in (11.43–23.37 cm)

Notes:

- This toad often hops, but will also walk. Watch for its tracks and scats where it has foraged at night, especially in sandy areas and on dirt roads.
- Where the range of this toad overlaps with the western toad (Anaxyrus boreas), this species appears in more arid habitats.
- A related species, the Fowler's toad (*Anaxyrus fowleri*) is found farther east all the way to the Atlantic coast and up into parts of New England. The two species overlap in range in eastern Texas, eastern Oklahoma, and in most of Louisiana.

American Toad (*Anaxyrus americanus*)

This is the iconic toad of the Great Lakes, the northern Midwest, and the northeastern United States. Adults grow from 2 to 5 inches in total length. This toad is typically brown, olive, or brick red, with a light stripe down the middle of the back. It has brownish spots and red or brown warts. The belly is typically spotted, unlike the Fowler's toad (*Anaxyrus fowleri*), which shares some of the same range. The cranial crest of adults is separated from the elongated parotoid glands.

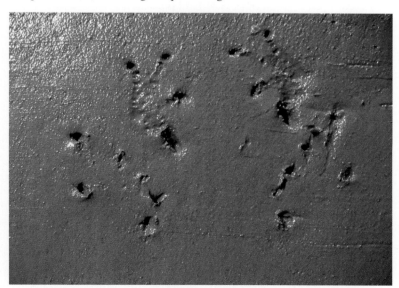

The track group of an American toad showing the details on the underside of the feet. JOE LETSCHE

Another track group from an adult American toad. DAN GARDOQUI

You can typically find this species in habitats such as woodlands, open deciduous forests, and grasslands, but also in suburban parks and even in yards, making the American toad one of the most commonly encountered amphibians.

Adults use both active and ambush styles of foraging for prey. They will eat almost any small animal they can capture, but especially invertebrates such as beetles, caterpillars, moths, and slugs. They benefit yards, gardens, and even agricultural fields as natural pest control.

Watch for the tracks of this species around the edges of ponds, sandy areas around creeks, and in dusty areas around suburban yards.

Track: Front: 0.8–1.3 in (2–3.3 cm) x 0.4–0.8 in (1–2 cm)

Tracks are medium in size. There are four toes and all tend to register clearly. The front tracks are K-shaped overall. Toe 4 is the shortest. Toe 2 is slightly shorter than toe 3. Toe 1 is the longest. No webbing is present on the front foot. The palmar tubercle is very large and teardrop shaped. The thenar tubercle is small, only one-quarter the size of the palmar tubercle. Toe tips are bluntly rounded. The subarticular tubercles are very large. A nuptial pad is present along the inner edge of toe 1 on adult males, and may become prominent enough to register in the track during the breeding season. Front tracks are oriented facing each other directly across the center line of the trail. The fronts may land to the inside of the hind tracks during a walk. When hopping rapidly, fronts tend to land ahead of the hind tracks, but still aligned somewhat closer to the center line of the trail than the hind tracks. The metacarpal area has granular skin.

Hind: 0.6–1.8 in (1.52–4.57 cm) x 0.3–1 in (0.76–2.54 cm)

Tracks are medium in size. There are five toes, and all tend to register clearly. Hind tracks are asymmetrical. Webbing is present between all toes. Toe 4 is the longest and webbing reaches less than halfway up this toe. Toe 1 is the shortest. Subarticular tubercles are very small and widely spaced. The inner metatarsal tubercle is large, raised, and ovoid in shape.

The inner metatarsal tubercle is located at the base of toe 5. The outer metatarsal tubercle is small, raised and only one-third the size of the inner metatarsal tubercle. It is located inset from the outer edge of the foot and in line with the base of toe 4. The outer metatarsal tubercle is positioned lower on the foot than the inner metatarsal tubercle. The metatarsal area has granular skin. The hind track is generally oriented directly forward or may pitch out up to 45 degrees away from the center line of the trail.

Track Group Size: 1.2–3 in (3–7.62 cm) x 1.4–3.5 in (3.56–8.89 cm)

Trail: Hop   Stride: 4–10 in (10.16–25.4 cm)
          Walk   Stride: 3.5–5 in (8.89–12.7 cm)

Notes:

• If you live in this toad's range and want to encourage this voracious pest eater, create some shelters for them to hide in during the day, and avoid spraying your yard or garden with pesticides or herbicides.

• This toad will frequently both hop and walk.

Great Plains Toad (*Anaxyrus cognatus*)

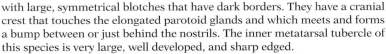

This is a large, very plump toad with relatively short limbs. It lives in grasslands and semiarid to arid scrublands in the Great Plains and southwestern United States. In the Great Plains, it depends on areas that include sand, so that they can dig down below the frost line to overwinter.

Adults grow between 2 and 4.5 inches in total length. Adults are tan, brown, olive green, or gray with large, symmetrical blotches that have dark borders. They have a cranial crest that touches the elongated parotoid glands and which meets and forms a bump between or just behind the nostrils. The inner metatarsal tubercle of this species is very large, well developed, and sharp edged.

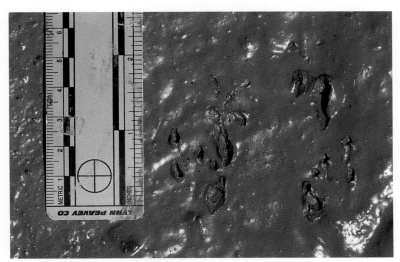

The track group of a Great Plains toad.

The walking trail of a Great Plains toad ending at a dig-in site.

Adults feed on a variety of invertebrates, and are known to be a major predator of cutworms, which damage food crops such as corn, beans, carrots, potatoes, and even turfgrass. This is another amphibian ally of the gardener and farmer.

Watch for the tracks of this species in sandy spots throughout its range. It tends to walk more frequently than it hops. Following a trail may lead to a dig-in spot, where the toad has burrowed down into loose soil for the day.

Track: Front: 1–1.5 in (2.54–3.81 cm) x 0.5–0.7 in (1.27–1.78 cm)

Tracks are medium in size. There are four toes and all tend to register clearly. The front tracks are K-shaped overall. Toe 4 is the shortest. Toe 2 is slightly shorter than toe 3. Toe 1 is the longest. No webbing is present on the front foot. The palmar tubercle is very large and teardrop shaped. The thenar tubercle is small, only one-quarter the size of the palmar tubercle. Toe tips are bluntly rounded. The subarticular tubercles are very large. A nuptial pad is present along the inner edge of toe 1 on adult males, and may become prominent enough to register in the track during the breeding season. Front tracks are oriented facing each other directly across the center line of the trail. The fronts may land to the inside of the hind tracks during a walk. When hopping rapidly, fronts tend to land ahead of the hind tracks, but still aligned somewhat closer to the center line of the trail than the hind tracks. The metacarpal area has granular skin.

Hind: 1–1.5 in (2.54–3.81 cm) x 0.5–0.8 in (1.27–2.03 cm)

Tracks are medium in size. There are five toes, and all tend to register clearly. Hind tracks are asymmetrical. Webbing is present between all toes. Toe 4 is the longest and webbing reaches less than halfway up this toe. Toe 1 is the shortest. Subarticular tubercles are very small and widely spaced. The inner metatarsal tubercle is very large, raised and slanted oval in shape. The inner metatarsal tubercle is located at the base of toe 5. The outer metatarsal tubercle is small, raised, and only one-third the size of the inner metatarsal tubercle. It is located inset from the outer edge of the foot and in line with the base of toe 4. The metatarsal area has granular skin. Hind tracks register with the toes pitching outward away from the center line of the trail up to about 45 degrees.

Track Group Size: 2.3–3 in (5.84–7.62 cm) x 2.3–3 in (5.84–7.62 cm)

Trail: Walk   Stride: 3–5.5 in (7.62–13.97 cm)

Hop   Stride: 2.5–6 in (6.35–15.24 cm)

Notes:

• The Great Plains toad is one of the plumpest toad species in the United States. Some individuals may be so fat that you can't see their legs when viewed from above. Plump adults may leave very wide belly impressions where they stop or when they hop.

Lizards

4

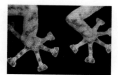

Lizards are the most diverse and, in many places, the most easily observed group of reptiles in North America. Ten families of lizards make up the population found north of Mexico: geckos (Geckkonidae), iguanids (Iguanidae), alligator lizards (Anguidae), Gila monsters (Helodermatidae), California legless lizards (Anniellidae), night lizards (Xantusidae), whiptails (Teiidae), typical Old World lizards (Lacertidae), skinks (Scincidae), and worm lizards (Amphisbaenids).

Scales cover the bodies of lizards, and they generally possess four limbs and a tail. Some species can shed their tail to avoid predators, an ability known as "autotomy." Many lizards have moveable eyelids, external ear openings, and skulls that are more solidly fused than those of snakes. Some lizards also have small bones underneath their scales called "osteoderms," which are a feature they share with crocodilians and which provide them with an added degree of physical protection.

Lizard Locomotion

Limbed lizards move in ways overall akin to mammals of a similar size. In lizard movement, however, the spine flexes side to side rather than up and down as seen in quadruped mammals. This tendency affects how lizards use their limbs, and therefore, how they leave tracks.

Lizards generally move in a quadrupedal walk or trot. The speed of the animal affects its foot placement. There are essentially three main types of walks and two main types of trots (from slowest to fastest): the understep walk, direct register walk, and overstep walk; direct register trot and overstep trot.

Walks

Lizards generally walk when they are relaxed, foraging, or deliberately trying to avoid attention from potential predators. Most lizards will use walks to some degree, though some species do so more than others. For example, more heavily built and wider-bodied lizards tend to walk more frequently.

During an understep walk, the lizard moves in such a way that the tracks of the hind feet register behind the tracks of the front feet. This is the slowest method of locomotion, and is not how most lizards typically move. Some exceptional lizards that frequently use this type of locomotion include: Gila monsters (*Heloderma suspectum*), banded geckos (*Coleonyx* ssp.), leaf-toed geckos (*Phyllodactylus* ssp.), and house geckos (*Hemidactylus* ssp.). Anoles (*Anolis* ssp.) will also commonly use this gait when climbing.

A slightly faster method of locomotion is the direct register walk. In this form of walking, the hind tracks end up partially or completely on top of the front tracks. Many lizard species commonly use this walk, and it is the baseline form of locomotion for all whiptails (*Aspidoscelis* ssp.).

The fastest type of walk is the overstep walk. Here, the hind tracks register just ahead of the front tracks, but not nearly as far as when the lizard does an overstep trot. Many lizards use the overstep walk, including species in the families Iguanidae and Phrynosomatidae. Skinks are likely to use this form of locomotion when moving at higher speeds in the open.

The understep walk of a desert banded gecko. SEB BARNETT

Trots

Lizards are unlikely to exhibit an understep trot, because they trot too fast for their hind feet to land behind their front feet. Even relatively slow trotting lizards, such as the large horned lizard (*Phrynosoma* ssp.), still end up doing either a direct register trot or overstep trot.

Direct register trotting is done by many lizards as they rapidly accelerate from a standstill or a walking gait into a high-speed overstep trot or

The overstep trot of a western fence lizard. SEB BARNETT

The bipedal running trail of an eastern collared lizard. SEB BARNETT

a bipedal run. In this form of locomotion, the hind tracks land completely or partially on top of the front tracks.

The overstep trot is one of the most commonly encountered forms of lizard locomotion. Most lizards are capable of using this method of movement, though some are more likely to sustain it for longer distances. Very fast lizards, such as zebra-tailed lizards (*Callisaurus draconoides*), fringe-toed lizards (Uma ssp.), and desert iguanas (*Dipsosaurus dorsalis*), often use this locomotion form to cross wide, open expanses.

Bipedal Run

Some species can also travel at high speeds in a bipedal run. This true run involves the lizard lifting its front feet completely clear of the substrate as it propels itself with only the power of its hind limbs. Studies conducted by Dr. Bruce Jayne using high-speed video have demonstrated that lizards running in this fashion actually keep their bodies relatively parallel to the ground, largely horizontal rather than sharply angled or vertical. Running using bipedal locomotion allows the lizards to extend their stride length and outpace potential predators or potential prey.

You may confuse this form of locomotion with the direct register trot. Note, however, that the trail width is wider and the stride shorter in the direct register trot.

Specialized Locomotion Forms

A select number of species use specialized locomotion forms. One example is called "sand swimming." This type of movement involves flexing the body in rhythmic sine waves as the lizard moves through loose soils. Lizards known for this are the Florida sand skink (*Neoseps reynoldsi*) and California legless lizard species complex (*Anniella ssp.*). Skinks may also use this form when in the cover of dense grass or leaf litter in order to more rapidly disappear.

Generally, sand swimming creates relatively even sine waves that resemble the trails of very small snakes. The trails, however, are not made on the surface, but just under

This is an example of a Colorado fringe-toed lizard (*Uma notata*) transitioning from a direct register walk to a direct register trot, and finally into a bipedal run. The lizard was moving from the bottom of the photo toward the top.

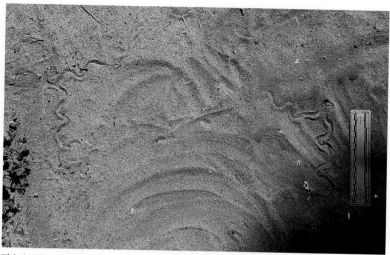

This is an example of a legless trail. The lizard started moving in the lower left, burrowing shallowly. Then, it went deeper and reemerged on the right side of the photo, moved just below the surface toward the top, and then traveled deeper back toward the right.

the surface. As the lizard moves, it creates a temporary tunnel that collapses just behind it. If the lizard moves further down into the substrate, little or no sign may be visible on the surface.

Legless lizard species, such as both California legless lizards and the glass lizards (*Ophiosaurus* ssp.) can also move on the surface of the substrate very similarly to snakes, using a form of lateral undulation. See the species account of the California legless lizards for a photographic example of this type of movement.

Tail and Body Drags

Tail drags are common for some species, and in some cases can be diagnostic. Certain lizards show tail drags most of the time, while others only occasionally register them. Whiptails nearly always leave a thin, continuous tail drag that appears perfectly placed on the center line of the trail. Desert iguanas leave consistent, but gently undulating tail drags. Other species, such as horned lizards, register tail drags most obviously when they use the direct register trot.

Many species drag their bodies as they display to each other, or when they come to a full stop and lower their bellies to the ground. A few lizards, including skinks and alligator lizards, regularly drag their bellies on the ground as they walk. Both of these groups of lizards create wide, obvious belly drags that also include a relatively wide tail drag. Belly drags of skinks tend to undulate, while those of alligator lizards tend to be relatively straight.

Foot Structure

Lizards show a diverse range of foot types and features, everything from the fan-shaped feet of leaf-toed geckos to the narrow and incredibly long hind feet of the zebra-tailed lizard. Most lizards have five toes on both front and hind feet. One exception is the Florida sand skink, which is very unusual in that it has only a single toe on each front limb, and only two toes on the hind limbs.

In order to simplify the description of toes in lizards, I have numbered them following the same formula used throughout this book. I start from the toe closest to the body, toe 1, and move outward to toe 5. Variations in size, length, and placement of toes on the foot are all helpful clues to sorting out different lizard species. The size, shape, and scalation of the metacarpal and metatarsal regions of the feet are also useful. Most lizard species have claws that will register in their tracks. Some have especially long claws or longer claws on certain toes.

Some lizards have specialized features on their feet. True geckos and many anole species possess expanded toe pads covered in microstructures called "lamellae." These toe structures allow them to climb sheer, smooth surfaces, such as waxy leaves and even panes of glass. Fringe-toed lizards have a fringe of scales on either side of the toes on both their front and hind feet. These act as a source of additional surface area that aids in locomotion on loose, sandy substrates. Chameleons have fused toes, with two toes on one side of their foot and three toes on the other. This gives them a very strong grip for holding onto branches, but makes them more awkward on the ground.

Track Morphology

The tracks themselves reflect all the different features of the feet. Even surprisingly subtle features, such as scalation, may be seen in clear tracks through fine substrates. The tracks, however, are not just mere stamps of the feet, but reflect the movement of the animal and its physical interaction with the landscape around it. With changes in speed or behavior, even tracks on a consistent substrate can show considerable variation. This is in part because the feet and their parts will move and change along with changes in behavior.

For example, a slowly walking fringe-toed lizard will leave a series of closely spaced tracks, a wide overall trail width, and will likely include a tail drag or even occasionally a body drag. Its hind feet are likely angled at between 45 and 90 degrees away from the center line of the trail. Most of the foot should register, including the metatarsal and metacarpal regions. These track details might be even more emphasized if the lizard is arching its back and displaying as it walks. As the lizard increases speed into a trot, several changes will occur. The trail width will narrow, the distance between tracks—the measurement known as the stride length—will increase, and a smaller portion of the feet will actually touch down and register. The tracks will be less distinct, which in loose substrate may no longer show easily distinguishable toes. The front track may appear as a single mark, while the hind track will appear as two marks made up of one mark of toes 1–4, and one mark made by toe 5.

The toes of lizards are also somewhat flexible and mobile, and may show some degree of variation in exact position in the tracks. There are consistent tendencies, however, even in species with very long toes. For instance, the tracks of trotting spiny lizards tend to show curvature at the toe tips of toes 1–4 angled toward the center line of the trail. The same can be said for the toe tips in the tracks made by walking collared and leopard lizards. The longest toe on the hind tracks of whiptail lizards often reaches significantly further than the other toes, and tends to curve near the tip and form a J shape. Sometimes, this toe may curve drastically, even to the extent of pointing backward. Look closely at the relationship of the toes, as well as the negative space between them. You can find information about the animal not only in the marks made directly by the animal's feet, but also in the space in between them.

Under certain conditions, distinguishing species by the tracks can be very challenging. In order to increase confidence in identification, an observer must study a variety of tracks of a single species under different conditions including different speeds and substrate conditions.

Ecological Clues and Interspecies Relationships

Although some lizards are more general in their habitat preferences, most are tied to specific ecological conditions, such as particular soil types or habitat structures and the shelter and prey they attract. Observing and recording these details can help you sort out what possible species are present in a given area.

As an example of using such macro- and microhabitat clues, consider a group of lizard species found together in parts of the Mojave Desert of

California. Picture a hillside strewn with granite boulders leveling out into a flat, open valley covered in creosote bushes. In this area, there may be anywhere from eight to thirteen species of lizard sharing one relatively small area of desert. Each can exist in this area because of a particular niche in the habitat, whether it's the relationship to a particular habitat element, or because the particular lizard is active during a certain time of day. You can find desert banded geckos (*Coleonyx variegatus*) on both the rocky hillside and in the valley below it. Although this species has less specific habitat preferences compared to other lizards in this same area, they are nocturnal, so they do not compete directly with any of the diurnal lizards and avoid completely those lizards that might make a meal of them. If large yucca or Joshua trees are present in this area, you may also find desert night lizards (*Xantusia vigilis*). They limit their activities almost exclusively to areas in and around the fallen limbs of yuccas, Joshua trees, and other large desert plants.

Another small lizard, the side-blotched lizard (*Uta stansburiana*) also calls the Mojave Desert home. Although a habitat generalist as well, it prefers areas that include at least small rock piles for thermoregulation. The long-tailed brush lizard (*Urosaurus graciosus*), a small relative of the side-blotched lizard, lives among the limbs of the bushes in the valley and goes largely unnoticed. This species prefers to stick close to desert shrubs, especially creosote bushes. It may occasionally dash from one bush to another, but you won't find it on the ground for very long or see it travel very far.

Also among the branches of the creosote bushes is a much larger lizard, the mostly herbivorous desert iguana (*Dipsosaurus dorsalis*). It is no threat to the brush lizard. It is comfortable on the ground, but will climb into desert shrubs to feed or seek shelter. Its wandering trails are commonly seen between the creosote bushes.

In small or large wind-blown dunes in this same valley, you might spot the medium-sized Mojave fringe-toed lizard (*Uma scoparia*). This lizard is very quick and alert, feeding on a variety of sand-dwelling invertebrates and occasionally small lizards. It hunts largely by sight.

At the edge of this sandy area and in the gravelly wash that comes down from the hillside, hunts the very fast zebra-tailed lizard (*Callisaurus draconoides*). This speedster is built to dash rapidly across the ground to escape predators or grab small prey. It is well acclimated to the open habitats, though it does not have the feet specially adapted for more open sand dunes as its relative, the fringe-toed lizard.

Near the base of the hill, watching from a large boulder pile is a chuckwalla (*Sauromalus ater*). This large, wide-bodied herbivore prefers not to stray far from the relative safety of its home here or in other rock crevices. Its habitat does, however, overlap somewhat with the desert iguana, in that they prefer similar plant species.

A bleached earless lizard warming in the light of the setting sun.

Sharing both the hillside and valley habitat is the tiger whiptail (*Aspidoscelis tigris tigris*), which walks around in a herky-jerky manner and seeks its prey both by sight and scent, unlike the other species mentioned here. It stops frequently to dig out hidden prey buried just under the surface, which is invisible and unavailable to the other lizards sharing its habitat.

Among the boulders and nearby small trees, you'll find two species of spiny lizards: the western fence lizard (*Sceloporus occidentalis*) and the larger yellow-backed spiny lizard (*Sceloporus uniformis*). Both might be found sharing the same general area; however, the larger species is more dominant. The yellow-backed spiny lizard also ranges down into the valley, away from the boulders, where they are often found in areas with yuccas and Joshua trees.

Higher up on the hillside, a Great Basin collared lizard (*Crotaphytus bicinctores*) watches for the movement of prey in all directions from atop a boulder. This is one of the top predators of the lizard community, and will potentially eat any other lizard (or vertebrate) it can capture and fit in its mouth. Below, down in the valley, its relative—the long-nosed leopard lizard (*Gambelia wislizenii*)—plays a similar role, preying on any lizards or other prey it can fit in its mouth. If these two species come in contact, they will compete directly for prey, or one might end up eating the other. So, generally, they avoid each other by occupying different habitats.

Notice the interplay of species mentioned above. Each fits into a particular place, tends to seek out specific microhabitats within each place, or has a particular behavioral strategy for survival that is different from the others. Knowing this not only enriches your understanding of these animals, but also can help you identify their tracks with greater accuracy. In areas rich in lizard diversity, such as the warmer southern portions of the United States, knowing such details is essential to improving tracking knowledge.

Alligator Lizards: Family Gerrhonotinae

Lizards in this family have relatively long bodies and short limbs, with moderate to long tails. Their hides are covered in scales that barely overlap one another. The stiff, rectangular dorsal and ventral scales are supported by underlying small bones known as "osteoderms." You can easily identify these lizards by noting the presence of a distinctive fold running along either side of their bodies. This fold is full of small, granular scales and is a vital point of flexibility so that the bodies of these armored creatures can expand and contract as they breathe. The flexibility also allows room for large meals and for developing eggs.

If you've ever tried to pick up one of these lizards, you might already know they have some very effective antipredator defenses. The first is that they are capable of delivering a surprisingly tenacious bite. They also will often attempt to smear their captor with their own feces. Finally, their tail can break off, but it generally does not grow back to the original length. The broken tail will writhe and flail around, distracting the predator long enough for the rest of the lizard to make a discreet escape. Southern alligator lizards have been observed to grab their own tail in response to approaching predatory snakes. In this circular shape, the lizard is more difficult to swallow.

At least three species—the Texas alligator lizard (*Gerrhonotus infernalis*), Panamint alligator lizard (*Elgaria panamintina*), and southern alligator lizard (*Elgaria multicarinata*)—have been observed to use their long tails as an additional limb while climbing.

Alligator lizards generally feed on a variety of small animals, such as insects, arachnids, and other invertebrates. The larger specimens may also feed on other lizards and even small mammals. All species are capable swimmers, and some may seek prey in shallow pools.

Southern Alligator Lizard (*Elgaria multicarinata*)

This might be the most familiar alligator lizard to many, as it is the one species whose range includes some of the most densely populated areas on the West Coast of the United States. You can spot it relatively often even in the suburban and urban yards of Los Angeles and the Bay Area.

It also grows to be the largest and longest species of alligator lizard north of Mexico, with some adults growing up to about 16 inches in total length.

You can distinguish adult males from females by their head shape—males have a broader and more triangular head. This gives them a stronger bite, which you can easily experience if you handle them.

Track: Front: 0.4–0.6 in (1–1.52 cm) x 0.4–0.7 in (1–1.78 cm)
    Tracks are small. There are five toes present and generally all register in the track. The front track is fan shaped and relatively symmetrical overall. Toes 3 and 4 are nearly the same length. Toe 2 is shorter than toe 3. Toes 1 and 5 are also similar in length. Claws are present on all toes and moderate in size. The metacarpal region of the foot is covered in granular scales, which are largest toward the back of the foot and progressively smaller as they near the base of the toes. Scales on the underside of the toes are short and wide, and form lines across the toes. The front track registers closer to the center of the trail. These tracks tend to be oriented directly forward or may

The typical understep walking trail of an adult southern alligator lizard.

pitch slightly toward the center of the trail.

Hind: 0.7–1 in (1.78–2.54 cm) x 0.6–0.8 in (1.52–2.03 cm)

Tracks are small to medium. There are five toes present and generally all register in the track. The hind track is strongly asymmetrical. Toe 4 is the longest. Toe 1 is the shortest. Toes 5 and 3 are similar length. Toe 2 is shorter than toe 3. There is a significantly large negative space between toes 3 and 4— as well as between toes 4 and 5—when compared with the negative space between the other toes. This significant negative space may create an angle between toes 3 and 4, and toes 4 and 5, of between 45 and

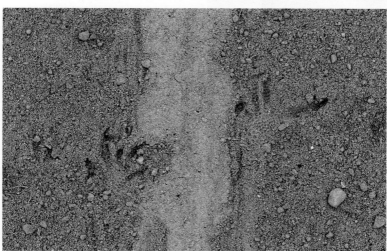

The close-up of a left front and right hind foot in the walking trail of a southern alligator lizard. Notice the scalation and the angles and proportions of the toes.

90 degrees. Claws are present on all toes and are of moderate size. The metatarsal region of the foot is covered in granular scales, which are largest toward the back of the foot and get progressively smaller as they near the base of the toes. Scales on the underside of the toes are short and wide, and form lines across the toes. The hind track registers more to the outside of the trail, and tends to be oriented directly forward or pitch slightly outward away from the center of the trail.

Trail: Understep Walk   Trail Width: 1.5–2 in (3.81–5.08 cm)

                            Stride: 3–3.5 in (7.62–8.89 cm)

          Belly Drag: 0.5–1 in (1.27–2.54 cm)

          Tail Drag: 0.25–0.5 in (0.63–1.27 cm)

Notes:

- This common lizard leaves very distinct trails that often include significant tail and belly drag. They are usually slow-moving animals, and travel often with an understep walk. If exposed they may speed up into awkward overstep trot. They might also choose to tuck their legs and use a kind of lateral undulation movement when rapidly sliding into cover.

- The delicate details of scalation in the tracks of this species can be observed in very fine substrates.

- This feisty species is quite capable of fending off a variety of predators. Along with the behaviors mentioned earlier, this alligator lizard species may grab its tail to form a loop when attacked by a snake. This causes it to become an awkward mouthful that might be too large for a snake to handle. If the snake

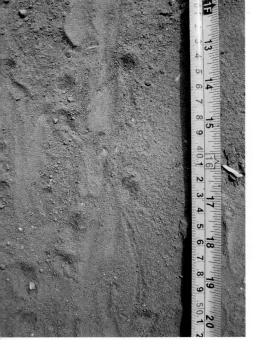

continues to walk its mouth around the lizard, it might just end up with only the tail breaking off in its mouth, allowing the lizard to escape.

- This species includes three recognized subspecies north of Mexico: the Oregon alligator lizard (*E. multicarinata scincicauda*), California alligator lizard (*E. multicarinata multicarinata*), and San Diego alligator lizard (*E. multicarinata webbii*).

Northern Alligator Lizard (*Elgaria coerulea*)

This lizard species ranges far north in North America. It ranges from the Sierra Nevada and the Bay Area in northern California to south-central British Columbia and southern Vancouver Island. The northern alligator lizard can tolerate colder temperatures than most other lizards, with

The walking trail of a southern alligator lizard with only three legs.

individuals having been observed being surface active on cool, moist days with air temperatures at 50 degrees.

Like the southern alligator lizard, adult males can be distinguished from females by their broader and more triangular head.

Track: Front: 0.25–0.5 in (0.63–1.27 cm) x 0.25–0.5 in (0.63–1.27 cm)

Tracks are small. There are five toes present and all generally register in the track. The front track is fan shaped and relatively symmetrical overall. Toes 3 and 4 are nearly the same length. Toe 2 is shorter than toe 3. Toes 1 and 5 are also similar in length. Claws are present on all toes and are moderate in size. The metacarpal region of the foot is covered in granular scales, which are largest toward the back of the foot and progressively smaller as they near the base of the toes. Scales on the underside of the toes are short and wide, and form lines across the toes. The front track registers closer to the center of the trail. These tracks tend to be oriented directly forward or may pitch slightly toward the center of the trail.

Hind: 0.3–0.5 in (0.76–1.27 cm) x 0.3–0.45 in (0.76–1.14 cm)

Tracks are small to medium. There are five toes present and generally all register in the track. The hind track is strongly asymmetrical. Toe 4 is the

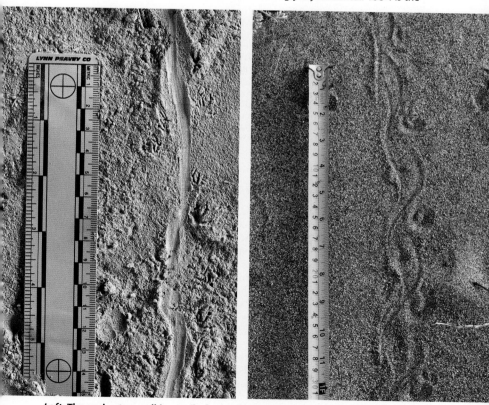

Left: The understep walking trail of a northern alligator lizard. *Right:* The rapid, direct register walk of a northern alligator lizard creating a beautiful sinuous trail.

longest. Toe 1 is the shortest. Toes 3 and 5 are similar length. Toe 2 is shorter than toe 3. There is a significantly large negative space between toes 3 and 4—as well as between toes 4 and 5—when compared with the negative space between the other toes. This significant negative space may create an angle between toes 3 and 4, and toes 4 toe 5, of between 45 and 90 degrees. Claws are present on all toes and are of moderate size. The metatarsal region of the foot is covered in granular scales, which are largest toward the back of the foot and progressively smaller as they near the base of the toes. Scales on the underside of the toes are short and wide, and form lines across the toes. The hind track registers more to the outside of the trail, and tends to orient directly forward or pitch slightly outward away from the center of the trail.

Trail: Understep Walk   Trail Width: 1–1.5 in (2.54–3.81 cm)
                                   Stride: 1.8–2.5 in (4.57–6.35 cm)
       Direct Register Walk   Trail Width: 1.6–2.1 in (4.06–5.33 cm)
                                   Stride: 2.5–3.7 in (6.35–9.4 cm)
       Belly Drag: 0.3–0.6 in (0.76–1.52 cm)
       Tail Drag: 0.2–0.4 in (0.5–1 cm)

Notes:

• This may be the only lizard species you come across in some parts of its range in the cool, moist, and forested Pacific Northwest regions.

• This species is rather shy, and is often quick to slip away from and out of sight of observers. You can observe it most easily on warm, sunny days around objects that act as heat traps, such as stone walls, rock piles, or other similar types of cover. You can invite it to yards in its range by supplying warm cover.

• This alligator lizard species does not lay eggs; rather, it carries its eggs inside its body and the young emerge fully developed upon birth—this is known as "ovipary." This provides a reproductive edge, especially in the cool habitats in much of its range. Since the female carries young in her body, she can actively seek out patches of warmth and not have to rely purely on the temperature conditions at a fixed nesting site.

Family Dactyloidae

Northern Green Anole (*Anolis carolinensis*)

This is a small to medium-sized lizard, green or brownish in color, and familiar to many. Its native range covers eastern Texas east to the Florida Keys, and north to the Carolinas. Some small populations have been introduced outside of its native range, including several in southern California.

These lizards have long, pointed heads, moderate bodies, and long, thin tails. The scales of these lizards are small and granular. Males and females are similar in appearance, but males have an extendable pink dewlap. This dewlap is used in combination with head bobbing and push-ups to intimidate rivals or impress females.

Green anoles can rapidly change color from green to brown, or vice versa. Fighting males often turn bright green and may have dark patches behind their eyes. These color changes may occur due to fluctuations in humidity, mood, temperature, and health. Because this species can rapidly change color, the misnomer of "American chameleon" has stuck. Anoles are not, however, closely related to chameleons.

This lizard is a popular species in the pet trade, and this is likely how new populations have spread outside of its native range. Given the proper conditions, it is relatively easy to keep them in captivity. This species is also easily stressed, however, and tends to be flighty, so it's not a good pet choice for those who like to frequently handle their lizards.

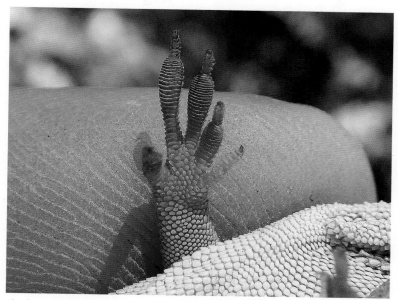

The front foot of the northern green anole. JASON KNIGHT

The hind foot of the northern green anole. JASON KNIGHT

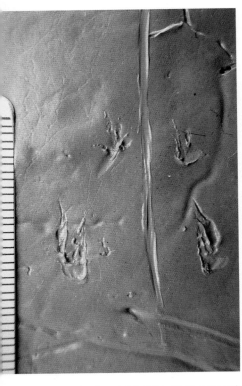

The track group of a hopping green anole, including the tail drag.
JASON KNIGHT

Anoles feed on invertebrates such as spiders, flies, moths, beetles, and even sometimes small crabs. They seek out their prey most often among foliage of shrubs, trees, and on walls. These lizards are able to climb even sheer vertical surfaces because lamellae are present on expanded pads located on each toe. These pads appear one-third of the way down from the toe tip rather than at the tip of the toes. Like some gecko species, these anoles can climb vertically or upside down, and can even scramble up smooth surfaces like panes of glass.

This species generally prefers areas that include dense, shady vegetation and a moist microclimate. Watch for their tracks in dusty or sandy areas near such vegetated sites. If you are lucky and find a set of tracks from this lizard crossing an area of fine, moist clay, you might be able to even see an impression of the lamellae.

Track: Front: 0.3–0.5 in (0.76–1.27 cm) x 0.2–0.4 in (0.5–1 cm)

Tracks are small. There are five toes present and all register in the track. In firm substrates, only the three middle toes tend to register. Front tracks are asymmetrical. Toe 4 is the longest toe, and along with toe 3, has the largest scansor pads. Toe 3 is slightly shorter in length. Toes 2 and 5 are similar in length, though toe 2 has a larger scansor pad. Toe 1 is the shortest and possesses little or no lamellae and no enlarged scansor pad. Toes 1 and 5 are attached on roughly the same plane on the front foot. Claws are small, relatively short, and may register immediately at the toe tips as tiny dots or very thin, short lines. The metacarpal region of the foot is small, and covered in small, granular scales. The front tracks tend to register oriented forward, in the direction of travel and parallel to the center of the trail or pitched slightly toward the center. They fall closer to the inside of the trail than the hind tracks.

Hind: 0.6–0.8 in (1.52–2 cm) x 0.2–0.5 in (0.5–1.27 cm)

Tracks are small. There are five toes present and all register in the track. The hind track is asymmetrical. Toe 4 is the longest, and along with toe 3, has the largest scansor pads. Toes 2 and 5 are similar in length, though toe 5 has a wider area of lamellae, and the same area on toe 2 containing lamellae is longer. Toe 1 is the shortest and possesses little or no lamellae and no enlarged scansor pad. There is a significant gap between toes 4 and 5. Toe 5 is the lowest toe on the hind foot. Claws are small, relatively short, and may register immediately at the toe tips as tiny dots or very thin, short lines. The metatarsal region of the foot is moderate, narrow,

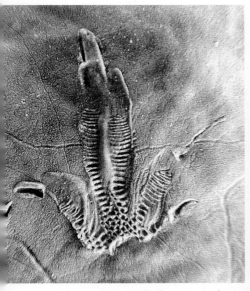

Left: The close-up of a front track of a green anole showing the scale details. JASON KNIGHT

Right: The close-up of a hind track of a green anole. JASON KNIGHT

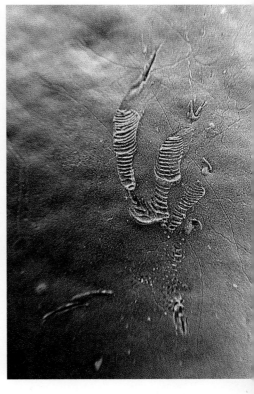

and covered in small, granular scales. The hind tracks tend to register oriented outward at about 45 degrees away from the center line of the trail. They fall further from the inside of the trail than the front tracks.

Trail: Trot   Trail Width: 1.2–2.4 in (3–6 cm)
                Stride: 1.4–3.4 in (3.56–8.64 cm)
       Hop   Track Group Size: 1.3–1.5 in (3.3–3.81 cm) x 1.3–1.5 in (3.3–3.81 cm)
                Stride: 1.2 in, 3.5 in, 10 in, 4.3 in
       Tail Drag: 0.1–0.3 in (0.25–0.76 cm)

       This species possesses a stiff tail, so the tail drag tends to be thin and straight and most obvious when the lizard is walking or hopping, but it will also register to a lesser degree during a trotting gait. During a trot, the tail will leave a series of broken segments composed of relatively straight lines.

Notes:

• Brown anoles have very few lamellae on their toes, compared to the native green anole (*Anolis carolinensis*). Therefore, the tracks of the Cuban brown anole (*Anolis sagrei*) can be distinguished by their narrower, more tapered toes.

• Green anoles are also active, display to each other frequently, and are impressively acrobatic. They are a great choice of species to observe for lizard watchers and naturalists in general.

Banded Geckos (Genus *Coleonyx*)

This small group of lizards is limited to the southwestern United States, and includes four different species. They have large eyes with vertical pupils. Genus *Coleonyx* belongs to the family Eublepharidae, which differs from geckos (family Geckkonidae) in that they lack adhesive toe pads and also have true eyelids that allow them to blink and close their eyes (Jones and Lovich 2009).

Unlike many lizards that live in arid and semiarid regions, banded geckos are largely nocturnal. They have soft, delicate skin covered in small granular scales. Two species, the reticulate banded gecko (*Coleonyx reticulatus*) and the Switak's banded gecko (*Coleonyx switaki*), have enlarged tubercles interspersed among their smaller, more uniform granular scales.

The legs of these geckos are long and thin, and the feet have slender toes that end in small claws. The fifth toe on the hind foot is angled sharply backward. The relatively fat tail has a constriction at the base, which aids in allowing the lizards to shed their tails in the face of predators. They are able to grow back their tails, but the new tails will never include the same pattern of markings as the original.

Although banded geckos are capable of short bursts of speed, they are generally slow moving when compared to many diurnal lizards. You can often observe them waving their tails when they walk or sometimes even when they stalk prey. Their prey includes invertebrates such as spiders, scorpions, and ground insects.

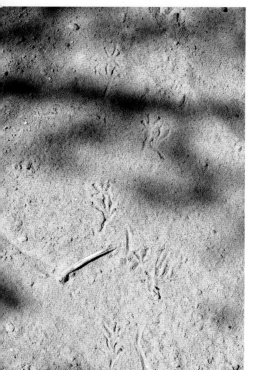

Banded geckos can vocalize, and sometimes squeak or hiss softly when handled or distressed. They also might suddenly leap into the air from your hand if you're holding them. You should handle them near the ground to reduce the risk of injury or stress that might cause them to lose their tails.

Despite banded geckos being largely restricted to desert regions in the United States, they are also prone to high rates of water loss through evaporation. They avoid this by being active at night when humidity is higher and temperatures are cooler. They seek shelter during the day in burrows, rock piles, and other types of

The close-up understep walk of a desert banded gecko.

cover that contain more humid conditions. These lizards may gather in pairs or small groups of up to eight, which some research suggests may help reduce evaporative water loss (Jones and Lovich 2009).

Western Banded Gecko (*Coleonyx variegatus*)

This is the most widespread, and most commonly encountered, of the banded gecko species. It ranges east from the southern coast of California through the Mojave Desert, and on through the Colorado and Sonoran Deserts. Then it ranges as far east as southeastern Arizona and extreme southwestern New Mexico.

This species has a variety of dorsal patterns of dark to somewhat diffuse

Another understep walking trail of a desert banded gecko.

Banded gecko understep walking trail.
SEB BARNETT

bands, blotches and spots. Observing several individuals from a relatively small area of desert can reveal a surprising variety of patterns. Generally, however, most individuals will have a background color that closely matches the ground color of the dominant substrate in the area.

Some of the best places to look for their tracks are in sandy areas between large rock formations in desert washes or in foothills. In sand dune habitats, you can most likely observe their tracks in flat, less exposed areas between dunes. Their meandering trail might lead you along the path of their night-time hunting foray, or back to one of a variety of daytime retreats.

Track: Front: 0.2–0.3 in (0.5–0.76 cm) x 0.3–0.4 in (0.76–1.02 cm)

Tracks are very small. There are five toes present and all register in the track. Toe 3 is the longest. Toe 1 is the shortest. Toes 1 and 5 directly oppose each other on the foot at about 180 degrees. Toes 2 and 4 are similar in length. Front tracks are relatively symmetrical and fan shaped. Claws are present and register as enlarged areas that are part of the toe tip. The metacarpal region of the foot is very small, and covered with fine, granular scales. Front tracks tend to orient directly in line with the direction of travel or may pitch slightly inside toward the center line of the trail. Front tracks also tend to fall more to the inside of the trail relative to the hind tracks.

Hind: 0.4–0.5 in (1.02–1.27 cm) x 0.4–0.5 in (1.02–1.27 cm)

Tracks are small. There are five toes present and all register in the track. Toes 4 and 5 are nearly the same length and are the longest toes. Toe 5 is positioned lowest on the foot, and generally registers pointing backward relative to the direction of travel. Toe 3 is slightly shorter than toe 4. Toe 2 is slightly shorter than toe 3. Toe 1 is the shortest. Claws are present and register as enlarged areas that are part of the toe tip. The metatarsal region is relatively small and covered in fine, granular scales. Hind tracks tend to land more to the outside of the trail relative to the fronts. Hind tracks are oriented in the direction of travel or pitch slightly away from the center line of the trail.

Trail: Understep Walk   Trail Width: 1.2–1.4 in (3.0–3.55 cm)
Stride: 1.8–2.1 in (4.57–5.33 cm)

This species general uses an understep walk as a preferred type of locomotion. Banded geckos are capable, though, of short bursts at faster gaits, such as a direct register trot, and they can also hop.

Notes:

• This species is arguably one of the most beautiful desert life forms, with its strongly contrasting banded and blotched patterns, delicate and semi-translucent skin, and mesmerizing eyes. It is also relatively common—you can see them on warm nights in their preferred habitats, or during a careful drive on a desert roadway. They can appear almost white against a blacktop road at night.

• Banded geckos typically lay a pair of oblong, leathery eggs in late spring to late summer. Females may lay up to three clutches.

• Banded gecko males and females are somewhat sexually dimorphic. At the base of their tails, male banded geckos have short cloacal spurs, one on each side. Switak's banded gecko adult males also develop bright yellow body coloration during the breeding season.

Family Anniellidae

This is the North American legless lizard family. It includes a former species called the California legless lizard (*Anniella pulchara*) which has since, through genetic research, been reclassified as a species *complex* that includes five similar-looking species. Further research may uncover more species to add to the complex. At the time of this writing, the five recognized

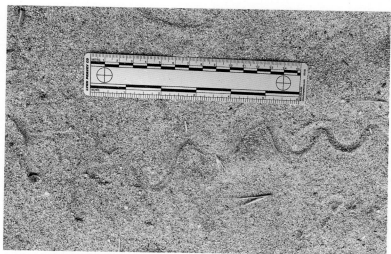

The slow-moving trail of a southern California legless lizard.

The tunneling trail of a southern California legless lizard.

The lateral undulation trail made by a rapidly escaping southern California legless lizard.

species are: the northern California legless lizard (*Anniella pulchara*), Temblor legless lizard (*Anniella alexanderae*), southern Sierra legless lizard (*Anniella campi*), Bakersfield legless lizard (*Anniella grinnellii*), and southern California legless lizard (*Anniella stebbinsi*).

The information given here is based on observations of specimens in the wild from the southern California legless lizard (*Anniella stebbinsi*). Given the general similarity of physical build and behavior shared by all of the species in this complex, it is likely that they all create a nearly identical trail pattern.

These legless lizards are small, slender, cylindrical lizards of 7 to 10 inches in total length. This species has a wedge-shaped head, no external ear openings, small eyes, and a countersunk lower jaw.

Their bodies are covered in small, rounded, tightly fitting and incredibly smooth scales. The dorsal surface is tan, gray, silvery, or blackish. A thin black line runs from the back of the head down the center of the back to near the tip of the tail. One or several equally thin black lines run along the sides and divide the dorsal surface from the yellowish ventral surface. The tail is moderate in length and blunted at the tip. It can be shed easily when grabbed by a predator or handled by a human.

Though these lizards can appear to look like very small snakes, several features make them distinctly different. They can blink their eyes. They also have no enlarged belly scales (scutes), which makes them different from all but the blind snakes.

Legless lizards eat insects, especially termites, beetles, caterpillars, and some spiders. They live in subterranean environments, and are only found on the surface under cover objects. They require warm, loose soil for their burrowing activity. When in loose soil and near the surface, they leave easily recognizable trails that are created by the collapsing tunnels as they pass.

Since their trails are very similar to those of snakes, they are most easily described and measured using the same techniques.

Trail: Sine Wave Tunnel   Overall Trail Width: 1–1.5 in (2.54–3.81 cm)
   Wave Length: 1–2.6 in (2.54–6.6 cm)
   Inner Trail Width: 0.3–0.5 in (0.76–1.27 cm)

The trail of this species is small, often showing a repeating sine wave pattern. The inner trail width is proportionately wide and fairly consistent, as the trail is not created by the same process as a snake traveling on the surface. Rather, the trail is formed by a collapsing tunnel; snakes push down and sideways or backward.

Lateral Undulation   Overall Trail Width: 2–2.3 in (5.08–5.84 cm)
Wave Length: 3–3.2 in (7.62–8.13 cm)
Inner Trail Width: 0.3–0.8 in (0.76–2 cm)

These legless lizard species can also use a form of lateral undulation when they have to move across a relatively firm surface that does not allow them to tunnel. This is not typical, but you may occasionally see this activity on the surface of an area with very firm sand. Note that this animal is not as efficient at this movement as a snake, and will likely switch to tunneling as soon as it has the option to do so.

Notes:

• Watch for the tracks of these species along the edges of shrubs in areas of sandy soil. They might even appear in large suburban yards with the appropriate soils and some adjoining native habitat. In this context they should be encouraged, as they feed on insects that might act as garden pests.

Family Iguanidae

Desert Iguana (*Dipsosaurus dorsalis*)

This lizard is found in some of the most extremely arid and hot desert regions in the western United States. It can tolerate incredibly high temperatures, and its activity has been observed at temperatures well over 110 degrees Fahrenheit. Robert C. Stebbins recorded body temperatures of greater than 116 degrees Fahrenheit, which is the highest temperature ever recorded for a wild terrestrial vertebrate.

This species is largely herbivorial, feeding on a variety of desert plants, especially creosote bushes (*Larrea tridentata*). These iguanas will also eat some insects, feces, and carrion.

Desert iguanas are moderately large, and pale gray to whitish in color. The body is covered in pale dots, and is scrawled lengthwise with small, reddish-brown lines. Small rows of dots encircle the long, round tail. The pattern and color of this species helps it blend perfectly with the mottled shadows underneath and within the cover of creosote bushes.

These lizards are territorial during the breeding season, but social and tolerant outside of it. Aggressive males chase and bite each other. They also engage in battles during which they face opposite directions and lash each other with their long tails.

They generally move in a direct register or overstep walk. When they need to, these lizards can move rapidly to escape danger. They are also skilled climbers, and may be found several feet off the ground in shrubs, feeding on leaves and flowers.

Track: Front: 0.6–1 in (1.52–2.54 cm) x 0.6–7 in (1.52–1.78 cm)

Tracks are medium in size. There are five toes present and all register in the track at slower speeds. At higher speeds and in loose substrates, only toes 2, 3, and 4 may register, or all toes might register as a single track. Tracks are not quite symmetrical, with toes 3 and 4 being the longest. Toe 1 is the shortest. Toes 2 and 5 are about equal in length. Claws are well developed, and are proportionately long. The claws tend to register most clearly at slow speeds. The metacarpal region of the foot is covered in

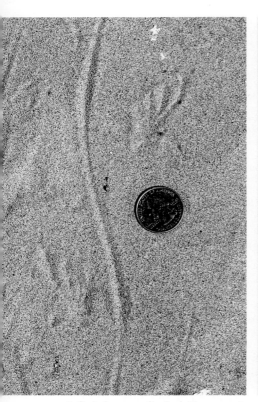 

Left: A close-up of a desert iguana using a direct register walk. *Right:* The bipedal running trail of a desert iguana.

small, keeled, diamond-shaped scales. The front track registers directly forward, or may occasionally pitch slightly toward the center of the trail.

Hind: 1.4–2 in (3.56–5.08 cm) x 0.9–1.3 in (2.29–3.3 cm)

Tracks are medium in size. There are five toes present and all tend to register in the track. Tracks are asymmetrical. Toe 4 is the longest. Toe 3 is about one-third shorter than toe 4. Toes 2 and 5 are about the same length. Toe 1 is the shortest. Claws are well developed and present on all the toes. The longest claw is present on toe 5. The metatarsal region is relatively long, narrow, and covered in keeled, diamond-shaped scales. The largest scales are at the base of the metatarsal area and the smallest are at the base of the toes. The hind tracks register directly forward, or may pitch outward away from the center of the trail up to 45 degrees or even up to 90 degrees at very slow speeds.

Trail: Overstep Trot Trail Width: 2–3.1 in (5.08–7.87 cm)
                        Stride: 10–11 in (25.4–27.94 cm)
       Overstep Walk Trail Width: 2.7–3.5 in (6.86–8.89 cm)
                        Stride: 3–6 in (7.62–15.24 cm)
       Tail Drag: 0.3–0.9 in (0.76–2.29 cm)

Tail drags of this species tend to gently curve, as they walk with the tail in a twisting motion. Tail drags are generally only visible during a walking gait, as the animal lifts it clear of the substrate during more rapid locomotion.

Notes:

- This is a common lizard of open, arid landscapes, and may be the most abundant lizard species where conditions are optimal. Look for them and their sign in sandy spots around the bases of creosote bushes.
- This lizard species is capable of bipedal running to escape potential predators.
- The desert iguana also has a strong relationship to its burrows, seeking shelter in them from heat, cold, and predators. It will either dig itself or adapt from other animals such as kangaroo rats. When near a burrow, it will often run to its entrance and pause there before finally disappearing underground.

Chuckwalla (*Sauromalus ater*)

These impressively large lizards are a familiar sight to many who have spent time around rock outcrops in the Mojave or Sonoran Deserts of the southwestern United States. Males are territorial, and actively defend their preferred sites against other intruding males. A single male's territory will overlap with that of multiple females. Males tend to tolerate females and young. You can often observe this social species in family groups.

This species tends to color match the habitats where it is found; you'll see darker individuals in lava bed areas, and lighter individuals in granite and sandstone outcrops.

Track: Front: 0.9–1.5 in (2.28–3.81 cm) x 1–1.5 in (2.54–3.81 cm)

Tracks are medium in size. There are five toes present and all tend to register in the track. The front track is relatively symmetrical and fan shaped overall. Toe 3 is the longest, and toe 4 is only slightly shorter. Toes 2 and 5 are roughly the same length. Toe 1 is the shortest and positioned lowest on the foot. Claws are well developed, moderately long, and can be blunted by constant wear on abrasive rocks. They register well in tracks, and might be the most obvious feature in the tracks in firmer substrates.

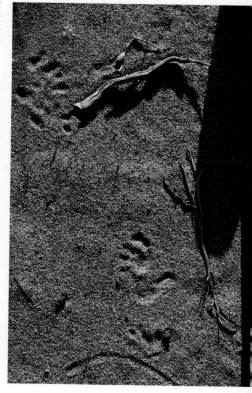

The overstep trotting tracks of a large adult male chuckwalla.

Another set of overstep trotting tracks of a chuckwalla.

Chuckwalla over-step walk trail.
SEB BARNETT

The metacarpal area is relatively large and covered in diamond-shaped scales. This whole region tends to register most obviously at slower speeds. Front tracks tend to pitch overall toward the inside of the trail at up to about 45 degrees.

Hind: 1.5–2 in (3.81–5.08 cm) x 1.1–1.6 in (2.8–4.06 cm)

Tracks are medium in size. There are five toes present and all tend to register in the track. The hind track is asymmetrical and similar in shape to other members of the family Iguanidae. Toe 4 is the longest, while toe 3 is only slightly shorter. Toe 2 is slightly shorter than toe 3. Toes 1 and 5 are similar in length. Toe 5 is positioned lowest on the foot. Claws are well developed, moderately long, and can be blunted by constant wear on abrasive rocks. The claws register well in the tracks, and might be the most obvious feature in the tracks in firmer substrates. The metatarsal area is relatively large, and is covered in diamond-shaped scales. This whole region tends to register most obviously at slower speeds and in softer

A windblown walking trail of a large chuckwalla showing the wide tail drag.

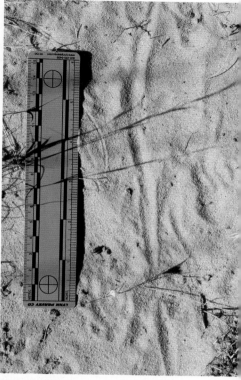

substrates. Hind tracks tend to face directly forward or pitched slightly to the outside away from the center of the trail.

Trail: Overstep Trot Trail Width: 4–5.6 in (10.16–14.22 cm)
Stride: 6–12.5 in (15.24–31.75 cm)

Tail Drag: 0.5–1 in

Tail drag may or may not be present. The tail drag of walking chuckwallas generally appears as a continuous line down the center of the trail. In gravelly soils, the continuous tail drag might be the only obvious part of the trail visible. When trotting, the tail is usually lifted clear of the substrate, though it might occasionally make contact off to one side of the trail.

Notes:

• This rock-dwelling species is rarely found far from its rock piles and nearby crevice safe havens.

• Chuckwallas generally move slowly, and even at top speed, they're much slower than many other desert dwellers that share their range and

The feeding sign of a chuckwalla on the buds of a desert shrub.

habitats. Watch for them basking on top of rock piles on warm summer days. If a predator approaches too closely, they will dash into a crevice and inflate their bodies, making them very difficult for predators to extract.

- You should look for chuckwalla tracks in areas where you find fine sand between rock piles or cliff-face crevices. Abundant scat is easier to spot in these locales, and concentrations of it indicate a preferred haunt of this charismatic species. Look for areas with fresher, greener scats to find these lizards. Old scats often desiccate to a pale white in the desert sun.
- This species is largely herbivorous and can sometimes be observed feeding actively several feet up in a shrub.

Family Crotaphytidae

Crotaphytidae includes both collared lizards and leopard lizards. North of Mexico, this includes eight species: the Great Basin collared lizard (*Crotaphytus bicinctores*), eastern collared lizard (*Crotaphytus collaris*), Sonoran collared lizard (*Crotaphytus nebrius*), Baja California collared lizard (*Crotaphytus vestigium*), reticulated collared lizard (*Crotaphytus reticulatus*), blunt-nosed leopard lizard (*Gambelia sila*), Cope's leopard lizard (*Gambelia copeii*), and long-nosed leopard lizard (*Gambelia wislizenii*). All have granular scales covering their bodies, and their color patterns match their surroundings.

Collared lizards are named for the two black rings that appear around the neck, and which frame a pale sandy or white ring. Baja collared lizards may have only a partial or totally absent posterior collar. Leopard lizards lack collars completely, and are named for the black spots and stripes found on their dorsal surface.

Relative to their bodies, both groups of lizards have proportionally large heads with strong, muscular jaws. Males have visibly larger and more muscular heads. The large jaw muscles allow both sexes to tackle a wide variety of prey species, including beetles, grasshoppers, crickets, spiders, and wasps. They also eat vertebrates, especially small lizards (sometimes including their own species), rodents, birds, and small amounts of vegetable matter such as flowers and leaves. Lizards belonging to this group tend to be sit-and-wait predators, watching from a perch for the approach of potential prey. Then they swiftly dash in to capture prey in their muscular jaws. They are the top lizard predators in the locations where they occur.

All Crotaphytidae species have very long tails, long legs, and well-developed feet. The muscular rear legs allow them to move with surprising speed, and they can switch from a rapid four-legged trot to an even faster bipedal run. This rapid run allows them to efficiently capture prey and to escape potential predators.

Approaching these lizards for close observation may be very difficult, as some will flee while an observer is still fifty to one hundred feet away. In contrast, a few individuals may appear incredibly comfortable or indifferent to close approach, even to within one or two feet.

When walking, these lizards tend to create a thin, relatively straight tail drag. They walk to explore, most often close to some kind of cover, such as rock piles or shrubs. They tend to trot when moving across large open areas, when feeling threatened, or when chasing potential competitors.

Eastern Collared Lizard (*Crotaphytus collaris*)

This collared lizard is the most wide ranging—you can find it anywhere from Arkansas and southern Missouri west to western Arizona. This species is also a popular pet.

These collared lizards can be impressively colorful in some regions, particularly the males, which can have brilliantly green bodies, legs, and tails crossed by yellowish stripes and spots. Females tend to be a more subdued green and tan. During the breeding season, females show reddishorange markings on their neck and sides. This might be an indicator that they are gravid.

Track: Front: 0.5–1 in (1.27–2.54 cm) x 0.5–0.8 in (1.27–2.03 cm)

Tracks are medium in size. There are five toes present and all register in the track at slower speeds. At higher speeds and in loose substrates, only

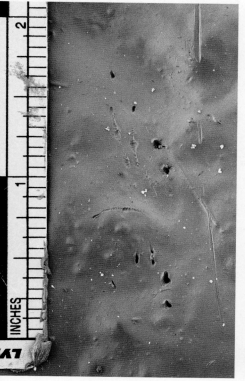

Left: The front and hind tracks of an eastern collared lizard in mud showing the fine scalation. *Right:* The direct register walk of an adult eastern collared lizard.

toes 2, 3, and 4 may register, or all toes might register as a single track. Tracks are not quite symmetrical, with toes 3 and 4 being the longest. Toe 1 is the shortest. Toes 2 and 5 are about equal in length. Claws are well developed, and are longest on toes 3 and 4. The claws tend to register most clearly at slow speeds. The metacarpal region of the foot is covered in small, keeled, diamond-shaped scales. The front track tends to register facing directly forward, or may occasionally pitch slightly toward the center of the trail.

Hind: 1.3–2 in (3.3–5.08 cm) x 0.7–1 in (1.78–2.54 cm)

Tracks are medium in size. There are five toes present and all toes tend to register in the track. Tracks are asymmetrical. Toe 4 is the longest. Toe 3 is about one-third shorter than toe 4. Toes 2 and 5 are about the same length. Toe 1 is the shortest. Claws are well developed and present on all the toes. The longest claw is present on toe 5. The metatarsal region is relatively long, narrow, and covered in keeled, diamond-shaped scales. The largest scales are at the base of the metatarsal area and the smallest are at the base of the toes. The hind tracks tend to register facing directly forward, or may pitch outward away from the center of the trail up to 45 degrees.

Left: The bipedal running trail of an eastern collared lizard. *Right:* The ventral view of an eastern collared lizard showing the foot structure and line of femoral pores.

Trail: Direct Register Trot Trail Width: 2.1–2.8 in (5.33–7.1 cm)
                                    Stride: 12–14 in (30.48–35.56 cm)
             Direct Register Walk Trail Width: 2.6–3.2 in (6.6–8.13 cm)
                                    Stride: 4.5–6.6 in (11.43–16.76 cm)
             Bipedal Run   Trail Width: 2–2.3 in (5.08–5.84 cm)
                                    Stride: 16–20 in (40.64–50.8 cm)
Tail Drag: 0.2–0.4 in (0.5–1 cm)
Tail drags of this species tend to be narrow and straight, and fall in the center of the trail. You'll generally only encounter tail drags during a walking gait, as the tails are lifted clear of the substrate during more rapid locomotion.

Great Basin Collared Lizard (*Crotaphytus bicinctores*)

This species of collard lizard is found furthest west and furthest north. It ranges into Kern County, California, and north into northeastern Oregon and southwestern Idaho. It also lives in the Mojave and Great Basin desert regions.

As with other collared lizard species, males tend to be more colorful than females. They tend to be greenish olive to grayish, with scattered pale spots that coalesce together on the hind legs and tail, creating a pale background with brownish or greenish spots. Some males show bands of yellow across the dorsal surface of the body. Females tend to be paler, in shades of tan or greenish gray.

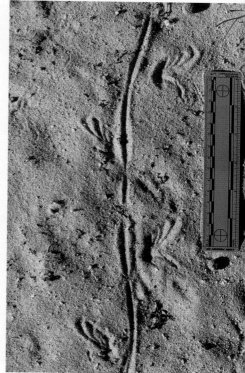

Track: Front: 0.5–0.8 in (1.27–2.03 cm) x 0.4–0.7 in (1–1.78 cm)

Tracks are medium in size. There are five toes present and all register in the track at slower speeds. At higher speeds and in loose substrates, only toes 2, 3, and 4 may register, or all toes might register as a single track. Tracks are not quite symmetrical, with toes 3 and 4 being the longest. Toe 1 is the shortest. Toes 2 and 5 are about equal in length. Claws are well developed, and are longest on toes 3 and 4. The claws tend to register most clearly at slow speeds. The metacarpal region of the foot is covered in small, keeled, diamond-shaped scales. The front track tends to register facing directly forward, or may occasionally pitch slightly toward the center of the trail.

The direct register walk of a Great Basin collared lizard.

Hind: 1.3–2 in (3.3–5.08 cm) x 0.7–1.5 in (1.78–3.81 cm)

Tracks are medium in size. There are five toes present and all tend to register in the track. Tracks are asymmetrical. Toe 4 is the longest. Toe 3 is about one-third shorter than toe 4. Toes 2 and 5 are about the same length. Toe 1 is the shortest. Claws are well developed and present on all the toes. The longest claw is present on toe 5. The metatarsal region is relatively long, narrow, and covered in keeled, diamond-shaped scales. The largest scales are at the base of the metatarsal area and the smallest are at the base of the toes. The hind tracks tend to register facing directly forward, or may pitch outward away from the center of the trail up to 45 degrees.

Trail: Direct Register Trot   Trail Width: 2.1–2.8 in (5.33–7.1 cm)
                                  Stride: 12–14 in (30.48–35.56 cm)
          Direct Register Walk   Trail Width: 3.2–3.7 in (8.13–9.4 cm)
                                  Stride: 6–7.3 in (15.24–18.54 cm)

Tail Drag: 0.2–0.4 in (0.5–1 cm)

Tail drags of this species tend to be narrow and straight, and fall in the center of the trail. Tail drags are generally only visible during a walking gait, as they are lifted clear of the substrate during more rapid locomotion

Notes:

• This species lives in some of the most arid landscapes in the western United States. You'll most likely encounter it in areas with rocky slopes that contain an abundance of prey, such as grasshoppers or small lizards.

• I have observed this species dig shallow burrows that serve as places to escape when pursued by predators, or perhaps to get away from extreme heat. The burrows were located underneath boulders, mostly horizontal in orientation, and were generally only about 12 inches in total length.

• In areas where this species occurs on black lava beds, it may be very dark in color. In general, the color and pattern will match the local substrate.

• This species has a tendency to be particularly wary of close approach. Advance in a slow, indirect path while keeping one eye on their body position at all times. With a little persistence, you'll be able to closely observe your specimen.

Long-nosed Leopard Lizard (*Gambelia wislizenii*)

This species generally lives in wide-open, sparsely vegetated, and relatively flat habitats. It appears in all four major desert regions in the United States.

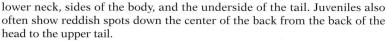

They tend to be pale whitish tan or gray overall. On top of that are darker spots and pale lines that cross in transverse orientation across the body, legs, and tail. Males and females are similar in color except during the breeding season. When gravid, females show reddish spots or blotches along the lower neck, sides of the body, and the underside of the tail. Juveniles also often show reddish spots down the center of the back from the back of the head to the upper tail.

Track: Front: 0.6–0.9 in (1.52–2.29 cm) x 0.6–0.8 in (1.52–2 cm)

Tracks are medium in size. There are five toes present and all five register in the track at slower speeds. At higher speeds and in loose substrates, only toes 2, 3, and 4 may register, or all toes might register as a single track. Tracks are not quite symmetrical, with toes 3 and 4 being the longest. Toe 1 is the shortest. Toes 2 and 5 are about equal in length. Claws are well

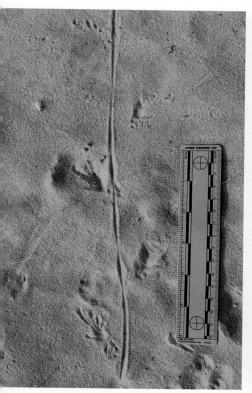

Left: **The direct register walk of a long-nosed leopard lizard.** *Right:* **The overstep trotting trail of long-nosed leopard lizard transitioning to a bipedal run near the middle of the image.**

developed, and are longest on toes 3 and 4. The claws tend to register most clearly at slow speeds. The metacarpal region of the foot is covered in small, keeled, diamond-shaped scales. The front track tends to register facing directly forward, or may occasionally pitch slightly toward the center of the trail.

Hind: 1.2–1.5 in (3–3.81 cm) x 0.8–1 in (2–2.54 cm)

Tracks are medium in size. There are five toes present and all five toes tend to register in the track. Tracks are asymmetrical. Toe 4 is the longest. Toe 3 is about one-third shorter than toe 4. Toe 2 and toe 5 are about the same length. Toe 1 is the shortest. Claws are well developed and present on all the toes. The longest claw is present on toe 5. The metatarsal region is relatively long, narrow, and covered in keeled, diamond-shaped scales. The largest scales are at the base of the metatarsal area and the smallest are at the base of the toes. The hind tracks tend to register facing directly forward, or may pitch outward away from the center of the trail up to 45 degrees.

Trail: Overstep Trot   Trail Width: 1.8–2 in (4.57–5.08 cm)
                              Stride: 10–12 in (25.4–30.48 cm)

Direct Register Walk Trail Width: 2.3–3 in (5.84–7.62 cm)
Stride: 5.4–6.2 in (13.72–15.75 cm)
Overstep Walk Trail Width: 2.5–3.3 in (6.35–8.38 cm)
Stride: 7.5–7.8 in (19–19.8 cm)
Bipedal Run Trail Width: 1.5–2 in (3.81–5.08 cm)
Stride: 14–19 in (35.56–48.26 cm)

Tail Drag: 0.1–0.3 in (0.25–0.76 cm)

Tail drags of this species tend to be narrow and straight, and fall in the center of the trail. Tail drags are generally only visible during a walking gait, as they are lifted clear of the substrate during more rapid locomotion

Notes:

- Leopard lizards are fast, wary animals that usually don't allow close approach. They tend to rapidly run for cover and hide in deeply shadowed areas under shrubs. Their broken, pale pattern with subtle dark spots and mottling make them nearly invisible in this setting.
- This particular lizard species has the incredible ability to eat lizards of nearly its own size. This is helped by its long head and powerful jaws.
- Unlike their collared lizard relatives, leopard lizards do not appear to be territorial. Rather, they seem to share overlapping home ranges (Jones and Lovich 2009).

Family Helodermatidae

This family includes the only two highly venomous lizards in North America: the beaded lizard (*Heloderma horridum*) and the Gila monster (*Heloderma suspectum*). These large lizards have bead-like, bone-reinforced scales, large heads with well-developed jaw muscles, and venom glands in the lower jaw. They have thick, fat-storing tails and well-developed limbs with fan-shaped feet. These generally slow animals tend to walk most of the time.

Gila Monster (*Heloderma suspectum*)

This famous lizard is the only venomous lizard found in the United States. It is the largest native lizard species, growing up to around 22 inches in total length. The Gila monster has an easily recognizable appearance, with a bright orange to yellow background broken by patterns of black bands and blotches on its body, tail, and legs. The mouth, nose, and cheeks are black while the top of the head is the pinkish, yellow, or orange, similar to the general color of the body. The highly contrasting coloration is likely aposematic, acting to warn predators about this species' potent venomous bite.

The head, body, limbs, and tail are covered in large granular scales that resemble beads. These are reinforced under the skin with small bones called "osteoderms," providing armor-like protection against would-be predators.

Gila monsters move slowly and will not act defensively unless harassed or picked up. Their strong venom is potentially dangerous to humans. These rather shy and beautiful animals should be treated with respect and left unmolested.

The range of the Gila monster includes much of southern and western Arizona, the southern tip of Nevada, a few mountain ranges in southeastern

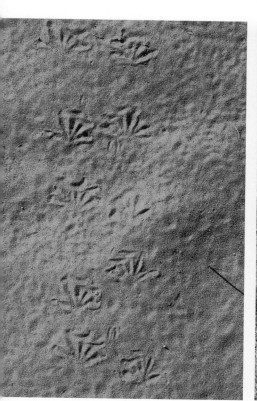

Left: The distinct overstep walking trail of a Gila monster. CAMERON ROGNAN *Right:* The walking trail of a Gila monster showing a wide tail drag.

The front and hind track of a Gila monster showing the scalation pattern.

California, the extreme southwestern tip of Utah, and southwestern New Mexico. They actively forage on bird and reptile eggs, as well as juvenile mammals such as rabbits and squirrels. The venomous bite aids in subduing mammalian prey.

Gila monsters have a relatively high rate of water loss to evaporation, especially for a desert-dwelling lizard. Like desert tortoises (*Gopherus agassizii*), these lizards can store water in their urinary bladder to help delay dehydration (Jones and Lovich 2009). They may also shift to being more active at night during the hotter times of the year, or especially during periods of high humidity before, during, and after rain events. They spend inactive periods in burrows, deep in rock crevices, or in other similar shelters.

These large, slow-plodding lizards leave relatively large tracks that you can easily recognize. Their front and hind feet are fan shaped, somewhat similar to a human hand. Their trails usually show tracks arranged in pairs, with a front track from one side and a hind from the other side, similar to those paired tracks left by a raccoon (*Procyon lotor*), using a 2 x 2 walking gait. Watch for their trails in areas of fine sand such as washes and riparian areas through desert canyons.

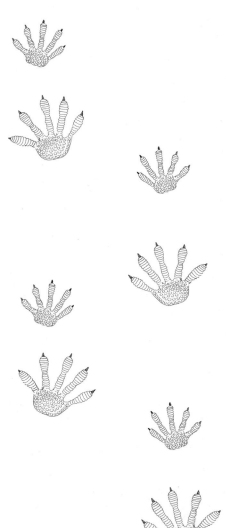

Gila monster understep walk trail.
SEB BARNETT

Track: Front: 1.44–1.6 in (3.66–4.06 cm) x 1.6–2.11 in (4.06–5.36 cm)

Tracks are medium in size. The front feet have five toes and all generally register clearly in the tracks. Toe 3 is the longest, but it is only slightly longer than toes 2 and 4. Toes 2 and 5 are nearly the same length. Toe 1 is the shortest and is on the same plane on the foot as toe 5. The claws are slightly longer on the front feet. The metacarpal area of the foot registers symmetrically, the widest point being directly behind toe 3. The front tracks generally angle directly forward in the direction of travel.

Hind: 1.34–1.48 in (3.40–3.76 cm) x 1.63–1.74 in (4.14–4.41 cm)

Tracks are medium in size. The hind feet have five toes and all tend to register clearly in the tracks. The longest toe is usually toe 4, but it is only slightly longer than toe 3. Toe 3 often reaches slightly further forward in the track. Toes 3 and 5 are about equal length. Toe 2 is slightly shorter than toe 3. Toe 1 is the shortest toe. Toe 5 is found lower down on the foot than toe 1, which helps distinguish hind tracks from front tracks. The claws are shorter on the hind feet than on the fronts. The metatarsal area registers asymmetrically, with the widest point being between toes 4 and 5. The hind tracks generally angle directly forward in the direction of travel.

Trail: Overstep Walk   Trail Width: 4–7 in (10.16–17.78 cm)
                     Stride: 5.81–6.48 in (14.76–16.46 cm)
        Tail Drag: 0.77–1.14 in (1.96–2.89 cm)

The tail drag may or may not be present, depending on both the depth of the substrate and whether or not the animal is elevating the tail horizontally. When observed, the tail drag may be located centrally in the trail, or off to one side. In older trails, the wide tail drag might be confused at first glance with the rectilinear trail left by a gopher snake or small rattlesnake.

Notes:

• This slow-plodding lizard species travels mostly using a consistent overstep walk. The hind tracks land just clear and only slightly ahead of the front tracks.

• It is always exciting to find the tracks of this elusive and beautiful lizard. Several trails of varying ages may indicate an area routinely used by the same individual. You may occasionally find the trails of several Gila monsters crossing the same area. This is especially likely during the breeding season.

• This species is under threat from poaching for private collections and from habitat loss. Please help protect this species in the wild by leaving it unmolested, and observe it from a respectful distance. These wonderful creatures deserve to have a chance to thrive in the wild.

Phrynosomatid Lizards: Family Phrynosomatidae

This family of lizards includes a diverse group of species with a variety of body forms. It comprises horned lizards, spiny lizards, earless lizards, and a handful of other species.

Earless Lizards

This group of lizards is named for their lack of an external ear opening. Lizards in this group are small to medium in size, with small, round heads that come to a point at the tip of the rostrum. The labial scales angle steeply like those in fringe-toed lizards (*Uma* ssp.) and zebra-tailed lizards (*Callisaurus* ssp.), likely helping them to burrow headfirst into loose soils.

These lizards have very long, thin toes on both the front and hind feet. Their tails are widest near the base, somewhat flattened from above, and become rounded and more tapered toward the tail tip.

These diurnal lizards are generally found in open, sandy areas, or at least in areas that include some loose soils in which to burrow during cold weather and for resting at night. All of them are very fast runners, and can potentially run using bipedal locomotion.

Zebra-tailed Lizard (*Callisaurus draconoides*)

This elegant, lean lizard species has very long front and hind limbs and feet.

Zebra-tailed lizards are capable of extremely rapid running, and can potentially run over eighteen miles per hour. At top speed, they will run bipedally on their long hind legs. This species often dashes rapidly with tail raised when initially approached, then stops suddenly and drops its tail. The tail is boldly patterned in black and white, giving the lizard its name. When this lizard drops its tail at the end of its rapid trot, it can look like the animal has disappeared.

Track: Front: 0.5–0.8 in (1.27–2 cm) x 0.4–0.9 in (1–2.3 cm)

Tracks are small. There are five toes present and all tend to register in the track. In loose substrates, only one or two toes may show clearly. Toes 3 and 4 are about equal in length, and are the longest toes on the front track. Toes 2 and 5 are about the same length, though toe 2 reaches slightly further forward in the track. Toe 1 is the shortest. Claws are well

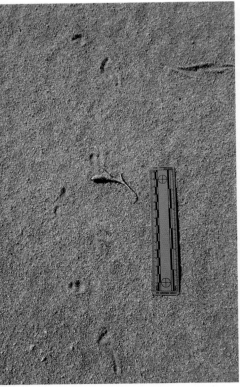

Left: The rapid overstep trot of a zebra-tailed lizard. *Right:* The bipedal run of a zebra-tailed lizard.

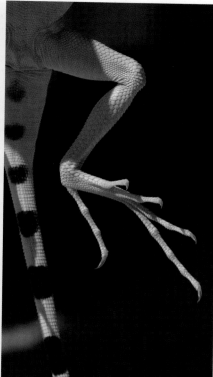

Left: The front foot of a zebra-tailed lizard. *Right:* The hind foot of a zebra-tailed lizard.

developed, relatively long, and present on all the toes. They tend to register in the track, though they may smudge together in loose substrate or at higher speeds. The metacarpal region of the foot is relatively small and covered in small scales. The largest scales are found nearest to the wrist. The front tracks register more to the inside of the trail than the hind tracks, and tend to orient vertically in line with the direction of travel.

Hind: 0.9–2 in (2.3–5 cm) x 0.8–1.5 in (2–3.8 cm)

Tracks are small to medium in size. There are five toes present and all tend to register in the track. All toes are proportionately very long and thin. Toe 4 is the longest. Toe 1 is the shortest. Toes 3 and 5 are about equal length. Toe 5 is the lowest on the foot and may orient between 90 and 180 degrees to the center of the trail. Claws are present, well developed, and relatively long. The metatarsal region is narrow, relatively long, and covered in small scales. Very little of the metatarsal region registers in the track. Hind tracks orient parallel to the center line of the trail.

Trail: Direct Register Walk   Trail Width: 3–3.2 in (7.62–8.13 cm)

                                          Stride: 5.5–6.6 in (13.97–16.76 cm)

Overstep Walk   Trail Width: 2.8–3 in (7.1–7.62cm)
                 Stride: 4–5 in (10.16–12.7 cm)

Overstep Trot   Trail Width: 2.5–2.7 in (6.35–6.86 cm)
                 Stride: 12–15 in (30.48–38.1 cm)

Bipedal Run   Trail Width: 1.7–2.5 in (4.32–6.35 cm)
                 Stride: 14–20 in (35.56–50.8 cm)

Tail Drag: 0.1–0.2 in (0.25–0.5 cm)

The tail drag of this species is generally only seen during slow walking gaits or briefly when it transitions from a walk into a rapid trot. When the lizard walks, the tail drag generally appears very thin, and fairly straight. When it transitions into a trot, it whips its tail wildly from one side to the other. This creates a zigzagging line between the tracks.

Notes:

- The zebra-tailed lizard is one of the fastest in the United States, and it's incredible to observe running at full speed. It can go from standing still to a full-tilt run in the blink of an eye.

- You'll generally find this species in open gravel or sand flats with sparse, widely spaced vegetative cover. It's occasionally found on the edges of open dunes close to gravelly or hard-packed ground. You won't usually find it farther into open dune habitats, which are the more likely the haunts of the similar fringe-toed lizard (*Uma* ssp.).

- You may observe this lizard basking on small rocks, low rises, or even dried cow patties. This species generally does not climb very often, and appears most comfortable on the ground where it can make a rapid escape at the approach of potential predators.

- Zebra-tailed lizards are very heat-tolerant, and may remain active on some of the hottest summer days when other reptiles have sought shelter from the intense heat.

- This species often sits and observes the landscape, waiting for potential prey to walk within range. It grabs prey in a quick dash. It may also consume some vegetative materials.

- Its predators include other lizard species (such as collared and leopard lizards), coachwhip snakes, foxes, and birds of prey.

Common Lesser Earless Lizard (*Holbrookia maculata*)

This small, active lizard has relatively long limbs and a short head. Their tails are somewhat flattened at the base, becoming round toward the tip. Typical individuals have two dark lateral bars on each side of their bodies just ahead of the midbody. Males typically have darker, more obvious lateral bars. The dorsal coloration varies with the dominant substrate in the area, but often includes blotches arranged in two rows from the neck down the mid-back to the base of the tail.

With the exception of some large scales on their heads, these lizards are covered in small, granular scales. This species is known to have about six subspecies, though the taxonomy is still under some debate.

One subspecies of this widespread lizard, the bleached earless lizard (*Holbrookia maculara ruthveni*), has become specially adapted to the white sand

A close-up of an overstep trotting trail of a bleached earless lizard.

dunes where it lives. It is largely white and tan in color, and its speckles appear much paler than those of more typical populations. It is the iconic species of lizard of White Sands National Monument in New Mexico.

This species commonly feeds on small invertebrates such as spiders, small beetles, and ants, and may also occasionally eat very small lizards. They usually catch their prey with a short dash after spotting them.

Watch for its tracks and scat in open, sandy areas.

Track: Front: 0.4–0.6 in (1–1.52 cm) x 0.3–0.5 in (0.76–1.27 cm)

Tracks are small. There are five toes present and all register in the track. Sometimes only the two or three lead toes will register at higher speeds in loose substrates. Track is asymmetrical. Toe 4 is the longest. Toe 3 is slightly shorter than toe 4. Toes 2 and 5 are about the same length. Toe 1 is the shortest. Toes 1 and 5 are attached on the same plane of the foot. Well-developed and moderately long claws are present on all toes, and register immediately at the end of the toe tip. The metacarpal region of the foot is small and covered in granular scales, and often does not register in the track. The front tracks tend to register directly in the line of travel, parallel to the center line of the trail. They often land closer to the inside of the trail than the hinds. At higher speeds—especially in loose substrate—the front tracks appear as narrow ovals.

Hind: 0.6–1 in (1.52–2.54 cm) x 0.3–0.6 in (0.76–1.52 cm)

Tracks are small. There are five toes present and all register in the track. At higher speeds in loose substrates sometimes only three of the toes will register clearly. Therefore, the hind tracks tend to appear considerably narrower at higher speeds. This generally includes toes 3, 4, and 5. Track is asymmetrical. Toe 4 is the longest. Toe 3 is slightly shorter than toe 4. Toes 2 and 5 are about the same length. Toe 1 is the shortest. Toe 5 is attached lowest on the foot. Well-developed and moderately long claws are present on all toes, and register immediately at the end of the toe tip. The metatarsal region of the foot is small, narrow, covered in granular scale, and often does not register in the track. The hind tracks tend to register directly in the line of travel or pitch outward at generally less than 45 degrees. At slower speeds, the hind track might be pitched outward

Left: The close-up of a direct register walking trail of a bleached earless lizard. *Right:* The overstep trotting trail of a bleached earless lizard. The trot becomes a bipedal run as the trail moves toward the upper left.

more dramatically at closer to 90 degrees. In a slow walk, toes 1 through 4 make a crescent-shaped arc away from the center line of the trail.

Trail: Overstep Trot   Trail Width: 1.3–2 in (3.3–5 cm)
                                    Stride: 5.5–7.5 in (13.97–19 cm)

         Direct Register Walk   Trail Width: 1.7–2.5 in (4.32–6.35 cm)
                                    Stride: 2.9–3.6 in (7.37–9.4 cm)

Tail Drag: 0.1–0.2 in (0.25–0.5 cm)

Tail drags are uncommon as generally this species carries the tail clear of the substrate when walking or trotting. When present, tail drags are usually very narrow. They will also drag their cloacal region immediately after defecating, which creates a relatively wide drag mark for several inches or occasionally for more than a foot. This drag line can be 0.3–0.8 in (0.76–2 cm) wide.

Notes:
• In the white sand dune habitat where this species' range and habitat overlap with that of the southwestern fence lizard (*Sceloporus cowlesi*), you may confuse the tracks of the two. Scrutinize the tracks closely to determine which

species you're observing. Note that there is a tendency for the front and hind tracks of southwestern fence lizards to be proportionally shorter and somewhat wider. Also, the front tracks of southwestern fence lizards tend to pitch inward toward the center line of the trail more dramatically than those of bleached earless lizards when they travel at higher speeds.

- Bleached earless lizards are a protected species, and much of their range is limited to military reservations and nationally protected federal land. Please do not collect these lizards from the wild. Take photos and enjoy them in their native habitat.

Northern Keeled Earless Lizard (*Holbrookia propinqua*)

This lizard species is similar to the common lesser earless lizard, although they do not have overlapping ranges. The keeled earless lizard is named for the keeled scales—scales with a ridge down their middle—with that are found dorsally on its body. The keels can only be seen at extremely close range, and generally with the help of a hand lens since the scales themselves are very small.

This species feeds on small invertebrates it finds in the sandy areas it inhabits, including small spiders, ants, beetles, and flies. Closely inspecting their scats can reveal more specifically what they are consuming and whether they seem to prefer a particular prey item in that area during that particular season.

The tracks of this species may be abundant in sandy areas, especially along the coastal areas of Texas.

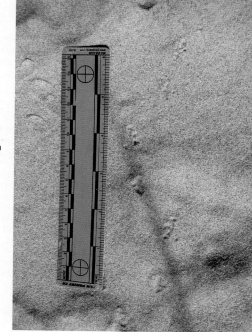

Track   Front: 0.3–0.4 in (0.76–1 cm) x 0.2–0.4 in (0.5–1 cm)

Tracks are small. There are five toes present and all register in the track. At higher speeds in loose substrates sometimes only the two or three lead toes will register. Track is asymmetrical. Toe 4 is the longest. Toe 3 is slightly shorter than toe 4. Toes 2 and 5 are about the same length. Toe 1 is the shortest. Toe 1 and toe 5 are attached on the foot on the same plane. Well-developed and moderately long claws are present on all toes, and register immediately at the end of the toe tip. The metacarpal region of the foot is small, covered in granular scales, and often does not register in the track. The front tracks tend to register directly in the line of travel, parallel to the center line of the trail. They tend to land closer to

The overstep trot of a northern keeled earless lizard. JASON KNIGHT

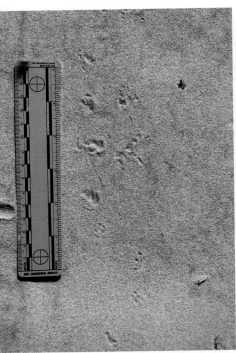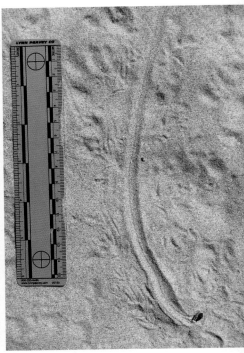

Left: The trotting trail of a northern keeled earless lizard coming to a stop, then resuming a trotting gait. JASON KNIGHT *Right:* A northern keeled earless lizard deposited a scat, then dragged its cloacal region through the sand. JASON KNIGHT

the inside of the trail than the hinds. At higher speeds—especially in loose substrate—the front tracks appear as narrow ovals.

Hind: 0.5–0.9 in (1.27–2.29 cm) x 0.3–0.5 in (0.76–1.27 cm)

Tracks are small. There are five toes present and all register in the track. At higher speeds in loose substrates, sometimes only three of the toes—generally toes 3, 4, and 5—will register clearly. Therefore, the hind tracks appear considerably narrower at higher speeds. Track is asymmetrical. Toe 4 is the longest. Toe 3 is slightly shorter than toe 4. Toes 2 and 5 are about the same length. Toe 1 is the shortest. Toe 5 is attached lowest on the foot. Well-developed and moderately long claws are present on all toes, and register immediately at the end of the toe tip. The metatarsal region of the foot is small, narrow, and covered in granular scales, and often does not register in the track. The hind tracks tend to register directly in the line of travel or pitch outward at generally less than 45 degrees. At slower speeds, the hind track might be pitched outward more dramatically at closer to 90 degrees. In a slow walk, toes 1 through 4 make a crescent-shaped arc away from the center line of the trail.

Trail:	Direct Register Trot	Trail Width: 1.7–2 in (4.32–5.08 cm)
		Stride: 2.6–3.5 in (6.6–8.9 cm)
	Overstep Trot	Trail Width: 1.1–1.3 in (2.79–3.3 cm)
		Stride: 4.7–5 in (11.94–12.7 cm)

Direct Register Walk Trail Width: 1.8–2 in (4.57–5.08 cm)
                                       Stride: 1.5–1.8 in (3.81–4.57 cm)
Tail Drag: 0.1–0.2 in (0.25–0.5 cm)
Tail drags are uncommon, as this species generally carries the tail clear of the substrate when walking or trotting. When present, tail drags are usually very narrow. Keeled earless lizards will also drag their cloacal region immediately after defecating, which creates a wider drag mark for several inches or occasionally for more than a foot. This drag line can be 0.3–0.6 in (0.76–1.52 cm) wide.

Notes:
- Watch for the tracks of this species in sandy areas such as inland dunes and barrier beaches. They are one of the most abundant lizard species on the barrier islands of the southern coast of Texas.
- The relatively long legs and proportionately long feet of these lizards allow them to move rapidly across the surface of sandy environments. Once they fully warm up, it's challenging to keep up with them.

Greater Earless Lizard (*Cophosaurus texanus*)

This widespread earless lizard species is similar in size and build to the zebra-tailed lizard (*Callisaurus draconoides*). Adult females have relatively drab brown, gray, or pale green dorsal regions with faint crescent-shaped blotches on each side. Gravid females may develop orange-pink on the chest, sides, or throat. Adult  males are brightly colored, with a base body color of gray but with orange spots on the upper forelimbs and upper back. Bright yellow infuses the dorsal region of the lower back, and meets with bright greenish blue coming up from the belly. Two thick black bars reach up from the ventral surface and onto the sides just past the midbody. The underside of the tail has bold black bands, which are bolder in males.

This species inhabits the Sonoran and Chihuahuan Deserts of the southwestern United States. They commonly inhabit flats, sandy washes, and nearby rocky hillsides. Similar to other earless lizard species, they will sit and watch for prey, then run out and grab it. They typically feed on invertebrates such as spiders, flies, beetles, and bees. Look for their tracks in patches of dust or fine sand in washes and on nearby trails.

Track: Front: 0.4–0.6 in (1–1.52 cm) x 0.3–0.5 in (0.76–1.27 cm)
           Tracks are small. There are five toes present and all register in the track. At higher speeds in loose substrates, only the two or three lead toes may register. Track is asymmetrical. Toe 4 is the longest. Toe 3 is slightly shorter than toe 4. Toes 2 and 5 are about the same length. Toe 1 is the shortest. Toes 1 and 5 are attached to the foot on the same plane. Well-developed and moderately long claws are present on all toes, which tend to register immediately at the end of the toe tip. The metacarpal region of the foot is small, covered in granular scales, and often does not register in the track. The front tracks tend to register directly in the line of travel, parallel to the center line of the trail. They tend to land closer to the inside of the trail than the hinds. At higher speeds—especially in loose substrate—the front tracks appear as narrow ovals.
           Hind: 0.8–1.2 in (2–3 cm) x 0.4–0.6 in (1–1.52 cm)

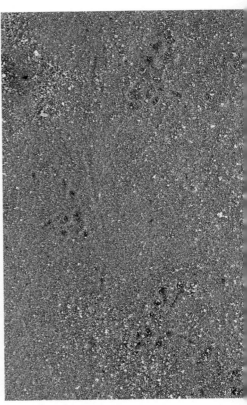

Left: The understep walking trail of a greater earless lizard showing significant tail dragging. *Right:* The subtle tracks of a greater earless lizard using a direct register walk.

Tracks are small. There are five toes present and all five toes register in the track. At higher speeds in loose substrates sometimes only three of the toes will register clearly. Therefore, the hind tracks tend to appear considerably narrower at higher speeds. This generally includes toe 5, toe 4, and toe 3. Track is asymmetrical. Toe 4 is the longest. Toe 3 is slightly shorter than toe 4. Toe 2 and toe 5 are about the same length. Toe 1 is the shortest. Toe 5 is attached lowest on the foot. Well-developed and moderately long claws are present on all toes, which tend to register immediately at the end of the toe tip. The metatarsal region of the foot is small, narrow, covered in granular scales, and often does not register in the track. The hind tracks tend to register directly in the line of travel or pitch outward at generally less than 45 degrees. At slower speeds, the hind track might be pitched outward more dramatically at closer to 90 degrees. In a slow walk, toes 1 through 4 make a crescent-shaped arc away from the center line of the trail.

Trail:	Understep Walk	Trail Width: 1.7–2.3 in (4.32–5.84 cm)
		Stride: 2.2–2.8 in (5.58–7.1 cm)
	Overstep Trot	Trail Width: 1.1–1.5 in (2.79–3.81 cm)
		Stride: 4.8–6 in (11.94–15.24 cm)
	Tail Drag: 0.1–0.2 in (0.25–0.5 cm)	

Notes:

- You should be able to easily spot this lizard, since it typically perches on low rocks along the edges of trails and washes. It frequently leaves its small scats on these low perches.
- When frightened, this species will break into a bipedal run. Just before and after running, the lizard with lift and wave the underside of its tail similar to the tail-wagging behavior of zebra-tailed lizards. This likely serves to distract predators.
- This is one of the most beautifully colored lizards in North America, and is a real pleasure to see up close in the wild.

Fringe-toed Lizards: Genus *Uma*

This small group of lizard species has a medium build, with large limbs and feet. The tail is relatively thick at the base, and also somewhat compressed, becoming rounder toward the tip. The specially adapted feet allow these lizards to move quickly and efficiently across loose sand.

Both the front and hind feet have horizontally projected scales that form fringes on the outer edges of each toe. These act to create a larger surface area, which reduces sinking in the sand in the same fashion a snowshoe reduces sinking into powdery snow. Similar sand adaptations have been observed in unrelated lizard species in extensively sandy regions in Africa and the Middle East.

Four species are commonly recognized in this group: the Coachella fringe-toed lizard (*Uma inornata*), Colorado Desert fringe-toed lizard (*Uma notata*), Mojave fringe-toed lizard (*Uma scoparia*), and Yuman fringe-toed lizard (*Uma rufopunctata*). All four are similar in appearance, with mostly subtle

The hind foot of a Mojave fringe-toed lizard showing the namesake fringe of scales.

The front foot of fringe-toed lizards also possesses a fringe of scales on the toes.

differences in markings. They are most easily told apart by their ranges.

According to Robert Stebbins, fringe-toed lizard species tend to avoid sand areas with grains much larger than 0.04 inch. Sand larger than this inhibits rapid burial under the surface. Fringe-toed lizards depend on the ability to bury themselves to avoid potential predators, to thermoregulate during temperature extremes, and to escape aggressive rivals.

The wedge-shaped head of this group of lizards helps them to more effectively bury themselves. They have countersunk lower jaws and enlarged labial scales, which also assist with this process by helping reduce drag on their head. They are able to disappear under the surface of the sand in seconds. They accomplish this by diving headlong into the loose sand and rapidly vibrating the head, body, and tail. This vibration makes the sand more like a liquid, which allows for easier penetration of the substrate, as well as making the sand flow over and cover up the animal even when it is buried just a few centimeters below the surface.

During hot periods of the year, the best time to observe these active lizards in the field is during midmorning or late afternoon. In the spring, you can see them during midday.

Fringe-toed lizards are territorial during the breeding season, but only males tend to aggressively defend their territories. One male's territory is several times larger than that of a female's, and tends to include the territories of several females within its boundaries.

Mojave Fringe-toed Lizard (*Uma scoparia*)

This is the most widespread fringe-toed species, found in many parts of the Mojave Desert of southeastern California. This medium-sized lizard is fast, alert, and generally wary of close approach. The most frequent view you'll usually get of this lizard is seeing it disappear in a puff of sand over the crest of the nearest dune. It's best to view them early in the morning, while they are still warming up.

This heat-loving species can tolerate a body temperature up to about 111 degrees Fahrenheit (Jones and Lovich 2009). Juveniles feed on a variety of invertebrates, including scorpions, beetles, and ants. Surprisingly, the adults are often largely herbivorous and feed on grasses, buds, flowers, and seeds.

The best places to see this species are areas of expanses of windblown dunes in open valleys or sandy edges of dry lake beds. Being active and terri-

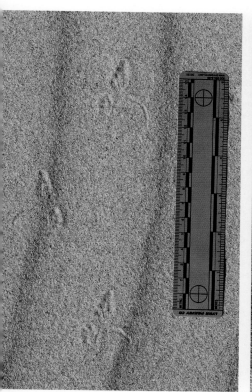

Left: The direct register walking trail of Mojave fringe-toed lizard. *Right:* The over-step trotting trail of a Mojave fringe-toed lizard.

torial, these lizards often leave interesting trails that show interactions between members of the same species.

Track: Front: 0.4–0.9 in (1–2.29 cm) x 0.38–0.6 in (0.96–1.52 cm)

Front tracks are small. There are five toes present and all register in the track at slow speeds. At more rapid speeds, only the two lead toes are likely to show up, or the entire track will show up as a single impression. Toe 4 is the longest, but only slightly longer than toe 3. Toes 2 and 5 are similar in length. Toe 1 is the shortest and is as low on the foot as toe 5. Toes 2, 3, and 4 have the most significant fringe of scales. Meanwhile, toes 1 and 5 show greatly reduced or absent fringing. The well-developed claws are long on all toes, and tend to register just ahead of the toe tip. The metacarpal region of the foot is small, covered in fine granular scales, and frequently does not register. The front tracks tend to register directly in the line of travel or pitched slightly toward the center line of the trail, and are closer to the center than the hind tracks.

Hind: 0.79–1.6 in (2–4 cm) x 0.35–1.3 in (0.88–3.3 cm)

Hind tracks are small to medium in size. There are five toes present and all register in the track at slower speeds. At more rapid speeds toes 1 through

4 make a single impression, and toe 5 makes its own mark as a dot or short line just to the outside and slightly behind the mark made by the other toes. Toe 4 is the longest. Toes 3 and 5 are the same length. Toe 2 is shorter than toe 3. Toe 1 is the shortest. Toe 5 is the lowest on the foot. Toes 2, 3, and 4 have the most significant fringe of scales. Meanwhile, toes 1 and 5 show greatly reduced or absent fringing. The well-developed claws are long on all toes, and tend to register just ahead of the toe tip. Sometimes the claw on toe 5 can appear oversized when it is dragged during slower movements. The metatarsal region is moderate, narrow, and covered in granular scales. It registers during very slow walking gaits, but generally does not register at all at higher speeds. The hind tracks tend to be oriented around 45 degrees at slow speeds, but the hind track becomes narrower and more oriented in the direction of travel with increased speed.

Trail: Understep Walk   Trail Width: 2.3–3 in (5.84–7.62 cm)
                             Stride: 2.1–3.6 in (5.33–9.14 cm)

        Direct Register Trot   Trail Width: 2.1–4.3 in (5.33–10.92 cm)
                             Stride: 4.7–6.8 in (11.94–17.27 cm)

        Overstep Trot   Trail Width: 1.6–2.5 in (4–6.35 cm)
                             Stride: 5.1–6.7 in (12.95–17 cm)

This species spends most of its time moving around either in a walk or overstep trot. It only uses the direct register trot when transitioning from a walk to a rapid overstep trot, and even then only for a short distance.

Tail Drag: 0.1–0.2 in (0.25–0.5 cm)

The tail drag of this species only registers during slow walking gaits and briefly as it transitions from a walk to a trot. During walks, tail drag appears as a slightly undulating and inconsistent line. As it transitions to a trot, the tail is whipped rapidly from one side to the other, creating a zigzagging line between the sets of tracks.

Notes:

• Mojave fringe-toed lizards prefer to live in areas of open sand dunes with sparse, scattered vegetation.

• They are very fast runners, and sometimes you will only see a small puff of sand when one runs across a sand dune.

• All fringe-toed lizards are able to bury themselves using a technique known as "sand swimming" in about one to three seconds (Jayne 2008). They will sometimes use this technique when pursued by a predator across an open sand dune. They also use this behavior to seek cooler sand during extremely hot conditions.

• This active species may leave some of the most abundant sign on the surface of sand dunes in certain areas.

Colorado Desert Fringe-toed Lizard (*Uma notata*)

This species of fringe-toed lizard is limited to lower elevations in extreme southeastern California, from the Colorado River west to the base of the Borrego Mountains. In this area—known as the Colorado Desert—this species prefers habitat with fine, windblown sands.

Just east of its range in Arizona and northern Baja California is the very similar Yuma fringe-toed lizard (*Uma rufopunctata*).

You can observe this species on warm spring and summer days foraging for insects such as ants, moths, sand roaches, grasshoppers, and caterpillars. They will also feed on seasonal flowers, buds, and leaves of certain desert shrubs.

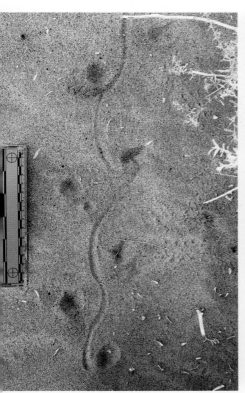

Left: The direct register trot of a Colorado Desert fringe-toed lizard. *Right:* The overstep walk of a Colorado Desert fringe-toed lizard.

They may share habitat with the flat-tailed horned lizard (*Phrynosoma mcallii*). Look close at the tracks to help distinguish these two species. The tracks of the Colorado Desert fringe-toed lizard tend to show a greater difference in size between the hind feet and the smaller front feet. Alternatively, those of the flat-tailed horned lizard are nearly the same size. Also, look closely at the morphological features, as the hind tracks of flat-tailed horned lizards have proportionately shorter toes. Keep an eye out for their long, smooth, and pill-shaped scats. The scat contents of the horned lizard are nearly always ant parts, whereas those of the fringe-toed lizard are more variable in shape and content.

Track: Front: 0.27–0.5 in (0.68–1.27 cm) x 0.27–0.45 in (0.68–1.14 cm)
Front tracks are small. There are five toes present and all register in the track at slow speeds. At more rapid speeds, only the two lead toes are likely to show up or the entire track will show up as a single impression. Toe 4 is the longest, but only slightly longer than toe 3. Toes 2 and 5 are similar in length. Toe 1 is the shortest and is as low on the foot as toe 5. Toes 2, 3, and 4 have the most significant fringe of scales. Meanwhile,

Here is an example of a slower over-step walk of the Colorado Desert fringe-toed lizard. Notice how at this slower speed the toes of the hind feet are visible in the track, although the fronts still don't register very clearly.

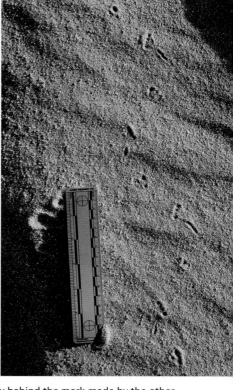

toes 1 and 5 show greatly reduced or absent fringing. The well-developed claws are long on all toes, and tend to register just ahead of the toe tip. The metacarpal region of the foot is small, covered in fine granular scales, and frequently does not register. The front tracks tend to register directly in the line of travel or pitched slightly toward the center line of the trail and are closer to the center than the hind tracks.

Hind: 0.8–1.2 in (2–3 cm) x 0.44–0.88 in (1.1–2.23 cm)

Hind tracks are small to medium in size. There are five toes present and all register in the track at slower speeds. At more rapid speeds toes 1 through 4 make a single impression, and toe 5 makes its own mark as a dot or short line just to the outside and slightly behind the mark made by the other toes. Toe 4 is the longest. Toes 3 and 5 are the same length. Toe 2 is shorter than toe 3. Toe 1 is the shortest. Toe 5 is the lowest on the foot. Toes 2, 3, and 4 have the most significant fringe of scales. Meanwhile, toes 1 and 5 show greatly reduced or absent fringing. The well-developed claws are long on all toes, and tend to register just ahead of the toe tip. Sometimes the claw on toe 5 can appear oversized when it is dragged during slower movements. The metatarsal region is moderate, narrow, and covered in granular scales. It registers during very slow walking gaits, but generally does not register at all at higher speeds. The hind tracks tend to be oriented around 45 degrees at slow speeds, but the hind track becomes narrower and orients more in the direction of travel with increased speed.

Trail: Direct Register Walk Trail Width: 2.5–3 in (6.35–7.62 cm)
Stride: 4–5.6 in (10.16–14.22 cm)

Direct Register Trot Trail Width: 1.9–3.2 in (4.83–8.13 cm)
Stride: 4–7.6 in (10.16–19.3 cm)

Overstep Trot Trail Width: 1.7–2.3 in (4.32–5.84 cm)
Stride: 5.6–7.6 in (14.22–19.3 cm)

This species spends most of its time moving around either in a walk or overstep trot. It only uses the direct register trot when transitioning from a walk to a rapid overstep trot, and even then only for a short distance.

Tail Drag: 0.1–0.2 in (0.25–0.5 cm)

The tail drag of this species only registers during slow walking gaits and briefly as it transitions from a walk to a trot. During walks, tail drag appears as a slightly undulating and inconsistent line. As the lizard transitions to a trot, it whips its tail rapidly from one side to the other, creating a zigzagging line between the sets of tracks.

Notes:

- This lizard moves very fast; it's built to run at high speeds on the fine, loose substrates that it calls home.
- This species and its tracks and sign can be most easily observed in sandy areas at the edges of open dunes. Although you can find it far from cover in open dunes, it tends to be more abundant closer to the cover of sparse shrubs at the edges of dune systems, or in sparsely vegetated depressions between dunes.
- For close observation, approach the lizard slowly and with care, as this species tends to make a rapid dash out of sight when it feels threatened.

Side-blotched lizard (*Uta stansburiana*)

This small, common lizard lives in much of the warmer parts of the western United States, and is related to the spiny lizards. Adults measure between 4 and 6 inches in total length. This lizard species is found in semiarid and arid habitats including desert scrub, chaparral, grassland, open woodlands, and the edges of sand dunes.

This lizard has small, weakly keeled dorsal scales that are cycloid or granular. Typically males are more colorful than females, and often have some blue scales on the dorsal surface. Both sexes have a fold of skin at the throat called a "gular fold." Both also have a very dark, bluish-black blotch on their side just behind the forelimb. Patterns vary significantly from location to location, and even from individual to individual. They can be striped, blotched, spotted, or a combination of all three. This species may also be largely uniform in color, without markings. Typical body color is tan, gray, or brown, often matching the dominant soil color of the area. Stripes are typically pale tan or yellowish, while blotches are dark brown or pale tan.

Males employ three kinds of mating strategies, and each type of male can be recognized by its throat color. Males with blue throats defend small territories, but usurp the territories of other males. Orange-throated males are the most dominant and control the most high-quality (and often larger) territories. These will also usurp the territories of other males. Males with yellow throats are the most similar to females in appearance and invade the territories of other males to sneak copulation. Yellow-throated males hold no territories. Barry Sinervo, a professor at the University of California, Santa Cruz, has studied and described this complex approach in great detail.

Side-blotched lizards feed largely on invertebrates, including flies, spiders, grasshoppers, and small scorpions. They are in turn preyed upon by many species, including larger lizards and snakes. They are generally the first species active in the mornings, and one of the last day-active lizards to be seen in the evenings. This likely helps them avoid at least some predators.

These small, lightweight lizards require fine substrates in which to leave tracks, such as dust, fine sand, or silt. Watch for their tracks in dusty trails, in washes, or on the edge of windblown sand dunes. Their tracks are similar to and can be confused with those of juvenile spiny lizards (*Sceloporus* ssp.).

Track: Front: 0.2–0.3 in (0.5–0.76 cm) x 0.2–0.3 in (0.5–0.76 cm)

Tracks are small. Front feet have five toes, and all toes tend to register at slower speeds. During very rapid trots, only the middle 2 or 3 toes might register. Toes 2 and 3 are the longest, and are about equal length. Toes 1 and 4 are about equal length. Toe 5 is the shortest. Toes 1 and 5 tend to angle strongly to the side. Track is fan shaped and symmetrical overall. Claws are short and present on all the toes, and may register as tiny dots at the toe tips. Metacarpal area is small and only registers at slower speeds. Front tracks tend to face directly forward or pitch slightly toward the inside of the trail.

Hind: 0.5–0.7 in (1.27–1.78 cm) x 0.4–0.5 in (1–1.27 cm)

Tracks are small to medium in size. Hind feet have five toes, all of which tend to register at slow to moderate speeds. During very rapid trots, toe 5 may not register, and neither may toe 1. Toe 4 is the longest and may register in line with toe 3 or might arch away slightly. Toe 5 is the lowest on the foot. It is well spaced away from toe 4 and generally angled about

Left: The typical overstep walk of a large side-blotched lizard, showing a thin tail drag. *Right:* The direct register walking trail of a side-blotched lizard in loose sand. The paired tracks next to the trail are from a kangaroo rat.

Left: Overstep trot of a side-blotched lizard in loose sand. The zipper-like trails on either side are those of several beetle species. *Right:* Another example of an over-step walk of a side-blotched lizard, this one lacking any tail drag.

90 degrees outward away from toe 4. Claws are short, present on all toes, and may register as tiny dots at the toe tips. Hind tracks tend to be pitched slightly to the outside. During a slow walking gait, the hind tracks can be angled outward more dramatically, between 45 and 90 degrees away from facing directly forward.

Trail: Overstep Trot   Trail Width: 1.3–1.5 in (3.3–3.81 cm)
                                Stride: 4–6.5 in (10.16–16.51 cm)
         Overstep Walk   Trail Width: 1.5–1.7 in (3.81–4.32 cm)
                                Stride: 1.6–3.3 in (4–8.38 cm)
         Direct Register Walk Trail Width: 1.4–1.6 in (3.55–4 cm)
                                Stride: 0.6–0.8 in (1.52–2 cm)
Tail Drag: 0.1–0.2 in (0.25–0.5 cm)

Tail drags are infrequent, as side-blotched lizards tend to stiffly carry their tails directly behind them. It might register for a short distance as the animal speeds up from a walk into a rapid trot. It might also register in particularly deep and loose substrate, especially during a slow walk. Belly marks might be visible where the lizard paused and dropped its body to the ground.

Notes:
- One of the most commonly encountered lizards in the drier portions of the western United States, this species is often quite visible and allows you a great opportunity to observe a variety of lizard behaviors.
- The scats of this species are very small, significantly smaller than those of the slightly larger common sagebrush lizard (*Sceloporus graciosus*).

Horned Lizards (Genus *Phrynosoma*)

This group is composed of squat, spike-covered lizards generally with short legs and broad bodies. The front and hind legs of horned lizards are similar lengths, and this combined with their wide bodies prevents them from running very fast. The short, squat appearance of the feet in the photos is typical of all the horned lizards.The trail left by a fast-running horned lizard looks much like the transitional trail of other, faster-moving species of lizards: the stride is rather short, and there is often a great deal of tail drag. The exception is the flat-tailed horned lizard, which tends to hold its tail parallel to the ground as it runs, and does not often leave much of a tail drag.

The front foot of a Texas horned lizard.

The hind foot of a Texas horned lizard.

Coast Horned Lizard (aka Blainville's Horned Lizard) (*Phrynosoma blainvillii*)

This iconic California lizard has experienced tremendous declines over much of its range. Adults measure 2.5 to 6.25 inches in total length. This lizard is often tan, gray, or reddish with paired dark brown or black blotches starting on the neck and running down the dorsal surface.

This species is typically found in chaparral, oak savannas, and open coniferous woodlands.

It feeds mostly on ants, especially harvester ants.

Track: Front: 0.3 in–0.5 in (0.83–1.35 cm) x 0.3 in–0.6 in (0.67–1.64 cm)

Tracks are small. There are five toes present and all tend to register in the track. Toe 3 is the longest, but toe 4 is nearly as long. Toe 2 is shorter. Toes 1 and 5 are nearly identical in length and are the shortest. The overall track is fan shaped and fairly symmetrical. Scales on toes are relatively large, triangular, and pointed. Well-developed and moderately long claws are present on all toes, which tend to register immediately at the end of the toe tip. Only the claws will register clearly in firmer substrates. The metacarpal region of the foot is small, short, wide, and covered by relatively large scales. The front track orients mostly forward or may pitch slightly toward the center line of the trail.

Hind: 0.4 in–0.7 in (1.03–1.75 cm) x 0.4 in–0.6 in (0.97–1.65 cm)

Tracks are small. There are five toes present and all tend to register in the track. Toe 4 is the longest. From toes 3 to 1, the lengths get progressively shorter. Toe 5 is angled relatively forward, and not significantly separated from the other toes.

The direct register trotting trail of an adult coast horned lizard. JEFF NORDLAND

Also, toe 5 is proportionately short. The metatarsal region is moderately wide and long, and is covered with relatively large scales. Scales on toes are relatively large, triangular, and pointed. Well-developed and moderately long claws are present on all toes, which tend to register immediately at the end of the toe tip. Only the claws will register clearly in firmer substrates. The hind track orients mostly forward or may pitch slightly outward from the center line of the trail, generally much less than 45 degrees.

Trail: Trot Trail Width: 0.5 in–2.1 in (1.2–5.25 cm)
Stride: 1.2–1.8 in (3–4.57 cm)

Tail Drag: 0.2–0.5 in (0.5–1.27 cm)

Tail drags are generally absent, except for where the lizard has rapidly sped up from a standstill to a trot. Here, a few diagonal marks will be left as the tail touches down a few times in response to the side-to-side arching of the body.

Notes:

- Coast horned lizards walk when foraging. They use a relatively slow, waddling trot when fleeing predators or chasing rivals. These lizards rely heavily on cryptic coloration and large spines to avoid predation.
- This species prefers coastal sage scrub habitat along the coast. Further inland, it also inhabits dense chaparral stands, mixed woodlands, riparian areas, and sand dunes. Some evidence shows that it can survive on the edges of grape vineyards and in remnant grassland areas in the Central Valley of California.
- The coast horned lizard has suffered serious decline from development and intensive agricultural conversion of its habitat. The introduction of invasive ant species, such as the Argentine ant (*Linepithema humile*) has also contributed to the decline of this species. Coast horned lizards avoid this ant species, and additionally these ants displace the native ant species that are vital to the diet of this lizard species.
- This is a federal and state threatened species. Observe them at a distance, and do not collect them. Like all horned lizard species, these will die a slow death via starvation in captivity.

Flat-tailed Horned Lizard (*Phrynosoma mcallii*)

This lizard has the most limited range in the United States of all the horned lizards. It limits its activity to sandy areas in flats and on sand dunes. Its range is largely restricted to the Colorado Desert in southeastern California, as well as in a small area at the extreme southwest of Arizona just east of the Colorado River.

It is also the most sand adapted of all of the members of the genus *Phrynosoma*. Its body is modified in several ways to facilitate more successful burrowing into fine sand: it has a flattened tail, reduced spines on its body, and even pointed cranial horns aligned with the long axis of the head that do not interfere with forward movement through sand (Stebbins 2012). This lizard shares its habitat with two species of fringe-toed lizards, another sand-adapted species.

Flat-tailed horned lizards tend to respond to potential threats in one of two ways: sitting still or dashing off at high speed and diving headfirst into loose sand. Despite the relatively open, sparsely vegetated dune habitats they inhabit, this species is incredibly cryptic. In fact, human observers must be very careful not to step on these lizards as they may sit tight and rely on their excellent camouflage for protection even with the threat of an incoming human foot.

When this species does decide to run, it will do so rapidly. With quick shakes of its head, body, and tail, this lizard can disappear very rapidly out of sight in loose sand. Fresh tracks might lead to a seeming "dead end" to the casual observer, but close inspection will reveal the subtle disturbance left behind by the burying action of the invisible lizard.

These lizards sun themselves by sticking only their heads out from the sand in the early morning or during cooler periods. Like other horned lizards, this can allow the flat-tailed horned lizard to warm up adequately to be surface active.

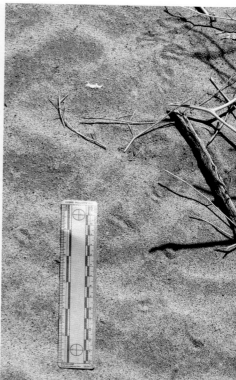

Left: The overstep trotting trail of a flat-tailed horned lizard. *Right:* The subtle walking trail leading to the hidden flat-tailed horned lizard.

Track: Front: 0.32–0.56 in (0.81–1.43 cm) x 0.32–0.53 in (0.81–1.35 cm)

Tracks are small. There are five toes present and all tend to register in the track. Toe 3 is the longest, but toe 4 is nearly as long. Toe 2 is shorter. Toes 1 and 5 are the shortest and nearly identical in length. The overall track is fan shaped and fairly symmetrical. Scales on toes are relatively large, triangular, and pointed. Well-developed and moderately long claws are present on all toes, which tend to register immediately at the end of the toe tips. Only the claws will register clearly in firmer substrates. The metacarpal region of the foot is small, short, wide, and covered by relatively large scales. Fronts land slightly to the inside of the trail compared to hinds. Front tracks mostly tend to angle directly forward in both a walk and a trot.

Hind: 0.48–0.729 in (1.21–1.85 cm) x 0.5–0.74 in (1.27–1.88 cm)

Tracks are small. There are five toes present and all tend to register in the track. Toe 4 is the longest. From toes 3 to 1, toes are progressively shorter. Toe 5 is angled relatively forward, and not significantly separated from other toes. Also, toe 5 is proportionately short. The metatarsal region is moderately wide and long, and is covered with relatively large scales. Scales on toes are relatively large, triangular, and pointed. Well-developed and moderately long claws are present on all toes, which tend to register immediately at the end of the toe tips at slower speeds. Only the claws will register clearly in firmer substrates. At higher speeds in loose substrates, toe marks combine into one or two major regions. Hind toes have a small fringe of scales similar to fringe-toed lizards that project horizontally and help add surface area to the foot. Hind tracks tend to angle toward the outside of the trail in a walk, but pitch slightly toward the center line of the trail in a trot.

Trail: Overstep Trot   Trail Width: 1.98–2.54 in (5.02–6.45 cm)
                       Stride: 3.9–4.6 in (9.9–11.68 cm)

This species creates tail drags much less frequently than other horned lizards. The tail is generally carried clear of the substrate during trots. You'll most likely observe tail drags in the areas where the animal creates a burrow, exits a burrow, or sets its body down and makes full contact with the substrate while thermoregulating.

Notes:
- The flat-tailed horned lizard walks when foraging. It often trots, but you won't see much detail in loose sand. Details of the tracks are most visible where the animal is walking while feeding, interacting with others of its kind, or near a basking or burrowing site. Use the similarity between the size of their front and hind tracks and trail pattern to help distinguish it from similar species.
- A narrower body and long legs allow flat-tailed horned lizards to move more quickly and bury themselves more swiftly than any other horned lizard.
- They share their sandy habitat with fringe-toed lizards and leave similar trails. Unlike fringe-toed lizards, the front and hind tracks of flat-tailed horned lizards are very similar in length and width.
- Look for distinctive scats along fresh trails.

Desert Horned Lizard (*Phrynosoma platyrhinos*)

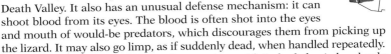

This common species is found throughout the Mojave, Great Basin, and parts of the Sonoran desert regions. As is true for many horned lizard species, this one can be very cryptic in coloration and behavior. This species has adapted well to very dry conditions, and can persist in incredibly arid places such as Death Valley. It also has an unusual defense mechanism: it can shoot blood from its eyes. The blood is often shot into the eyes and mouth of would-be predators, which discourages them from picking up the lizard. It may also go limp, as if suddenly dead, when handled repeatedly.

This species will often bask in the mornings with its body buried and only its head above the surface. As with other horned lizards, this one uses this technique to effectively heat up its body temperature before emerging to begin a day of foraging on the surface. They are much harder to locate as they are thermoregulating.

Track: Front: 0.4–1 in (1–2.54 cm) x 0.4–0.66 in (1–1.67 cm)

Tracks are small. There are five toes present and all tend to register in the track. Toe 3 is the longest, but toe 4 is nearly as long. Toe 2 is shorter. Toes 1 and 5 are also nearly identical in length and are the shortest. The overall track is fan shaped and fairly symmetrical. Scales on toes are relatively large, triangular, and pointed. Well-developed and moderately long claws are present on all toes, which tend to register immediately at the end of the toe tips. Only the claws will register clearly in firmer

The overstep trot of an adult desert horned lizard.

substrates. The metacarpal region of the foot is small, short, wide, and covered by relatively large scales. The front track mostly faces forward or may be pitched slightly toward the center line of the trail.

Hind: 0.67–1 in (1.7–2.54 cm) x 0.55–0.84 in (1.4–2.13 cm)

Tracks are small. There are five toes present and all tend to register in the track. Toe 4 is the longest. From toe 3 to 1, toes are progressively shorter. Toe 5 is angled forward, and not significantly separated from other toes. Also, toe 5 is proportionately short. The metatarsal region is moderately wide and long, and is covered with relatively large scales. Scales on toes are relatively large, triangular, and pointed. Well-developed and moderately long claws are present on all toes, which tend to register immediately at the end of the toe tips. Only the claws will register clearly in firmer substrates. The hind track mostly faces forward or may pitch slightly outward from the center line of the trail, generally much less than 45 degrees.

Trail: Overstep Trot   Trail Width: 2.3–2.6 in (5.84–6.6 cm)
Stride: 5–9.7 in (12.7–24.64 cm)

Tail Drag: 0.2–.5 in (0.5–1.27 cm)

Tail drag may be present or absent. It is most likely to be visible when the animal is trotting, as this action lifts the body and tips the tail down. In shallow substrate the tail drag may be broken up into diagonal lines.

Notes:

• This species tends to be found in desert areas that are sparsely vegetated and have some loose or sandy soils. Their color patterns tend to match the soil patterns beautifully.

• The desert horned lizard is not a particularly fast runner. It relies much more on cryptic behavior and coloration to elude predators.

• Watch for this species basking on small rocks or small rises in the landscape, especially in the mornings and evenings.

• This ant-loving species appears to prefer harvester ants, but will consume other species of invertebrates along with occasional vegetative matter.

• In rocky areas where their tracks may be difficult to find, watch for their distinct scats composed largely of ant parts.

Pygmy Short-horned Lizard (*Phrynosoma douglasii*)

This is the smallest horned lizard species, as well as the species that ranges the farthest north. Its range is largely restricted to the shrub-steppe habitats of the northern Great Basin and Columbia Basin regions.

Being so small, this species generally does not leave very clear tracks. The trails, however, are often fairly distinct and easy to identify in appropriate substrates. You can most easily observe trails on sunny days after it has rained.

Track: Front: 0.3–0.4 (0.76–1 cm) x 0.3–0.4 (0.76–1 cm)

Tracks are small. There are five toes present and all tend to register in the track. Toe 3 is the longest, but toe 4 is nearly as long. Toe 2 is shorter. Toes 1 and 5 are also nearly identical in length and are the shortest. The overall track is fan shaped and fairly symmetrical. Scales on toes are relatively large, triangular, and pointed. Well-developed and moderately long claws are present on all toes, which tend to register immediately at the end of the toe tips. Only the claws will register clearly in firmer substrates. The metacarpal region of the foot is small, short, wide, and covered by relatively large scales. The front track mostly faces forward or may pitch slightly toward the center line of the trail. Due to this species' tendency to move in a direct register gait, front tracks are not commonly observed.

Hind: 0.4–0.55 in (1–1.4 cm) x 0.4–5 in (1–1.27 cm)

Tracks are small. There are five toes present and all tend to register in the track. Toe 4 is the longest. From toe 3 to 1, toes are progressively shorter. Toe 5 angles relatively forward, and is not significantly separated from other toes. Also, toe 5 is proportionately short. The metatarsal region is moderately wide and long, and is covered with relatively large scales.

The walking trail of a pygmy short-horned lizard stands out against the rain-pocked sand.

The direct register trot of a male pygmy short-horned lizard.

Scales on toes are relatively large, triangular, and pointed. Well-developed and moderately long claws are present on all toes, which tend to register immediately at the end of the toe tips. Only the claws may register clearly in firmer substrates. The hind track mostly faces forward or may pitch slightly outward from the center line of the trail, generally much less than 45 degrees.

Trail: Direct Register Walk — Trail Width: 0.92–1.31 in (2.34–3.32 cm)
Stride: 0.36–1 in (0.91–2.54 cm)

Direct Register Trot — Trail Width: 1.2–1.7 in (3–4.32 cm)
Stride: 1.31–2.6 in (3.32–6.6 cm)

Tail Drag: 0.1–0.3 in (0.25–0.76 cm)

Tail drags are generally absent, except in places where the lizard has rapidly sped up from a standstill to a trot. This species can leave a more continuous mark when it travels in a trot. Trotting males make a continuous line that can show an angled, squared pattern. Trotting females drag their tails and tend to show a more smooth, sinuous pattern.

Notes:
- This species consumes ants, but also eats many other invertebrates. It sometimes even includes some vegetative matter in its diet.
- The scats of this species are relatively small, but still show the shape characteristic of horned lizards.
- This species has the smallest horns of all the horned lizards, and therefore relies most on cryptic coloration and behavior to avoid predation.
- This species may sometimes vocalize when harassed by a predator or picked up by a human.
- You can identify male pygmy short-horned lizards by their tail drags, which form almost squared-off curves with each undulation. Females leave a more subtle, straighter, sinuously curving line.

Texas Horned Lizard (*Phrynosoma cornutum*)

This large species of horned lizard, once found across a wide range in the southwestern United States, has greatly declined. Adults measure between 2.5 and 7 inches in total length. They are tan or grayish with a white line down the center of the back from the neck and down onto the tail. Paired pale-bordered, oval-shaped blotches are found on the back.

Track: Front: 0.5–0.8 in (1.27–2 cm) x 0.4–0.8 in (1–2 cm)

Tracks are small. There are five toes present and all tend to register in the track. Toe 3 is the longest, but toe 4 is nearly as long. Toe 2 is shorter. Toes 1 and 5 are also nearly identical in length and are the shortest. The overall track is fan shaped and fairly symmetrical. Scales on toes are relatively large, triangular, and pointed. Well-developed and moderately long claws are present on all toes, which tend to register immediately at the end of the toe tips. Only the claws will register clearly in firmer substrates. The metacarpal region of the foot is small, short, wide, and covered by relatively large scales. The front track mostly faces forward or may pitch slightly toward the center line of the trail.

Hind: 0.8–1.7 in (2–4.32 cm) x 0.6–0.8 in (1.52–2 cm)

Tracks are small. There are five toes present and all tend to

The evenly undulating direct register trotting trail of a Texas horned lizard.

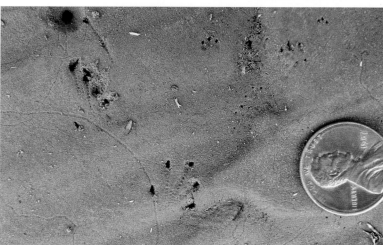

The front and hind track of a Texas horned lizard, with a set of pocket mouse tracks in the upper right. JONAH EVANS

register in the track. Toe 4 is the longest. From toe 3 to 1, toes are progressively shorter. Toe 5 is angled relatively forward, and not significantly separated from other toes. Also, toe 5 is proportionately short. The metatarsal region is moderately wide and long, and is covered with relatively large scales. Scales on toes are relatively large, triangular, and pointed. Well-developed and moderately long claws are present on all toes, which tend to register immediately at the end of the toe tips. Only the claws will register clearly in firmer substrates. The hind track mostly faces forward or may pitch slightly outward from the center line of the trail, generally much less than 45 degrees.

Trail: Overstep Trot   Trail Width: 3.4–4.2 in (8.64–10.67 cm)
                                   Stride: 5–6.5 in (12.7–16.51 cm)

This gait is not a dramatic overstep; rather the hind tracks fall just past the front tracks. This is the top speed for this lizard species.

Direct Register Trot   Trail Width: 3–4 in (7.62–10.16 cm)
                                   Stride: 4–6 in (10.16–15.24 cm)

Tail Drag: 0.2–0.8 in (0.5–2 cm)

Tail drag may be most visible during a trot. In shallow substrate the tail drag may be broken up into diagonal lines. In deeper substrates, it will appear as a continuous curving wave not unlike the trail made by a small snake.

Notes:
• This cryptic species is a joy to find in the field.

Regal Horned Lizard
(*Phrynosoma solare*)

This species has a well-developed crown of large cranial horns. Adult grow between 3.5 and 6.5 inches in total length.

Track: Front: 0.5–0.8 in (1.27–2 cm) x 0.4–0.8 in (1–2 cm)

Tracks are small. There are five toes present and all tend to register in the track. Toe 3 is the longest, but toe 4 is nearly as long. Toe 2 is shorter. Toes 1 and 5 also nearly identical length and are the shortest. Overall track is fan shaped and fairly symmetrical. Scales on toes are relatively large, triangular, and pointed. Well-

The zigzagging direct register trotting trail of a regal horned lizard.

developed and moderately long claws are present on all toes, which tend to register immediately at the end of the toe tips. Only the claws will register clearly in firmer substrates. The metacarpal region of the foot is small, short, wide, and covered by relatively large scales. The front track mostly faces forward or may pitch slightly toward the center line of the trail.

Hind: 0.8–1.7 in (2–4.32 cm) x 0.6–0.8 in (1.52–2 cm)

Tracks are small. There are five toes present and all tend to register in the track. Toe 4 is the longest. From toe 3 to toe 1, toes are progressively shorter. Toe 5 is angled relatively forward, and is not significantly separated from other toes. Also, toe 5 is proportionately short. The metatarsal region is moderately wide and long, and is covered with relatively large scales. Scales on toes are relatively large, triangular, and pointed. Well-developed and moderately long claws are present on all toes, which tend to register immediately at the end of the toe tips. Only the claws will register clearly in firmer substrates. The hind track mostly faces forward or may pitch slightly outward from the center line of the trail, generally much less than 45 degrees.

Trail: Overstep Trot   Trail Width: 3.4–4 in (8.64–10.16 cm)
                            Stride: 7–10 in (17.78–25.4 cm)

This gait is not a dramatic overstep; rather, the hind tracks fall just past the front tracks. This is the top speed for this lizard species.

Direct Register Trot   Trail Width: 2.5–3.5 in (6.35–8.89 cm)
                            Stride: 4–6 in (10.16–15.24 cm)

Tail Drag: 0.2–0.3 in (0.5–0.76 cm)

Tail drag may be present or absent. When the animal is trotting, it lifts the body and tips the tail down, likely making the tail drag appear. In shallow substrate the tail drag may be broken up into diagonal lines.

Notes:
- This horned lizard species appears to prefer harvester ants and does not seem to be affected by their venomous stings—at least when eating them.
- This species is difficult to see, so it might be easier to look for active harvester ant nests and then search around them for the lizard.

Family Iguanidae

In the United States, this family includes chuckwallas and true iguanas. There are about 40 species in this family, including the well-known and widely recognized green iguana of pet trade fame.

Green Iguana (*Iguana iguana*)

This nonnative and potentially giant lizard can grow up to 6 feet long from nose to tail tip. This is several feet longer than even the largest native lizard species.

The overstep trotting trail of a green iguana. LORI SHINDEL

The very large track of a mature adult male green iguana. JENNIFER TKACZYK

The green iguana's range in the United States is limited to parts of the southern tip of Florida, mainly the suburban edges of the Everglades and several islands in the Keys. This species was most likely introduced by irresponsible pet owners. Despite its common appearance in the pet trade, this species' needs can be difficult to meet by inexperienced keepers.

In its native range from southern Mexico through much of the northern half of South America, it is a common species. Adults are largely arboreal in their native habitat of tropical forest, and commonly dive out of the trees into water below.

This huge herbivore prefers leaves, flowers, and fruit. In its native range, the green iguana is commonly hunted for food by humans.

Watch for their tracks and trails in sandy areas in southern Florida.

Track: Front: 2.5–4 in (6.35–10.16 cm) x 3–4 in (7.62–10.16 cm)

Tracks are medium to large. There are five toes present and all register in the track at slower speeds. At higher speeds and in loose substrates, only toes 2, 3, and 4 may register or all toes might register as a single track. Tracks are not quite symmetrical, with toes 3 and 4 being the longest. Toe 1 is the shortest. Toes 2 and 5 are about equal in length. Claws are well developed and proportionately long. The claws tend to register most clearly at slow speeds. The metacarpal region of the foot is covered in small, keeled, diamond-shaped scales. The front track tends to register directly forward, or may occasionally pitch slightly toward the center of the trail.

Hind: 4.1–5 in (10.41–12.7 cm) x 3.1–5 in (7.87–12.7 cm)

Tracks are large. There are five toes present and all tend to register in the track. Tracks are asymmetrical. Toe 4 is the longest. Toe 3 is about one-third shorter than toe 4. Toe 2 and toe 5 are about the same length. Toe 1 is the shortest. Claws are well developed and present on all the toes. The longest claw is present on toe 5. The metatarsal region is relatively long, narrow, and covered in keeled, diamond-shaped scales. The largest scales are at the base of the metatarsal area and the smallest are at the base of the toes. The hind tracks tend to register directly forward, or may pitch outward away from the center of the trail up to 45 degrees or even up to 90 degrees at very slow speeds.

Trail: Direct Register Walk   Trail Width: 10–12 in (25.4–30.48 cm)
                                Stride: 10–13 in (25.4–33 cm)

Tail Drag: 0.4–1 in (1–2.54 cm)
Tail drags of this species tend to gently curve, as they walk with a twisting motion. Tail drags are generally only visible during a walking gait, as the animal lifts it clear of the substrate during more rapid locomotion.

Notes:
- Though it's restricted in range and not native, finding the tracks of this lizard can be a pleasure. They are so large and impressive that it can feel almost like looking at the tracks of small dinosaurs.

Night Lizards: Family Xantusiidae

Because of their small size and nocturnal nature, this is arguably the least known group of lizards in the United States. There are eight species in the United States, including Wiggins' night lizard (*Xantusia wigginsi*), the desert night lizard (*Xantusia vigilis*), Sierra night lizard (*Xantusia sierrae*), island night lizard (*Xantusia riversiana*), granite night lizard (*Xantusia henshawi*), sandstone night lizard (*Xantusia gracilis*), Bezy's night lizard (*Xantusia bezyi*), and Arizona night lizard (*Xantusia arizonae*).

Despite the name, these lizards are not all strictly nocturnal. The name is more likely based on the presence of vertical pupils, which many other nocturnal animals also possess. In this case, however, the pupil structure is likely a result of the lizards adapting to dark microhabitats—such as crevices under bark and between boulders—where they live.

Night lizards also possess several other physical features that help distinguish them from other lizards, including large, plate-like scales on their heads, rows of rectangular scales on their ventral surface, and a lack of moveable eyelids. They must clean the clear scale over their eyes with their tongues.

Desert Night Lizard (*Xantusia vigilis*)

This very small and rather delicate lizard is rarely found far from shelter and may spend much of its life under the same area of cover. Its incredibly reclusive nature caused early naturalists to believe it was simply very rare. Common forms of preferred cover include fallen yuccas, dead Joshua tree limbs, slabs of bark, and dead agaves. You may also find them under the cover of wooden boards or cardboard sheets around the edges of desert towns.

Desert night lizards are covered in fine, granular scales that extend on the limbs and cheeks. The tail has scales that are more elongated and rectangular, but still very small. You'll find enlarged scales on the top of the head and around the mouth. The belly has slightly enlarged rectangular scales that are arranged in very neat rows.

The subtle impression of the body of a desert night lizard, showing a hind track and the square scale pattern of the ventral surface.

Desert night lizard understep trail.

These lizards feed on very small invertebrates such as termites, ants, and beetles, which are found under the same cover where the lizards live.

This species is often found in groups of its own kind. Lizard researcher Alison R. Davis studied this species for several years and discovered that these aggregations are formed by genetically related lizards. Night lizards are viviparous, which is unusual for a desert-dwelling reptile. Young desert night lizards tend to stay with their siblings and their parents under the same cover object, sometimes for up to several years. One likely reason for these social groups is that the proximity reduces rapid heat loss during freezing temperatures (Davis 2009).

Although this is a relatively common lizard throughout its range and in appropriate habitat, much still remains to be learned about this species' hidden lives. Finding their tracks is a rare treat, and you'll likely find them only in areas of fine dust next to fallen yucca stalks or similar cover objects.

The toes of this species are very thin and delicate. They are even smaller and finer than the toes of the desert banded gecko (*Coleonyx variegatus*). The tracks above were recorded under controlled conditions in the field, but also in fine dust very near to where the desert night lizards were found. For reliable tracks, the substrate must be fine enough to clearly register the tracks of small invertebrates. The most common sign left by this species might be the fragments of shed skin found under cover objects.

Track: Front: 0.2–0.3 in (0.5–0.76 cm) x 0.2–0.3 in (0.5–0.76 cm)
Tracks are very small. The front feet have five toes, all of which tend to register in the track. All toes are very thin. The front track is fan shaped and relatively symmetrical overall. Toes 3 and 4 are the longest and project equally far in the track; toe 3 may project slightly further. Toes 2 and 5 are about the same length. Toe 1 is the shortest. Front track tends to pitch to the inside of the central line of the trail. All toes possess small claws which tend to register at the toe tips. The metacarpal region of the foot is small and covered in granular scales. Front tracks orient parallel to or pitch slightly in toward the center line, and are positioned closer to the inside of the trail than the hind tracks.

Hind: 0.3–0.4 in (0.76–1 cm) x 0.3–0.4 in (0.76–1 cm)
Tracks are very small. The hind feet have five toes, all of which tend to register in the track. Track is asymmetrical overall. Toe 4 is the longest and tends to register between 45 and 90 degrees outward away from the center line of the trail. Toes 3 and 5 are about the same length. Toe 5 is positioned lowest on the foot and is usually oriented at about 90 degrees from the center line of the trail. All toes possess small claws which tend to register at the toe tips. All toes are very thin. The metacarpal region is small and covered in granular scales. Hind tracks orient parallel to the center line of the trail.

Trail: Understep Walk   Trail Width: 0.5–0.9 in (1.27–2.29 cm)
Tail Drag: 0.1–0.2 in (0.25–0.5 cm)

Notes:
- This is one of the smallest lizard species in North America. Its tracks are tiny and comparable in size to those of large insects. The foot morphology is similar overall to desert banded geckos (*Coleonyx variegatus*), except that the night lizard's toes are skinnier and more tapered.
- It's rare to find the tracks of this species, given its tendency to remain at the site of one particular fallen Joshua tree branch or pile of debris. You'll most likely locate the tracks of this species in fine dust near such cover.
- If you find these lizards, please be sure to place their cover exactly as you found it, and return the animals to the exact location where they were found. Due to this species' specialized lifestyle, it is sensitive to disturbance despite being common at some locations.

Family Scincidae

Members of this family are commonly called "skinks." Skinks have a shiny, smooth body with overlapping (cycloid) scales. These scales are reinforced underneath with osteoderms, which give the body a relatively hard exterior. This group contains a diverse number of body forms, from fully limbless and snake-like bodies to stout forms with large limbs. The exact number of species that live in the United States is still somewhat uncertain.

On the underside of their tails, skinks have a line of enlarged scales not unlike the large ventral scutes on the belly of a snake. This might reduce drag during the typical skink locomotion of "belly-walking," which is a type of direct register walk in which at least a portion of the belly and much of the underside of the tail maintains contact with the substrates. Skinks hold their

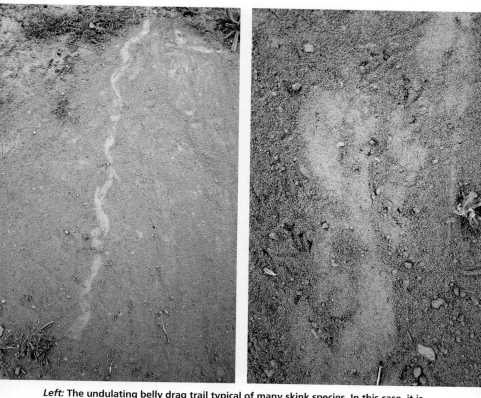

Left: The undulating belly drag trail typical of many skink species. In this case, it is the trail of a western skink. *Right:* A close-up of a western skink trail shows the subtle tracks on either side of the belly drag.

bodies in a sprawling posture, with legs held far out to the sides and the belly slung very low. As they move, they bend their bodies in repeating S-curves that are very similar to the lateral undulation of snakes.

Skinks have very skinny toes, which are relatively flexible. They arch up significantly toward the claws, and the front half of the toes generally does not touch down as the skink steps, except in deep, fine substrates. Their feet can also twist significantly as they rapidly belly-walk, which can obscure their tracks.

Some skinks will trot short distances, during which they can raise their bodies to a height where it no longer contacts the ground, and only the tail drag will be visible. This gait is atypical, and likely not energy efficient—the animals tend to rapidly switch to the belly walk after only a short distance.

It is vital to properly replace the cover objects where skinks are found. They seek out spots that hold moisture, so be sure to reseal the stone, log, or other object to help retain the beneficial qualities of these microhabitats.

Western Skink (*Plestiodon skiltonianus*)

This is a common skink found from southern British Columbia along the Pacific coast down to northern Baja California. It prefers habitats such as grasslands, chaparral, woodlands, and mixed forests. In drier areas, you'll find it most often found in rocky, well-vegetated spots near water.

This small, shiny lizard sometimes goes by the colloquial name "blue-tailed skink," as it sometimes has a blue tail, especially when younger. Adults are generally tan with three dark stripes. During the breeding season, reddish coloring appears on the adults around the mouth, tail, and occasionally on the sides.

It feeds on a variety of invertebrates such as beetles, caterpillars, grasshoppers, and crickets. It hunts for these by foraging underneath the cover of grass or fallen leaves. You'll most often be able to observe its trails when it temporarily slips out of cover to cross an open trail or ground squirrel throw mound.

Track: Front: 0.2–0.3 in (0.5–0.76 cm) x 0.2–0.3 in (0.5–0.76 cm)

Tracks are small. There are five toes present and all register in the track. The front track is somewhat asymmetrical. Toes 3 and 4 are equally long, and are the longest. Toe 1 is the shortest. Toes 2 and 5 are about equal in length. Toe 5 is positioned lowest on the foot. Claws are present on all toes, sharp, and moderately long. They register as tiny dots up ahead of the toes. The metacarpal region of the foot is covered in smooth, granular scales. Broad and smooth scales run across the toes and form lines. The front tracks generally register forward and parallel to the center line of the trail, or they pitch inward to the center of line. Front tracks land slightly more to the inside of the trail than hinds.

Hind: 0.5–0.8 in (1.27–2 cm) x 0.4–0.6 in (1–1.52 cm)

Tracks are small. There are five toes present and all register in the track. The hind tracks

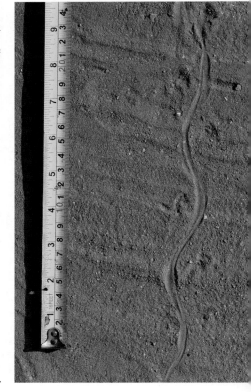

The direct register walking trail of a western skink showing less belly dragging and a more distinct tail drag.

are asymmetrical. Toe 4 is the longest. Toe 3 is only slightly shorter than toe 4. Toes 2 and 5 are about the same length. Toe 1 is the shortest. Toe 5 is positioned lowest on the foot. Claws are present on all toes, sharp, and moderately long. They register as tiny dots ahead of the toes. The metatarsal region of the foot is covered in smooth, granular scales. Broad and smooth scales run across the toes and form lines. The hind tracks generally register perpendicular to the center of the trail at about 90 degrees. Hind tracks land slightly more to the outside of the trail than fronts.

Trail: Direct Register Walk   Trail Width: 1–1.6 in (2.54–4.06 cm)

                                        Stride: 1.8–3.8 in (4.57–9.65 cm)

Tail Drag: 0.1–0.2 in (0.25–0.5 cm)

This species creates narrow tail drags, which are often much less obvious than the significantly wider marks left by the belly drag. The tail drag may follow inside the curving line made by the belly drag, or may register as another line curving in the opposite direction.

Belly Drag: 0.3–0.9 in (0.76–2.29 cm)

The belly drag of this species is wide and usually obvious. In loose substrates, belly drags might register more clearly than the marks made by the feet. Belly drags tend to undulate gently back and forth in waves, not unlike the trail of a slow-moving snake.

Notes:

- This species is the most familiar western skink species, occurring even in suburban neighborhoods of the major metropolitan areas.
- Watch for the tracks of this species crossing dusty trails in the summer, especially near riparian areas or shady, moisture-holding rock piles.
- The brilliant blue tail of this species is most obvious in juveniles. Adults tend to have more subdued blue on their tails. In some populations, the color may fade completely to tan at maturity.
- This species is capable of autotomy and will shed its tail when grabbed by potential predators or human hands.

Gilbert's Skink (*Plestiodon gilberti*)

Although its range and habitat overlap somewhat with the western skink (*Plestiodon skiltonianus*), Gilbert's skink tends to better tolerate more arid conditions. You can find them in more open areas than western skinks, when they are found together. You can even find this skink in the Mojave Desert along riparian corridors. It mainly lives in central and southern California, but also to a lesser extent in the southern tip of Nevada and western Arizona.

This species is sometimes referred to as the "red-tailed skink," as juveniles have a pinkish-purple or red tail. Some juveniles may also possess the reddish color on their limbs. As they age, they tend to fade in color, with adults generally being tan with three dark lateral stripes. The stripes do not reach very far onto the tail, unlike those of the western skink. Older males have very diffuse or absent striping and can appear tan overall, with orange-red coloring developing on the heads of older males.

This species is rarely seen in the open as it prefers to live and forage under cover. Its diet consists mostly of invertebrate prey, including small scorpions. Occasionally large Gilbert's skinks may eat juveniles of their own species.

Watch for their trails in dusty or sandy spots near areas of abundant ground debris. They tend to avoid densely brushy or forested areas.

The walking trail of an adult Gilbert's skink. JEFF NORDLAND

Track: Front: 0.3–0.5 in (0.76–1.27 cm) x 0.3–0.5 in (0.76–1.27 cm)

Tracks are small. There are five toes present and all register in the track. The front track is somewhat asymmetrical. Toes 3 and 4 are the longest, and are equal in length. Toe 1 is the shortest. Toes 2 and 5 are about equal in length. Toe 5 is positioned lowest on the foot. Claws are present on all toes, sharp, and moderately long. They register as tiny dots ahead of the toes. The metacarpal region of the foot is covered in smooth, granular scales. Broad and smooth scales run across the toes and form lines. The front tracks generally face forward, parallel to the center line of the trail, or pitch inward to the center of the line. Front tracks land slightly more to the inside of the trail than hinds.

Hind: 0.8–1.2 in (2–3.05 cm) x 0.6–0.8 in (1.54–2 cm)

Tracks are small. There are five toes present and all register in the track. The hind tracks are asymmetrical. Toe 4 is the longest. Toe 3 is only slightly shorter than toe 4. Toes 2 and 5 are about the same length. Toe 1 is the shortest. Toe 5 is positioned lowest on the foot. Claws are present on all toes, sharp, and moderately long. They register as tiny dots up ahead of the toes. The metatarsal region of the foot is covered in smooth, granular scales. Broad and smooth scales run across the toes and form lines. The hind tracks generally register perpendicular to the center of the trail at about 90 degrees. Hind tracks land slightly more to the outside of the trail than fronts.

Trail: Direct Register Walk   Trail Width: 2.2–2.5 in (5.59–6.35 cm)
Stride: 3.5–5 in (8.89–12.7 cm)

Tail Drag: 0.2–0.4 in (0.5–1 cm)

Tail drags created by this species are narrow and often much less obvious than the significantly wider marks left by the belly drag. The tail drag may follow inside the curving line made by the belly drag, or register as another line curving in the opposite direction.

Belly Drag: 0.7–1.1 in (1.78–2.79 cm)

The belly drag of this species is wide and usually obvious. In loose substrates, belly drags might be much clearer than the marks made by the feet. Belly drags tend to undulate gently back and forth in waves, not unlike the trail of a slow-moving snake.

Notes:
- This is a slightly larger species of skink than the western skink.
- There is some evidence to suggest that the Gilbert's skink is actually a species complex. Further research is needed.
- Observing this species in the wild is a treat. You will most likely encounter it during periods of high humidity or rain during warmer parts of the year.

Little Brown Skink (*Scincella lateralis*)

The habitat of this tiny brown skink stretches eastward from central and eastern Texas to Florida, and northward to southern New Jersey. It is certainly one of the smallest reptiles in the United States, with adult bodies reaching only about 2 inches long plus a tail of about one and half times that length.

This species is brown on top, with a dark lateral stripe running from the snout, passing through the eye, and extending along the sides to at least partway down the tail. Little brown skink juveniles do not possess a colorful tail.

You can find these skinks in moist woodland leaf litter, especially under oaks. They do not generally climb, but limit their foraging to underneath the litter, and, to a lesser degree, on top of it. Here they seek out small invertebrates such as spiders, flies, and beetles. Because they are so small, a wide variety of predators prey upon them, including other lizards, snakes, birds (even some songbirds), small mammalian carnivores, and even occasionally large spiders.

Watch for their subtle tracks in fine sandy areas such as sand roads or trails that pass through southeastern woodlands. You can catch these skinks by gently grabbing a handful of the leaf litter into which they have fled.

Track: Front: 0.1–0.2 in (0.25–5 cm) x 0.1–0.2 in (0.25–5 cm)

Tracks are very small. There are five toes present and all register in the track. The front track is somewhat asymmetrical. Toes 3 and 4 are the longest, and are equal in length. Toe 1 is the shortest. Toe 2 and toe 5 are about equal length. Toe 5 is positioned lowest on the foot. Claws are present on all toes, sharp, and moderately long. They register as tiny dots ahead of the toes. The metacarpal region of the foot is covered in smooth, granular scales. Broad and smooth scales run across the toes and form lines. The front tracks generally face forward parallel to the center line of the trail, or pitch inward to the center of the line. Front tracks land slightly more to the inside of the trail than hinds.

Hind: 0.3–0.4 in (0.76–1 cm) x 0.1–0.3 in (0.25–0.76 cm)

Tracks are small. There are five toes present and all register in the track. The hind tracks are asymmetrical. Toe 4 is the

The minute walking trail of a little brown skink. JONAH EVANS

longest. Toe 3 is only slightly shorter than toe 4. Toes 2 and 5 are about the same length. Toe 1 is the shortest. Toe 5 is positioned lowest on the foot. Claws are present on all toes, sharp, and moderately long. They register as tiny dots ahead of the toes. The metatarsal region of the foot is covered in smooth, granular scales. Broad and smooth scales run across the toes and form lines. The hind tracks generally register perpendicular to the center of the trail at about 90 degrees. Hind tracks land slightly more to the outside of the trail than fronts.

Trail: Direct Register Walk   Trail Width: 0.6–0.8 in (1.52–2 cm)
                                              Stride: 0.7–1.3 in (1.78–3.3 cm)

Tail Drag: 0.1–0.2 in (0.25–0.5 cm)
Tail drags created by this species are narrow and often much less obvious than the significantly wider marks left by the belly drag. The tail drag may follow inside the curving line made by the belly drag, or it may appear as another line curving in the opposite direction. The tail drags of this species can be relatively straight and register largely along the center line of the trail.

Belly Drag: 0.2–0.4 in (0.5–1.27 cm)
The belly drag of this species is wide and usually obvious. In loose substrates, belly drags might register more clearly than the feet marks. Belly drags tend to undulate gently back and forth in waves, not unlike the trail of a slow-moving snake. This species can also make a relatively straight belly drag.

Notes:
• Where found, this species of skink is often easily the most common lizard.
• It has demonstrated to be somewhat resilient to human habitat alterations and can often be found in the leaf litter of suburban backyards.

Florida Sand Skink (*Neoseps reynoldsi*)

This is a unique and specialized North American lizard; its range is completely limited to the sand scrub habitat of central Florida. It rarely travels out in the open on the surface; rather, it tunnels mostly just below the surface in a similar manner to the California legless lizard (*Anniella* sp.). The Florida sand skink is only a few inches long, with pale cream or whitish coloration, including scattered black dots and a black line running from the tip of the snout through each eye.

Sand skinks have four tiny limbs, with a single toe on the forelimbs and two toes on the hind limbs. As the lizard moves in an undulating wave through the sand, it tucks its limbs against its body. These skinks also have wedge-shaped heads, small eyes, and a countersunk lower jaw, and they lack external ear openings. The ventral side is flattened, though the rest of the body is cylindrical in cross-section.

This lizard species feeds largely on invertebrates it finds as it tunnels through the sand, especially antlion larvae and beetle larvae.

Given the size of their limbs and their manner of locomotion, these skinks don't often leave tracks with their limbs; rather, they leave undulating trails created by the collapsing tunnels made as they "swim" through the sand. You'll most likely encounter the trails of this species in sandy soil near the edge of a line of shrubs that have dropped a lot of leaf litter.

Trail: Sine Wave Tunnel   Overall Trail Width: 0.6–1 in (1.52–2.54 cm)
                                            Wave Length: 1.4–2 in (3.56–5.08 cm)
                                            Inner Trail Width: 0.1–0.2 in (0.25–0.5 cm)

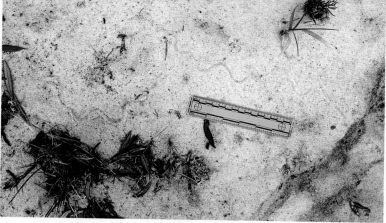

The evenly undulating trail of the Florida sand skink.

The trail of this species is small, often showing a repeating sine wave pattern. The inner trail width is proportionately narrow and fairly consistent, as the trail is formed by a collapsing tunnel rather than by the animal pushing down and sideways or backward as snakes do.

Notes:

- This species has lost a great deal of its native range to habitat loss and alteration. It should never be collected from the wild.
- Like other skinks, this species can lose its tail when grabbed by a predator or handled. Although the tail will grow back, it is best to prevent this traumatic event in the first place. It may even reduce the skink's likelihood of breeding successfully, and no doubt affects the energy efficiency of tunneling through the sand.

Spiny Lizards (Genus *Sceloporus*)

This large and diverse group of lizards found from Canada down to Panama includes common and widespread species, such as the western fence lizard (*Sceloporus occidentalis*) and eastern fence lizard (*Sceloporus undulatus*), both of which are commonly referred to colloquially as "blue bellies." The spiny lizard group includes some of the most visible lizards in the United States, as species in this genus like to bask in exposed, sunny locations.

Generally, spiny lizards are small to medium-sized lizards with keeled and pointed scales covering their bodies. They have short, blunt, and somewhat wedge-shaped heads. They are stocky

The front foot of a Yarrow's spiny lizard, showing the typical foot morphology.

Left: The hind foot of a Yarrow's spiny lizard, showing the typical foot morphology. *Right:* The walking trail of a granite spiny lizard, showing both the similarity to and difference from those of whiptail lizards.

creatures, and may be round or somewhat flattened in cross-section. Most species in the United States have tan, brown, or gray dorsal coloration. Blotches, chevrons, or stripes appear on their backs, tails, and sometimes their limbs.

Most species have some degree of blue color on the ventral surface, usually in patches on the sides and/or on the throat. The most dramatically blue species is the granite spiny lizard (*Sceloporus orcutti*), which has a solid blue ventral area often including all of the underside of the head, the entire belly, the underside of all the limbs, and even under the tail. Adult male granite spiny lizards in some populations may also have a great deal of blue coloration on the dorsal surface that fades to turquoise on the sides and ends in a wide purple area on the back. This is arguably one of the most fantastically colored lizards in North America.

Females of all species show reduced areas of blue on the ventral surface or may lack them completely.

Most spiny lizards climb on vertical surfaces, and very few are active on the ground. They climb quickly, able to dash up or across near vertical rock faces with apparent ease. This helps them escape many predators, as well as chase down prey out of the reach of ground-dwelling lizard species.

Spiny lizards feed on a variety of invertebrates including flies, beetles, ants, bees, wasps, true bugs, spiders, moths, butterflies, caterpillars, grasshoppers, and crickets. Some of the larger spiny lizards occasionally feed on vertebrate prey, including smaller lizards (sometimes even their own species), nestling birds, and other small vertebrates. Some species include vegetation in their diets such as flowers and leaves.

Spiny lizards hunt largely using visual cues, and may use a combination of active foraging methods and sit-and-wait techniques to capture prey. You can identify areas they frequent by the presence of their scat.

These animals are actively territorial, and you can often observe them posturing and chasing rivals. This conduct is especially true of dominant males. Common behaviors include push-ups, head bobs, and pushing out the throat or sides and standing stiff legged to show the colored ventral patches. Disputes can be over a log, tree, rock pile, or wall with an especially desirable basking area.

Most spiny lizard species lay eggs, and can have one or multiple clutches of eggs annually. Eggs are generally deposited in a shallow nest chamber dug by the female shortly before laying. Several species of spiny lizards are viviparous, including the Yarrow's spiny lizard (*Sceloporus jarrovii*) and crevice spiny lizard (*Sceloporus poinsettii*).

All share a similar foot morphology, which makes it relatively simple to distinguish them from other lizard species. The challenge, however, occurs in areas where several spiny lizard species coexist in the same habitat. The front feet have five distinct toes, with toes 3 and 4 being about equal length or toe 3 being slightly longer than toes 2 and 4. The metacarpal area of the front foot is moderately wide.

The hind feet have moderately long toes, with toe 4 being the longest. An obvious gap exists between toes 4 and 5 that forms a straight line. The negative space between toes 1, 2, 3, and 4 is variable. At higher speeds, there is often less negative space. Meanwhile, at slower speeds, the toes may spread distinctly. The most dramatic gap occurs between toes 3 and 4.

During high-speed gaits, such as overstep trots, the front tracks face directly forward in the direction of travel. The hinds tend to pitch out away from the center line of the trail slightly. In contrast, during slow gaits, such as an understep walk, the hind feet will angle between 45 and 90 degrees outward relative to the center line of the trail.

At this time you may confuse them with the tracks of a whiptail lizard (*Aspidoscelis* sp.), which has hinds that angle 90 degrees outward from the center line of the trail. A close inspection of the lengths of toes 4 and 5 in the tracks can help. Whiptail lizards have a particularly long toe 4 that reaches well beyond the much shorter toe 5. Also, in the tracks of whiptail lizards toe 4 often curves dramatically backward or forward. The shorter, stockier toe 4 of spiny lizards curves much less and is similar in length to toe 5 on the same foot.

Western Fence Lizard (*Sceloporus occidentalis*)

One of the most familiar lizards of the western United States, adult western fence lizards grow between 6 and 9.5 inches in total length. Adults vary in color and pattern somewhat, but appear tan or gray with a darker pair of blotches down the back with patches of blue scales. The deep blue throat and ventral surfaces

The front and hind track of an adult western fence lizard.

of this lizard, especially those of adult males, gives them their local name of "blue belly."

They live in a variety of habitats, such as woodlands, chaparral, oak woodlands, and suburban and urban areas. Typical microhabitats include fences, rock walls, or rock piles.

Track: Front: 0.4–0.8 in (1–2 cm) x 0.3–0.7 in (0.76–1.78 cm)

Tracks are small. Front feet have 5 toes, and all toes tend to register at slower speeds. During very rapid trots, only the middle two or three toes might register. Toes 2 and 3 are the longest, and are about equal length. Toes 1 and 4 are about equal length. Toe 5 is the shortest. Toes 1 and 5 tend to angle strongly to the side. The track is fan shaped and symmetrical overall. Claws are short, present on all toes, and may register as tiny dots at the toe tips. The metacarpal area is small and only registers at slower speeds. Front tracks tend to face directly forward or pitch slightly toward the inside of the trail.

Hind: 0.5–1.1 in (1.27–2.79 cm) x 0.5–1.1 in (1.27–2.79 cm)

Tracks are small to medium in size. Hind feet have five toes, all of which tend to register at slow to moderate speeds. During very rapid trots, toe 5 may not register, and sometimes toe 1 will not register either. Toe 4 is the longest and may register in line with toe 3 or might arch away slightly. Toe 5 is the lowest on the foot. It is spaced well away from toe 4 and generally angles about 90 degrees outward away from toe 4. Claws are short, present on all toes, and may register as tiny dots at the toe tips. Hind tracks tend to pitch slightly to the outside. During a slow walking gait, the hind tracks can angle outward more dramatically, between about 45 and 90 degrees away from facing directly forward.

Trail: Overstep Trot   Trail Width: 1.9–3.4 in (4.83–8.64 cm)

Stride: 4.1–6.8 in (10.41–17.27 cm)

Tail Drag: 0.2 in (0.5 cm)

You'll rarely see a tail drag in a fence lizard track, as they tend to stiffly carry their tails directly behind them. It might register for a short distance as the animal speeds up from a walk into a rapid trot. It might also register in particularly deep and loose substrate. Belly marks might be

The overstep trot of a western fence lizard.

visible where the lizard paused and dropped its body to the ground.

Notes:

- Tracks of this lizard species are some of the most frequent you'll encounter, since they are often found in suburban areas, parks, and green spaces frequented by humans. Look for their tracks in sandy spots or in the fine dust often found on well-used trails during drier periods.

- This is a wonderful species to study if you want to observe a variety of lizard behaviors. With a careful approach, you'll be able to observe this species, which often allows close observation. If you don't disturb it, the animal will go about its routines of hunting for insects and interacting with fellow fence lizards using a variety of body postures and movements, as well as sometimes chasing rivals.

- This "blue belly" is well known by many reptile fans. With the right kinds of structures, you can easily invite this species to live in your backyard, especially if you include warm, sunny basking sites, a water feature, and a variety of plants that can act both as cover and to attract insects for food. Any owner or tenant that includes these beautiful and lively lizards as residents in their yard should count themselves lucky.

- Western fence lizards are also extremely beneficial in the fight against Lyme disease. When a tick carrying the bacteria feeds on this lizard's blood, a protein in the lizard's body eliminates the bacteria. In areas where this lizard is found, only a small percentage of the ticks carries the disease-causing bacteria.

Desert Spiny Lizard (*Sceloporus magister*) and Yellow-backed Spiny Lizard (*Sceloporus uniformis*)

These two closely related species are a familiar presence in the deserts where they reside. They are both large, with robust bodies and impressively spiny scales. These two species consume a variety of invertebrates, especially beetles, grasshoppers, caterpillars, centipedes, and spiders. They also consume some leaves, flowers, and berries when in season. Males sometimes tackle large prey items including small lizards and nestling birds.

Track: Front: 0.7–1 in (1.78–2.54 cm) x 0.7–1 in (1.78–2.54 cm)

Tracks are medium in size. Front feet have five toes, and all toes tend to register at slower speeds. During very rapid trots, only the middle two or

The bipedal run of a yellow-backed spiny lizard.

three toes might register. Toes 2 and 3 are the longest, and are about equal length. Toes 1 and 4 are about equal length. Toe 5 is the shortest. Toes 1 and 5 tend to angle strongly to the side. The track is fan shaped and symmetrical overall. Claws are short, present on all toes, and may register as tiny dots at the toe tips. The metacarpal area is small and only registers at slower speeds. Front tracks tend to face directly forward or pitch slightly toward the inside of the trail.

Hind: 1.3–1.5 in (3.3–3.81 cm) x 0.7–1.3 (1.78–3.3 cm)

Tracks are medium in size. Hind feet have five toes, all of which tend to register at slow to moderate speeds. During very rapid trots, toe 5 may not register, and sometimes toe 1 will not register either. Toe 4 is the longest and may register in line with toe 3 or might arch away slightly. Toe 5 is the lowest on the foot. It is spaced well away from toe 4 and generally angles about 90 degrees outward away from toe 4. Claws are short, present on all toes, and may register as tiny dots at the top tips. Hind tracks tend to pitch slightly to the outside. During a slow walking gait, the hind tracks can angle outward more dramatically, between about 45 and 90 degrees away from facing directly forward.

Trail: Overstep Trot   Trail Width: 3–4.1 in (7.62–10.41 cm)
                                    Stride: 6–8 in (15.24–20.32 cm)

Tail Drag: 0.3–0.5 in (0.76–1.27 cm)

Tail drags are infrequent, as these large spiny lizards tend to stiffly carry their tails directly behind them. A drag might register for a short distance as the animal speeds up from a walk to a rapid trot. It might also register in particularly deep and loose substrate. Belly marks might be visible where the lizard paused and dropped its body to the ground.

Notes:

• These spiny lizards are often nearly twice the size of adult western fence lizards (*Sceloporus occidentalis*). They are often more wary than this smaller species and are more likely to take flight for the nearest cover while an approaching person is still many paces away. They act more tame when they appear in suburban fringe habitats or in busy campgrounds.

• This alert, quick-moving species can climb and run on open ground equally fast.

• Look for tracks crossing sandy washes between boulder-strewn hillsides and between trees in riparian areas. Also, watch for their scats on prominent snags and trailside boulders that they use for basking sites.

Sagebrush Lizard (*Sceloporus graciosus*)

This small spiny lizard favors mid- to high-elevation areas that include open and shrubby habitats. They are named for the sagebrush (*Artemisia* species) with which they are often found in the Columbia Basin and Great Basin regions. Sagebrush lizards also inhabit piñon and juniper woodlands, open pine forests, and open, sunny riparian areas. This lizard generally dwells on the ground, but will forage and seek refuge in the branches of low-growing shrubs.

When comparing its scales to the similar, but larger, western fence lizard, those of this species are smaller, and less dramatically spiny. This lizard generally has a grayish-brown coloration, and the pattern can vary, but it can include black and brown blotches and paler crossbars. Most individuals have a dark bar on the leading edge of the shoulder. There is also a broad, pale dorsolateral stripe. Adult males have a blue mottling on the throat or a uniform blue throat patch. They also have two long, broad blue patches on the belly. These are much paler or nearly absent in females. Gravid females often show some degree of deep orange-red on their chin, around their ears, and sometimes down onto their sides.

This species is generally shyer than the western fence lizard. They're easily disturbed, and run for cover in a shrub, burrow, or rock crevice. With a little patience and a slow approach, you can observe them hunting and displaying to each other. They are agile climbers, and feed on prey such as ants, flies, spiders, and beetles.

In many of the habitats in which they live, the soil tends to be coarse and rocky. Look for pockets of fine dust on trails or small sandy patches. In some parts of their range, they live in and around areas that include small sand dunes. One species of sagebrush lizard, the dunes sagebrush lizard (*Sceloporus arenicolus*) is an endangered species whose natural range is limited to the shinnery oak dune systems of eastern New Mexico and western Texas.

Track: Front: 0.3–0.5 in (0.76–1.27 cm) x 0.3–0.4 in (0.76–1 cm)

Tracks are small. Front feet have five toes, and all tend to register at slower speeds. During very rapid trots, only the middle two or three toes might register. Toes

The overstep trot of a sagebrush lizard.

2 and 3 are the longest, and are about equal length. Toes 1 and 4 are about equal length. Toe 5 is the shortest. Toes 1 and 5 tend to angle strongly to the side. The track is fan shaped and symmetrical overall. Claws are short, present on all toes, and may register as tiny dots at the toe tips. The metacarpal area is small and only registers at slower speeds. Front tracks tend to face directly forward or pitch slightly toward the inside of the trail.

Hind: 0.6–0.8 in (1.52–2 cm) x 0.4–0.6 in (1–1.52 cm)

Tracks are small to medium in size. Hind feet have five toes, all of which tend to register at slow to moderate speeds. During very rapid trots, toe 5 may not register, and sometimes toe 1 will not register either. Toe 4 is the longest and may register in line with toe 3, or it might arch away slightly. Toe 5 is the lowest on the foot. It is well spaced away from toe 4 and generally angles about 90 degrees outward away from toe 4. Claws are short, present on all toes, and may register as tiny dots at the toe tips. Hind tracks tend to pitch slightly to the outside. During a slow walking gait, the hind tracks can angle outward more dramatically, between about 45 and 90 degrees away from facing directly forward.

Trail: Overstep Trot Trail Width: 1.5–1.8 in (3.81–4.57 cm)
                       Stride: 4.2–6.5 in (10.67–16.51 cm)
        Overstep Walk Trail Width: 1.5–2.2 in (3.81–5.59 cm)
                       Stride: 3.5–3.8 in (8.89–9.9 cm)

Tail Drag: 0.1–0.2 in (0.25–0.5 cm)

Tail drags are infrequent as these small spiny lizards tend to stiffly carry their tails directly behind them. A drag might register for a short distance as the animal speeds up from a walk to a rapid trot. It might also register in particularly deep and loose substrate. Belly marks might be visible where the lizard paused and dropped its body to the ground.

Notes:
- This small spiny lizard ranges very high into some mountain ranges, and can even be found above 10,000 feet in some parts of the Sierra Nevada.
- In areas where substrate is not likely to register their tracks, watch for the scats of this lizard at favorite basking sites on exposed boulders, logs, or even large branches of low-growing shrubs.

Southwestern Fence Lizard (*Sceloporus cowlesi*)

This spiny lizard belongs to a confusing species complex once lumped under a single species: eastern fence lizard (*Sceloporus undulatus*). It was included with two other lizards, *Sceloporus consobrinus* and *Sceloporus tristichus*, under that single designation and referred to as subspecies. Continued genetic studies may end up clarifying the taxonomical challenges.

This particular lizard ranges through the southern two-thirds of New Mexico, southeastern Arizona, and west Texas. Its preferred habitat includes semidesert and grassland areas, woodlands, and some pine forests.

The scales of this species are strongly keeled and larger than those of the sagebrush lizard (*Sceloporus graciosus*). The overall color varies from tan or brown to nearly black or nearly white. It can appear much darker on colder days before the sun warms the air temperature throughout the day. The pattern is somewhat variable; the back is often an even gray, with a pale dorsolateral stripe that is sometimes interrupted, and often two rows of medium to small blotches down the back. Males have bright turquoise-blue patches on either side of the belly, as well as on either side of the throat. Females show little to no blue on the throat or belly.

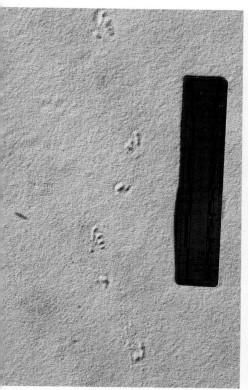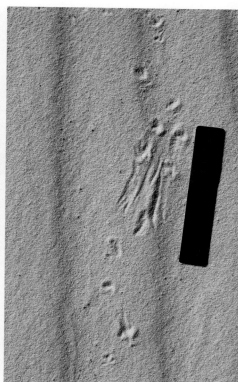

Left: The overstep trot of a southwestern fence lizard. *Right:* A trotting south-western fence lizard comes to an abrupt stop, then resumes trotting.

I have included this lizard in this book in part because its tracks are often found in association with those of the bleached earless lizard (*Holbrookia maculata ruthveni*) in the sand dune habitats they share. In this particular habitat, the lizard often favors dead yucca or fallen cottonwood branches as sites for thermoregulation. This species will also spend a considerable amount of time foraging on the ground.

This species feeds on a variety of small invertebrates such as ants, grasshoppers, and flies. They dart out to catch these insects as they come within a certain distance of the lizard's basking or resting site.

Watch for their tracks in sandy areas near large shrubs or stands of yucca. In areas where good substrate is not available for tracks, keep an eye out for scats on logs, fence posts, and debris piles.

Track: Front: 0.3–0.5 in (0.76–1.27 cm) x 0.3–0.4 in (0.76–1 cm)

Tracks are small. Front feet have five toes, and all tend to register at slower speeds. During very rapid trots, only the middle two or three toes might register. Toes 2 and 3 are the longest, and are about equal in length. Toes

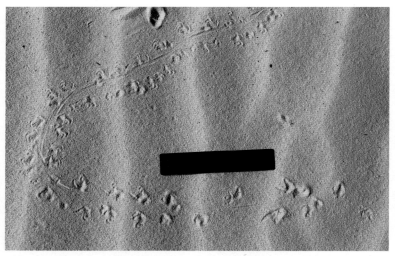

A slowly walking fence lizard turns toward the right and then speeds up into a trot.

1 and 4 are about equal in length. Toe 5 is the shortest. Toes 1 and 5 tend to angle strongly to the side. The track is fan shaped and symmetrical overall. Claws are short, present on all toes, and may register as tiny dots at the toe tips. The metacarpal area is small and only registers at slower speeds. Front tracks tend to face directly forward or pitch slightly toward the inside of the trail.

Hind: 0.5–0.8 in (1.27–2 cm) x 0.3–0.6 in (0.76–1.52 cm)

Tracks are small to medium. Hind feet have five toes, all of which tend to register at slow to moderate speeds. During very rapid trots, toe 5 may not register, and sometimes toe 1 will not register either. Toe 4 is the longest, and may register in line with toe 3 or arch away slightly. Toe 5 is the lowest on the foot. It is well spaced away from toe 4, and generally angles about 90 degrees outward away from toe 4. Claws are short, present on all toes, and may register as tiny dots at the toe tips. Hind tracks tend to pitch slightly to the outside. During a slow walking gait, the hind tracks can angle outward more dramatically, between about 45 and 90 degrees away from facing directly forward.

Trail: Overstep Trot   Trail Width: 1.3–1.6 in (3.3–4.06 cm)

                                  Stride: 6–6.8 in (15.24–17.27 cm)

       Direct Register Walk   Trail Width: 1.8–2.2 in (4.57–5.59 cm)

                                  Stride: 1–1.7 in (2.54–4.32 cm)

Tail Drag: 0.1–0.2 in (0.25–0.5 cm)

Tail drags are infrequent as these large spiny lizards tend to stiffly carry their tails directly behind them. A drag might register for a short distance as the animal speeds up from a walk to a rapid trot. It might also register in particularly deep and loose substrate. Belly marks might be visible where the lizard paused and dropped its body to the ground.

Notes:

• You'll likely best identify this species by carefully scrutinizing a combination of markings—including both dorsal and ventral—and by noting the specific range and habitat where you observe it.

- The faster this lizard moves, the more the hind tracks tend to point in the direction of travel. The slower it moves, the more perpendicular the hind tracks register to the center line of the trail. Knowing this about the lizard, as well as understanding its tendency to drag its tail at slower speeds, can be very helpful in interpreting its speed from the trail.

Family Geckkonidae

This is a large family, including about 1,200 species worldwide, although only a handful of species are found in the United States and only one of those is native: the peninsular leaf-toed gecko. Most geckos are nocturnal and all of them have enlarged, adhesive toe tips that allow them to climb even the smoothest surfaces. These geckos lack eyelids, and must clean their eyes with their tongues.

Peninsular leaf-toed gecko (*Phyllodactylus nocticolus*)

This unique and unusual lizard is named for its distinctive toe features. All members of the genus *Phyllodactylus* have twin scansor pads that make the toe tips appear leaf-like in shape. These lizards are able to easily run over and straight up, as well as go upside down and underneath, the surfaces of boulders where they live.

Their range is extremely limited, because north of Mexico they are only found along the eastern slopes of the Peninsular Range in southern California. Their color and pattern help them match the patterns on the boulders. Their skin is pinkish tan and covered in granular scales that are interspersed with larger keeled tubercles. Brown spots and blotches cover the head, limbs, and body. These tend to converge into thin bands on the tail. The gecko can easily shed its tail if a potential predator grabs it. Once the tail regenerates, it will have less of a bold pattern and may even be completely pattern free.

Like other small lizards, these geckos feed largely on small invertebrates, which they find in the crevices of their habitat, on the surface of boulders, or on the ground where they may forage at night.

This species, like other geckos in this family, does not have eyelids, so

The direct register walk of a leaf-toed gecko.

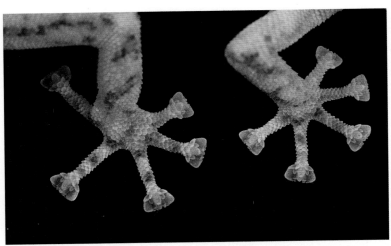

The front and hind feet of the leaf-toed gecko showing its namesake toe tips.

they must clean their eyes using their tongues. They are capable of vocalizing in squeaks and chirps.

Track: Front: 0.3–0.5 in (0.76–1.27 cm) x 0.3–0.5 in (0.76–1.27 cm)

Tracks are small. There are five toes present and all five register in the track. The front tracks are roughly symmetrical and fan shaped overall. All toes have paired scansor pads, divided by a claw. Toe 4 is the longest. Toe 3 is slightly shorter. Toes 2 and 5 are about the same length. Toe 1 is the shortest. Toes 1 and 5 are located directly across from each other on the foot. Claws are present but small, and are too short to register in the tracks. The metacarpal region of the foot is very small, covered in granular scales, and registers mostly in the small area between the toes. The front track pitches away from the center line of the track up to about 45 degrees. It lands closer to the inside of the trail than the hind track.

Hind: 0.5–0.6 in (1.27–1.52 cm) x 0.4–0.5 in (1–1.27 cm)

Tracks are small. There are five toes present and all 5 toes register in the track. The hind tracks are asymmetrical. All toes have paired scansor pads, divided by a claw. Toes 4 and 5 are about the same length. Toe 3 is slightly shorter than toe 4. Toe 2 is slightly shorter than toe 3. Toe 1 is the shortest. Toes 1 and 5 are located directly across from each other on the foot. Toe 5 points backward dramatically, between 180 and 200 degrees. Claws are present but are small and too short to register in the tracks. The metatarsal region of the foot is small, covered in granular scales, and registers mostly in the small area between the toes. The hind track pitches sharply away from the center line, landing perpendicular to the direction of travel at about or slightly beyond 90 degrees. It lands more to the outside of the trail than the front track.

Trail: Direct Register Walk   Trail Width: 1.6–1.9 in (4.06–4.83 cm)
                                  Stride: 1.8–2.5 in (4.75–6.35 cm)

Tail Drag: 0.2–0.3 in (0.5–0.76 cm)

Tail drags are generally absent, as this species lifts its tail clear of the substrate when it moves. The animal may leave tail marks when it comes to a full stop or dramatically changes direction.

Notes:
- Given this lizard's very limited range, please respect it and the California law protecting it, and do not collect these lizards from the wild.
- If you wish to observe these lizards, use a good flashlight or headlamp and move slowly and quietly through the habitat where they will most likely occur.
- Some evidence suggests that males of this species may mark their territories with fecal pellets (Jones and Lovich 2009). These observations were made in captivity and need to be verified in the field with further tracking studies.
- The tracks photographed for this book were made in a controlled setting. Look for them in the field in fine sand or dust between large boulder fields.

Whiptails and Their Allies: Family Teiidae

The small to medium-sized lizards in this family have long, tapered tails, narrow bodies, and pointed heads. Around twenty-three species are found north of the Mexican border, though this family includes about sixty species throughout the Americas.

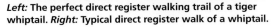

Left: The perfect direct register walking trail of a tiger whiptail. *Right:* Typical direct register walk of a whiptail.

This lizard group can be challenging to identify down to species, especially in areas where the ranges of multiple species overlap. The presence of hybrid species makes identification extra difficult, and can frustrate even experienced herpetologists. The genus *Cnemidophorus* was updated to *Aspidoscelis*. These changes are under some debate, so both are included here.

Whiptails are alert lizards, and are active foragers. They hunt using both visual and olfactory cues to locate prey, which they often find on the surface of the ground, under litter, or even underground. Whiptails appear to hunt incessantly and continue to move as they search for food. They prey on a variety of invertebrates including beetles, spiders, grasshoppers, crickets, and termites. The largest species of whiptail, the giant spotted whiptail, also sometimes includes small lizards in its diet.

The upper body of the whiptail is covered in granular scales, while the ventral surface is covered in large, squared scales. The leading edges of the front and rear legs feature enlarged scales. The rear legs are medium in length, with feet that have very long toes. Toe 4 is the longest on the hind feet. Front legs are shorter, with toe 3 being the longest.

Whiptails travel in a direct register walk most of the time, though they can sprint very fast over short distances, and are likely able to bipedally run at high speeds. They have a visually distinct, twitchy walk that can be recognized from a distance.

Tiger Whiptail (*Aspidoscelis tigris*)

Track: Front: 0.5–0.6 in (1.27–1.52 cm) x 0.4–0.6 in (1–1.52 cm)

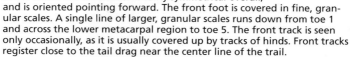

Tracks are small. The front foot has five toes, all of which tend to register in the track. The front track is relatively symmetrical and fan shaped overall. Toe 3 is the longest. Toes 2 and 4 are about the same length. Toes 1 and 5 are about same length. Claws are present and tend to register on all toes. The metacarpal region is small, relatively short, and often does not register well in the track. The front track is fairly symmetrical overall, and is oriented pointing forward. The front foot is covered in fine, granular scales. A single line of larger, granular scales runs down from toe 1 and across the lower metacarpal region to toe 5. The front track is seen only occasionally, as it is usually covered up by tracks of hinds. Front tracks register close to the tail drag near the center line of the trail.

Hind: 1–1.5 in (2.54–3.81 cm) x 0.8–1.3 in (2–3.3 cm)

Tracks are medium in size. The hind foot has five toes, all of which tend to register in the track. The hind track is asymmetrical. Toe 4 is the longest, and significantly longer than toe 3. Also, toe 4 will often curve, sometimes pointing back toward the center line of the trail. Toe 2 is shorter than toe 3, and toe 1 is shorter than toe 2. Toe 5 is far back on the foot and distinctly separate from other toes. Claws are present on all toes, moderate in size, and tend to register in the track. The metatarsal region is long, narrow, and covered in fine granular scales. It tends to only partially register in the track. The hind tracks tend to orient perpendicularly to the center line of the trail.

Trail: Direct Register Walk   Trail Width: 2.8–3.5 in (7.1–8.9 cm)

Stride: 2.8–4.4 in (7.1–11.2 cm)

Tail Drag: 0.1–0.3 in (0.25–0.76 cm)

The direct register walking trail of a tiger whiptail.

This species generally leaves a continuous, straight tail drag. Under certain substrate conditions, tail drag may be the only part of the trail that is obviously visible.

Notes:

- These lizards tend to use a direct register walk while foraging or moving from place to place. They can sprint when startled, running from a potential threat. They walk in a twitchy, jerky manner and start and stop frequently while foraging.

- Explore the landscape by poking under cover, into crevices, and investigating different spots out in the open. The tiger whiptail also makes frequent digs in areas of loose soil, likely after it locates buried prey by scent with a flicking of its tongue.

- This lizard will occasionally climb into low shrubs in search of potential prey.

- Because they are highly active, tiger whiptails may leave many trails in a small area as they travel in search of prey. In certain sandy areas, they may leave one of the most abundant trails of any vertebrate.

Little White Whiptail (*Aspidoscelis gypsi*)

Track: Front: 0.3–0.5 in (0.76–1.27 cm) x 0.3 in (0.76 cm)

Tracks are small. The front foot has five toes, all of which tend to register in the track. The front track is fan shaped overall. Toe 3 is the longest. Toes 2 and 4 are about the same length. Toes 1 and 5 are about same length. Claws are present and tend to register on all toes. The metacarpal region is small, relatively short, and often does not register well in the track. The front track is fairly symmetrical overall, and is oriented facing forward. The front foot is covered in fine, granular scales. A single line of larger, granular scales runs down from toe 1 and across the lower metacarpal region to toe 5. The front track is seen only occasionally, as it is usually covered up by hind tracks. The front tracks register close to the tail drag near the center line of the trail.

Hind: 0.9–1 in (2.29–2.54 cm) x 0.5–0.7 in (1.27–1.78 cm)

Tracks are medium in size. The hind foot has five toes, all of which tend to register in the track. The hind track is asymmetrical. Toe 4 is the longest, and significantly longer than toe 3. Also, toe 4 will often curve, sometimes pointing back toward the center line of the trail. Toe 2 is shorter than toe 3, and toe 1 is shorter than toe 2. Toe 5 is far back on the foot and distinctly separate from other toes. Claws are present on all toes, moder-

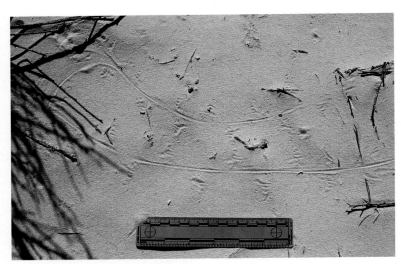

The direct register walk of a little white whiptail.

ate in size, and tend to register in the track. The metatarsal region is long, narrow, and is covered in fine granular scales. It tends to only partially register in the track. Overall, the hind tracks tend to orient perpendicularly to the center line of the trail.

Trail: Walk   Trail Width: 2–2.7 in (5–6.86 cm)

Stride: 1.3–3.6 in (3.3–9.14 cm)

Tail Drag: 0.1–0.2 in (0.25–0.5 cm)

The little white whiptail leaves a continuous, straight tail drag with tracks on either side. Under certain substrate conditions, tail drag may be the only part of the trail that is obviously visible.

Notes:

• This species is unique to white gypsum sand dunes of south central New Mexico. It is the only whiptail species found in this habitat.

• An active forager, the little white whiptail walks from place to place and makes frequent shallow digs in search of prey.

• It can sprint very quickly when it feels threatened.

• You may also occasionally observe this species climbing up into shrubs in search of prey. I observed one climbing several feet into a soap tree yucca.

Little Striped Whiptail (*Aspidoscelis inornata*)

Track: Front: 0.3–0.4 in (0.76–1 cm) x 0.2 (0.5 cm)

Tracks are small. The front foot has five toes, all of which tend to register in the track. The front track is fan shaped overall. Toe 3 is the longest. Toes 2 and 4 are about the same length. Toes 1 and 5 are about same length. Claws are present and tend to register on all toes. The metacarpal region is small, relatively short, and often does not register well in the track. The front track is fairly symmetrical overall, and is oriented facing forward. The front foot is covered in fine, granular scales. A single line of larger, granular scales runs down from toe 1 and across the lower metacarpal region to toe 5. The front track is seen only occasionally, as it is usually

The direct register walking trail of a little striped whiptail.

covered up by hind tracks. The front tracks register close to the tail drag near the center line of the trail.

Hind: 0.8–1.1 in (2–2.79 cm) x 0.4–0.9 in (1–2.29 cm)

Tracks are medium in size. The hind foot has five toes, all of which tend to register in the track. The hind track is asymmetrical. Toe 4 is the longest, and significantly longer than toe 3. Also, toe 4 will often curve, sometimes pointing back toward the center line of the trail. Toe 2 is shorter than toe 3, and toe 1 is shorter than toe 2. Toe 5 is far back on the foot and distinctly separate from other toes. Claws are present on all toes, moderate in size, and tend to register in the track. The metatarsal region is long, narrow, and covered in fine granular scales. The metatarsal region tends to only partially register in the track. Overall, the hind tracks tend to orient perpendicularly to the center line of the trail.

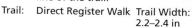

Trail: Direct Register Walk   Trail Width: 2.2–2.4 in

Stride: 2.1–3.2 in

Tail Drag: 0.1–0.3 in

Notes:

- Although the little striped whiptail is a close relative of the little white whiptail, it's found over a much larger range, especially in areas with sandy soils.
- You can use the measurements for this species for two other related species with adjoining ranges: the Arizona striped whiptail (*Aspidoscelis arizonae*) and Pai striped whiptail (*Aspidoscelis pai*).
- Whiptails are fascinating lizards to watch in the field and their trails expressively demonstrate their energy, alertness, and nearly continuous motion.

Crocodilians

5

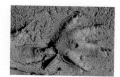

This group of large, well-armored reptiles is found around the world in the tropical and semitropical regions. They all possess powerful tails, long skulls with powerful jaw muscles, and four strong, well-developed limbs. Three species are found in United States: the American alligator (*Alligator mississippiensis*), the American crocodile (*Crocodylus acutus*), and the introduced spectacled caiman (*Caiman crocodilus*).

The last two species have very limited ranges, and I did not encounter their tracks or sign in the field while doing field work for this book.

Members of this group generally move around with an understep walk that involves lifting the body well off of the ground. They can also slide on steep mud banks by tucking their legs back and rapidly moving their tails. A few species are known to bound on open ground when they perceive immediate danger. This gait is sometimes mistakenly called a "gallop," which crocodilians are not physically capable of doing.

The large hind track of an American alligator.

Crocodilians have distinct foot morphology, with five toes on their front feet and only four toes on their hind feet. Their tracks can be quite large and their trails impressively wide. They also create other kinds of obvious sign such as digs, large nests, and scats.

American Alligator (*Alligator mississippiensis*)

This large, scaly crocodilian species is common in many parts of the southeastern United States. It was once endangered, but its population now numbers in the millions. It is the dominant predator of the swamps, marshes, and river systems where it dwells.

Male alligators can grow to well over twelve feet long (the largest on record was just over nineteen feet long), while females average between six and eight feet long. Large alligator specimens can weigh several hundred pounds. All crocodilians have skin reinforced with osteoderms. The most obvious ones are found under the major enlarged scales along the top of the neck and upper back.

American alligators eat a wide variety of prey, including fish, waterfowl, wading birds, snakes, turtles, small mammals, and even deer. Their powerful jaws can crush smaller prey, and their large teeth help them to hold on. These predators ambush their prey by sneaking up to within striking distance. Recent studies by Vladmir Dinets from the University of Tennessee, Knoxville, have shown that alligators sometimes use sticks as a kind of bait. The alligators place the sticks on their heads near active heronries, and wait for a nesting bird to come collect the material for nest building (Dinets et al 2013).

The bite force of large alligators is incredible, measured at around 2,000 pounds per square inch. They can be a potential threat to humans if harassed, and all alligators should be treated with respect.

Understep walking trail of an American alligator. SEB BARNETT

These are osteoderms, plates of bone under the large scales on the dorsal surface of an alligator. GEORGE L. HEINRICH

The cranium and three parts of the lower jaws of an adult American alligator.

Much of alligators' length is made up by their powerful tails. They swim using their tails, but they use their webbed feet at slower speeds. Like other water-dwelling reptiles, alligators spend a great deal of time thermoregulating by basking in the sun.

Track: Front: 3.73–5 in (9.4–12.7cm) x 3.41–4 in (8.7–10.2 cm).

Tracks are large to very large in size. Front feet have five toes, all of which tend to register in the track. Toe 3 is the longest. The foot often angles to the inside, so that toe 5 points either to the outside or back and to the outside of the trail at about 45 degrees. Claws are found on the first three toes. Webbing is present and most obvious between toes 2 and 3, and toes 3 and 4. There is no webbing between toes 4 and 5. Toes may show muscular segments not unlike a large bird's. A scale pattern may show in fine substrates, and is most obvious in the palm and heel of the track.

Hind: 6–8 in (15.2–20.3 cm) x 4–5.33 in (10.2–13.5cm)

Tracks are large to very large in size. Hind feet have four toes, all of which tend to register in the track. The foot generally points in the direction of travel. Toes 2 and 3 are nearly the same length. Claws are often visible, but there is no claw present on toe 4. Webbing may be visible in the track, and the largest is between toes 3 and 4. Toes may show muscular segments not

Left: The front and hind track of an American alligator. JONAH EVANS *Right:* A typical, trough-like slide from an alligator entering water in deep mud. GEORGE L. HEINRICH

The hide from the edge of the belly of an adult American alligator.

unlike a large bird's. A scale pattern may show in fine substrates, and is most obvious in palm and heel of the track.

Trail: Understep Walk   Trail Width: 11–24+ in (27.99–60.96 cm)

Stride: 18–28+ in (45.72–71.12 cm)

Trails often include a tail drag down the middle, and are generally found where alligators cross overland from one wetland area to another, or where they come out to bask. Alligators may also leave tail drags in the open or through well-vegetated areas. In the latter case, look for associated compressed vegetation.

The right front foot of an American alligator.

Notes:

- Alligators tend to use a high understep walk most of the time. In this case, the legs are placed underneath the body and raise it high off the ground as the animal walks. Alligators will occasionally move in a faster direct register walk. In extensive muddy areas, they may slide along on their bellies using the tail and hind legs.

- This is the largest native reptile species regularly seen throughout the south-eastern United States. Its close relative, the American crocodile (*Crocodylus acutus*), has much more limited range, and is found only in the southern tip of Florida.

- These huge reptiles are vital elements of an ecosystem, and large alligators act as ecological engineers by digging out large pools—called "gator holes" —that provide essential water-holding spots during drier times of the year. In the Everglades, large gator holes are found at the heart of nearly every cypress hummock.

Snakes

6

There are approximately 131 known species of snakes found in the United States and Canada. Snakes have such wonderfully different bodies from our own, having no limbs, no external ears, and eyes that can never blink. They are simple and beautifully elegant in design. Snakes are a diverse group of animals, demonstrating a wide range in color, pattern, body shape, and size. Snakes have adapted to practically every environment on the North American continent. Seven main families of snakes are found north of Mexico, including slender blind snakes (Leptotyphlopidae), blind snakes (Typhlopidae), boas (Boidae), pythons (Pythonidae), colubrid snakes (Colubridae), elapid snakes (Elapidae), and viperid snakes (Viperidae).

Snake Locomotion

The mesmerizing way in which snakes move has always fascinated humans. Their graceful, flowing form is truly a wonder of design and adaptation. Snake bodies are perfectly suited to move through the landscapes where they dwell, whether those landscapes are deserts, grasslands, wetlands, forests, jungles, or in the open ocean. Within each type of landscape, snakes that fill particular niches abide, and they have adapted physically and behaviorally to survive there.

Snakes can move in a variety of ways, depending on their body types and the substrate through which they move. Lateral undulation, rectilinear locomotion, slide-pushing, sidewinding, concertina locomotion, sliding, swimming, and vertical and horizontal climbing are all different major forms of movement.

Most snake species have enlarged belly scales called "ventral scales" or "scutes" that face backward. Those scales slide forward easily, but grip if pushed backward against a surface. They are the key that allows snakes to travel over any surface—including tree trunks, which they can climb—with relative ease. The two families of blind snakes are the only groups in North America that lack enlarged ventral scales. These snakes only spend limited amounts of time traveling on the surface, and are generally found underground where enlarged ventral scales are not advantageous.

In experiments studying snake locomotion, snakes that had their ventral scales covered by a jacket of cloth struggled to make any forward progress at all (Hu and Shelley 2012). This demonstrates how important these scales are to a snake's locomotion.

Most snake species can use several different methods of locomotion depending on conditions. Though a species can use all of the types mentioned here, some specialize in certain forms. Recognizing the type of locomotion present in a trail is your first clue to help tell groups of snakes apart.

Lateral Undulation

The most common and widely recognized form of locomotion is lateral undulation, also called "serpentine locomotion." To move this way, the animal sends a wave of motion down the length of the body, starting from the head and pushing backward. This gives the snake purchase through friction and allows it to move forward.

In the past, observers thought that snakes needed small objects on the soil surface in order to gain forward momentum. This did not explain,

An example of a lateral undulation trail, in this case the trail of a Great Basin gopher snake.

however, why many species could make forward progress on smooth surfaces. Later studies, especially those done by Dr. David Hu, demonstrated that snakes can push off of very small imperfections on smooth surfaces to move. Snake movement does not require a great deal of friction. The animals push downward as well as backward with certain areas of their bodies, using the ventral scales to grip and propel them. When moving in a lateral undulation, the snake's body is not evenly flat to the ground as it might at first seem to be. Instead, in order to reduce drag and increase friction of its belly scales, the snake lifts certain areas while pushing down and back with others. This is difficult to see from overhead, but you can observe this movement of the snake with your head at ground level or through video footage. Also, you can observe this activity in the tracks, where gaps may appear in the continuity of the trail or where decreased contact leaves a significantly narrower portion to the trail.

Generally speaking, the slower the lateral undulation movement is, the narrower the trail will be and the less consistent the waves of locomotion will appear. The trail may show a relatively straight or gently curving line, interspersed with occasional waves of motion accompanied by small ridges of displaced substrate. As the snake moves more rapidly using lateral undulation, the waves become more consistent and the trail becomes wider. Snakes that travel extremely rapidly using lateral undulation, such as coachwhip snakes (*Coluber flagellum*) and patch-nosed snakes (*Salvadora* sp.), may show another step in this progression: when the snake is bursting forward, the trail may narrow somewhat. The snake whips waves of movement down the length of its body so rapidly that it forms only fragments of trail, registering clearly only at the points of body contact, spots that are usually associated with the most dramatic ridges of displaced substrate. Snakes only use such high-speed lateral undulation over short distances to pursue prey or escape danger.

Concertina Locomotion

In this form of movement, a snake's body goes through a repeating start-and-stop pattern. The snake usually does this when it is wedging the anterior part of its body against a narrow area such as between the sides of a tunnel, a narrow area between boulders, or while climbing between the fissures of tree bark. Part of the snake's body bunches into a set of tight waves that push against the sides of the narrow space. Once the snake applies this pressure, it holds this part of the body still while bringing the posterior part forward and straightening it out. Immediately after, it bunches together the posterior part that was just straightened out, and wedges it against the two surfaces to either side of the animal. Then it straightens the anterior portion and pulls it forward. Thus, different parts of the snake's body follow slightly different paths.

This type of movement is based on expansion and contraction, similar to the movement of a concertina, the accordion-like instrument after whch it is named.

A snake that is climbing across the surface of a smooth branch may also use a modified form of the concertina locomotion. In this case, the animal's body follows a similar start-and-stop fashion of forward movement. In this situation, though, the snake will grip some of the circumference of the branch with the part of the body that it holds still, rather than pushing sideways against two surfaces. On a branch, all parts of the snake's body follow a similar path to minimize the possibility of falling. If the branch is small enough, a snake may wrap its body several times around the diameter in order to maintain grip and reduce the likelihood of a fall. Studies by Dr. Greg Byrnes at the University of Cincinnati reveal surprising information about how hard snakes will grip a branch. His studies found that snakes apply between two and five times the pressure needed to simply hold on and maintain a grip. This suggests that snakes apply additional force in order to greatly reduce the potential for injury.

Sidewinding

This is perhaps the most unusual snake locomotion form. A variety of snake species use a sidewinding motion, but the famous sidewinder rattlesnake (*Crotalus cerastes*) demonstrates it most dramatically and beautifully in North America. Snakes locomote this way by lifting parts of

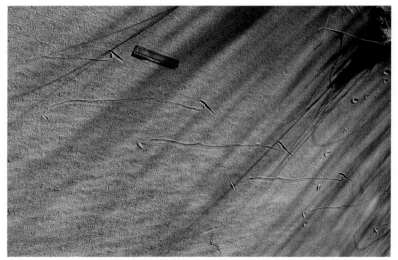

Here are some tracks left by the Mojave Desert sidewinder rattlesnake (*Crotalus cerastes cerastes*). This animal was moving from right to left.

their bodies clear off of the substrate, and sending a rolling S-curve down the body from the head to the tail. The body does not move directly forward, but rather in a diagonal direction, which leaves a row of repeated angled lines behind the animal.

With sidewinder rattlesnakes, these lines are generally cleanly separated by negative space, as they carry their tails well clear of the substrate. When longer and thinner snakes sidewind—such as colubrids like garter snakes—the diagonal lines are connected by a thin, wispy line left by the dragging tail. Also, colubrids leave more inconsistent and messy sidewinding trails, in contrast to the nearly perfect repeating pattern left by sidewinder rattlesnakes.

Although some moderately long-bodied snakes locomote this way under certain circumstances, short-bodied snakes moving on very loose, fine-grained substrate use this method most efficiently. Large and heavy-bodied snakes, such as pythons, are unlikely to sidewind because the movement requires that they repeatedly lift so much of their bodies clear of the ground.

Slide-pushing

Slide-pushing is a temporary form of locomotion used by snakes trying to get a grip on extremely smooth surfaces. Here, a snake sends a single wave of motion down the length of its entire body. The animal slides from one side to the other as it makes incremental forward progress. In a natural setting, snakes usually only do this on extremely loose, fine material, or on incredibly smooth surfaces such as fine, wet silt or fine-grained, water-soaked rock. Snakes can also slide-push when frightened, whipping wildly from side to side while they try to grip a surface in order to make an explosive forward movement.

This is not an energy-efficient gait, and it is not known to be used by any species as a preferred mode of travel.

Rectilinear Locomotion

Larger, heavier-bodied snakes use this form of locomotion while moving through open areas. Snakes achieve forward progress by making rippling, muscular waves that pull the snake forward in a relatively straight line. These ripples form small concentric undulations that lift, pull forward, drop, and push down the ventral scales. This rippling wave is similar to the method in which a millipede moves its legs. The key is the traction gained through the use of the enlarged ventral scales. The snake's body makes only slight sideways digressions as the animal moves forward. This means there isn't very much significant displacement of substrate along its trail, as there would be when a snake travels using lateral undulation. The subtle displacement that does happen is most visible as the snake changes direction while using rectilinear locomotion.

This impressively deep and wide trail was left by a large Burmese python in southern Florida. The animal was moving toward the camera.

This is a slow form of movement, but can be more efficient then lateral undulation for heavier-bodied snakes over short distances. It is often used by the larger pit vipers such as rattlesnakes, as well as by pythons and boas. This method implies that the snake is relatively relaxed, and is used by the snake to pass unnoticed in areas of thin cover when trying to remain unseen by a nearby predator. If the snake is threatened or excited, it would shift into one of the faster gaits.

Some snakes, including long, narrow-bodied species, will use this form of locomotion as they slow down and come to a stop in an area for basking. They will also use this type of locomotion as a way to travel slowly between two areas of cover without being detected. As you follow these trails, note how they show the snake entering or leaving such a spot using lateral undulation.

Identifying Snake Trails

The marks left by the passing of a snake's body are most accurately called "trails" since they are generally continuous, and are not made by feet. Distinguishing snakes down to particular species via their trails alone can be very difficult. As with tracking any other animal, you must carefully scrutinize all of the available evidence.

There are several main features to notice while looking at a snake trail: overall trail width, inner trail width, tail drag, and ridges. The overall trail width is the measurement at the widest portion of the trail. In other words, in a lateral undulation trail, you will measure from the highest point of one wave to the lowest point in the trough of that same wave. This is most accurately measured if you draw a straight line across the top of one wave and the bottom of the lowest point of that same wave, in the manner that you'd measure a sine wave on a graph. I have provided the wave length in this book as a useful measurement for each species of snake. The wave length is measured here from the apex of the peak of one wave to the apex of the peak of the adjacent wave.

The inner trail is the portion that is left by the belly of the snake. This can be challenging to measure accurately, as it can vary dramatically with speed and the amount of pressure the snake applies to the substrate along different parts of the trail. The best location to record this measurement is where the snake does not apply any sideways pressure and does not create an exaggerated sideways slide. In other words, where the snake is moving directly forward in a fairly straight line is the ideal spot to measure inner trail width to the highest degree of accuracy. This is not always possible, so measure the inner trail width in a portion of the trail that is not exaggerated by sideways sliding.

The tail drag is the mark left by the part of the body behind the cloaca, or vent, of the passing snake. Some snakes will carry their tails slightly or well off the ground. This is especially true of rattlesnakes. Therefore, rattlesnake trails often don't show drags made by the tail tip, but tend to show very thick lines down the middle of the trail where the tail base contacts the ground. Occasionally, one may drag its rattle through the substrate for a short time, but this is very unusual.

Ridges are those piles of substrate displaced by the passing of a snake. They tend to indicate the spots where the snake applied the most downward and backward or sideways pressure. The pattern and location of these ridges can help you understand the speed of the snake. They can

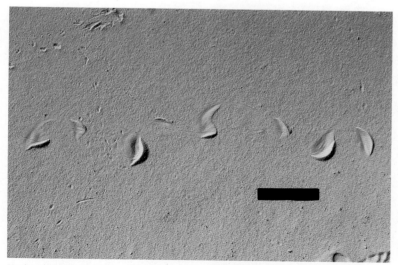

Notice the ridges in this photo of a gopher snake trail. The direction of travel in this case is from right to left.

also help to narrow down the possible species that may have created the trail. These ridges are also the key to quickly assessing which direction the snake was traveling, since snakes tend to push the substrate back and to the side in order to move forward.

Changing Gaits

A single snake species may use several different forms of locomotion as it shifts from one substrate to another, or as it reacts to the presence of prey or a potential predator. For example, a garter snake moving through an open area of loose sand may attempt to use a modified sidewinding movement to make forward progress. As soon as that same snake touches firmer substrate, it will switch to a lateral undulation. If it sees an approaching predator nearby while still on the loose sand, it may lash its body back and forth in a slide-push as it tries desperately to move more quickly, but struggles to get a grip. This same snake may approach

a wetland area where it has sensed the presence of frogs in a modified rectilinear form as it stalks closer, then switch into a rapid lateral undulation as it attempts to capture its prey. This can all happen within a relatively short distance, and is a major part of the challenge of identifying snakes via tracks and trails. Unlike the clear gait transitions made by four-legged vertebrates such as mammals, snakes show their transitions through the changing form of lines in the substrate.

Just as we can go from a crawl, to a walk, to a jog, and then a run, snakes too can shift from one form of locomotion to another. Some snakes may be better able to use certain types of locomotion, as their bodies may be particularly long and skinny, or relatively short and very

A rapidly moving snake slows down as it comes toward the camera.

thickly built. They may have a preferred way of moving that they will tend to use most of the time. For a garter snake, this is likely to be lateral undulation. Given the opportunity, the snake will use the form that is most energy efficient for it.

Each of the forms of locomotion discussed earlier can be considered very similar to the "gaits" used by limbed creatures such as mammals. A mammal may travel through a landscape and smoothly switch from a walk to a trot, then a lope and into a full-out gallop. This same mammal may have a preferred or "baseline" gait that it uses as its most energy-efficient method for moving as it travels. Snakes too will seek to move in a manner with which they are most comfortable and takes the least effort whenever possible.

The preferred means of locomotion used by various species of snakes may be largely based on not just their environment, but also on their physical build. It can be really helpful to look at snakes as having one of several body forms. Certain bodies are better equipped to move efficiently in one or more gaits, but may be more hampered when they are forced to use another. A good example of this is a very slim snake such as coachwhip. These are fast, relatively thin snakes that live in warm parts of the southern United States. They are very efficient at using lateral undulation to move around the landscape, and travel this way at high speeds. They are, however, so long that sidewinding effectively is very challenging for them. Not so for the short, stout sidewinder rattlesnake. This species is extremely efficient at sidewinding over loose sand, and can do it at a fairly rapid rate.

Here are several basic body types observed in snakes: short and stout, medium thickness and moderately long, long and skinny, and short and relatively skinny. Short and stout snakes tend to be most likely to use gaits such as rectilinear locomotion. Sidewinders are much more likely to be using side-winding locomotion than many other short and stoutly built snakes, but even they will use rectilinear locomotion frequently. Short and stout builds are found among hog-nosed snakes, pit vipers, water snakes, and boas.

Additionally, extremely large and heavy snakes will tend to use rectilinear locomotion method often. This includes such snakes as the introduced Burmese pythons.

Snakes of medium build and moderately to very long bodies might be more likely to use lateral undulation as well as sometimes using rectilinear locomotion. This includes snake such as gopher snakes, pine snakes, rat snakes, and indigo snakes.

Snakes that are long and skinny are often fast, active animals. This body type is found in racers, coachwhips, patch-nosed snakes, whip-snakes, ribbon snakes, rough green snakes, and vine snakes. The last two species are largely arboreal, and use this long, skinny build to help bridge gaps between branches. Some of these fast-moving snakes can travel in such a rapid lateral undulation that much of their body is held off the ground as they zip forward quickly.

The Challenge of Identification

Because of a general similarity in physiology as well as a lack of limbs, snakes provide a real challenge when it comes to the process of identifying their trails. Get to know the area that you're tracking—a solid understanding of species distribution and habitat preference in that area is hugely beneficial. You'll be able to considerably shorten the list of potential species if you're armed with this information.

It is vital that you try to observe snake trails that you wish to identify on a relatively flat substrate. Snakes will alter their gaits considerably when climbing or descending a slope. Also, seek out trails that represent a baseline form of locomotion for the particular snake species, one that represents the snake moving at a comfortable, normal speed. Do not use the trails of a snake acting out a particular behavior such as capturing prey, dashing momentarily to avoid a predator, or another such trail as an example of baseline locomotion.

Look closely at the accounts of snake species included in the following section of the book to better understand the subtle clues that help distinguish them via their trails. If you encounter a snake trail not included in this book, take the opportunity to photograph as much of the trail as you can and record detailed notes of your observations, so that if you encounter that species' trail again in the future, you will have a database of recorded knowledge you can refer back to. Ask yourself, what patterns continue to appear in this species' trail under similar conditions? Are there unique physical features or patterns within the trail? What are they? Are there features or patterns similar to those that appear in the trails of other species? What species? Continue to study snake trails and use the notes and information found here to serve as a starting place.

The beautiful scarlet kingsnake resting on a palm frond.

Boas: Family Boidae

These snakes are the relatives of the better-known—and much larger—tropical species like the boa constrictor (*B. constrictor*). American boas are much smaller, averaging between one and three feet in length. They all have proportionately small eyes, numerous body scales, and narrow ventral scales.

Rubber Boa (*Charina bottae*)

This species has been split into two closely related species: the northern rubber boa (*Charina bottae*) and the southern rubber boa (*Charina umbratica*). The first species ranges more widely, and is found from southern British Columbia down to mountains of central California. The second, smaller species is limited to a few mountain ranges in southern California. Rubber boas grow between 14 and 33 inches in total length. Adult males are generally smaller than females and have larger spurs on either side of their cloacas.

These boas prefer coastal coniferous forests, oak woodlands and savannas, and even moist areas in sagebrush steppe. They are not as heat tolerant as other snakes and will seek cover underground or in dense, moist areas during periods of hot weather. This species is frequently nocturnal.

Both species feed largely on small mammals, and frequently raid the nests of deer mice, voles, and similar small mammals so they can feed on the young. To a lesser extent they also feed on nesting birds and lizards. They kill their prey using constriction. Scars frequently appear on their heads, bodies, and especially tails from the bites of small rodents. These boas are born with round, blunted tails that are reinforced with bone. Often, due to increased scarring, their tails become more blunted with age.

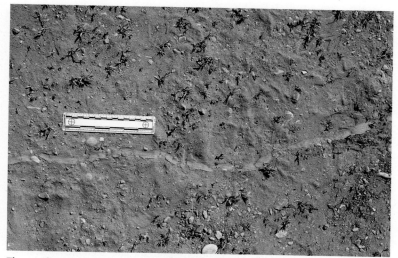

The rectilinear trail of a rubber boa moving from right to left.

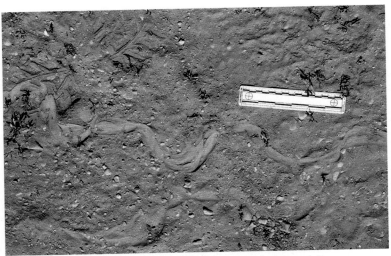

The lateral undulation trail of a rubber boa moving from right to left.

Scars on the end of the tail may also be from bites from potential preda-tors. Rubber boas sometimes cover their heads with their coils and raise their blunt tail up like a false head. This likely tricks predators into biting the hard and more expendable tail tip rather than the true head.

Rubber boas are generally olive green or greenish gray; the southern species is often pinkish orange or brown. Many individuals of both species have yellowish-colored ventral surfaces. A few individuals in both species have dark, even black, eyes while most have green, amber, or brownish eyes.

This snake gets its name from the loose skin on its body that looks rubbery, and even folds when the snake sharply bends. It generally moves slowly and is quite gentle. This species is not a good choice as a pet unless it is bred in captivity, as those caught in the wild will refuse to feed. They also live quite long—the oldest rubber boa on record lived between fifty and seventy years (Hoyer and Hoyer).

These boas spend significant amounts of time underground and under cover objects on the surface. They are amazingly cold tolerant, and have been observed on cool rainy nights.

You'll often encounter the trails of rubber boas as they cross dirt roads in spring and early summer.

Trail: Lateral Undulation   Overall Trail Width: 2.4–3.9 in (6–9.9 cm)
Inner Trail Width: 0.6–1.3 in (1.52–3.3 cm)
Wave Length: 4.3–7.2 in (10.9–18.28 cm)

The lateral undulation trails of rubber boas are often messy compared to the even trails of other snakes. The waves left in this trail type tend to vary significantly, and are often interspersed with relatively straight sections of trail. Compared to the trails of other snakes, the lateral undulation trails have a rather narrow overall trail width (or they could be described as having a generally low wave height). This demonstrates the tendency for this snake to move slowly, even at its typical "top speed" via the lateral undulation form of locomotion.

Rectilinear Locomotion   Overall Trail Width: 0.7–1.1 in (1.78–2.79 cm)
                         Inner Trail Width: 0.5–0.9 in (1.27–2.29 cm)

This form of locomotion is commonly used by this species when moving at a relaxed pace, often when it moves close to or in an area of cover.

Tail Drag: 0.2–0.5 in (0.5–1.27 cm)

The tail drags of this species frequently appear in trails and are similar to rattlesnake tail drags in width and consistency. The bluntness of the tail varies from individual to individual and changes over time due to increased scarring.

Notes:

- This incredibly gentle and slow-moving snake is a pleasure to find in the wild. Along with the rosy boa, this species is an ideal snake to introduce to children or to anyone who is working to get over their fear of snakes.

- Rubber boas are unique because they are the only boas found in the cool northern environments. They are often elusive, and despite ranging widely across much of the Pacific states, they often go about their lives unobserved by human eyes.

Rosy Boa (*Lichanura trivirgata*)

This beautiful, slow-moving snake is familiar to many in the pet trade. It prefers rocky areas in coastal sage-scrub, chaparral, foothills, canyons, and washes in semiarid and arid environments. In arid environments, you'll observe it more frequently near riparian areas. It is more active at night. Adults grow to between 17 and 44 inches in total length.

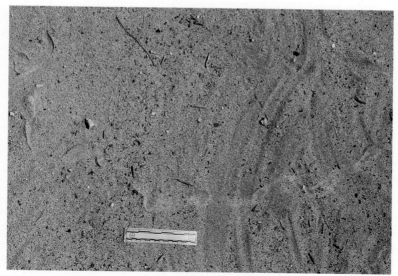

The lateral undulation trail of a rosy boa moving left to right.

The northern three-lined boa (*Lichanura orcutti*) was once thought to be part of the rosy boa species, but is now considered separate. Some contention exists about the exact taxonomy of the rosy boa. Some researchers place rosy boas in the genus *Charina*, the same as the rubber boa.

Northern three-lined boas have two different pattern classes: *roseofusca* and *gracia*. *Roseofuscas* have three irregular, poorly defined reddish-colored lines that extend down the length of the dorsal surface. The majority of the grayish-colored body shows reddish dots or small blotches of the same color as the three stripes. *Gracias* have a much cleaner pattern, with three distinct, well-defined lines running the length of the dorsal surface. Unlike *roseofuscas*, *gracias* show few or no dots or small blotches on the body between or around the main stripes.

The rosy boa (*Lichanura trivirgata*) has three poorly defined stripes that run the length of the dorsal surface, with profuse spots or small blotches of the same color as the stripes. The pattern is very faint, and some populations appear to be a single pinkish-gray or olive-gray color.

In general, rosy boas (*Lichanura* ssp.) are heavily built, and very gentle in nature, like the rubber boas (*Charina* ssp.). Rosy boas feed on a variety of small prey, including lizards, mammals, frogs, snakes, and nesting birds. They constrict to kill their prey.

Given the rosy boa's tendency to live in rocky habitats, the most common sign you'll likely find is their shed skins. The trails of rosy boas are very similar to those of rubber boas.

Trail: Lateral Undulation   Overall Trail Width: 2.5–4.5 in (6.35–11.43 cm)
Inner Trail Width: 0.8–2 in (2–5 cm)
Wave Length: 5.8–8 in (14.7–20.32 cm)
Similar to the lateral undulation trails of rubber boas, those of rosy boas are often messy compared to the even, repeatedly undulating trails of

The rectilinear trail of a rosy boa moving left to right.

other snakes. The undulating waves left by this species tend to vary significantly, and are often interspersed with relatively straight sections of trail.
Compared to the lateral undulation trails of other snakes, these have a
rather narrow overall trail width (or they could be described as having a
generally low wave height). This demonstrates the tendency for this snake
to move slowly, even at its typical top speed.

Tail Drag: 0.2–0.4 in (0.5–1 cm)

The tail drag of this species is frequently seen and relatively wide. Its drag
is less wide proportionately than that of rubber boas or rattlesnakes.

Notes:

- This shy, slow-moving snake is a pleasure to see in the wild. It is often hidden
under surface objects, but you can observe its surface action more frequently
than that of rubber boas. They use either slow lateral undulation or rectilinear
locomotion.

- You'll most likely observe this snake's tracks crossing dusty trails through prime
rocky habitats.

- Wild rosy boas should not be collected. There is a large captive population,
and it is important to leave healthy, breeding animals in wild populations. Like
rubber boas, they are a long-lived species, with some snakes living up to thirty
years in captivity. Continued collecting pressures on wild populations should
be seriously considered.

Pythons: Family Pythonidae

This family of snakes is native to Africa, Asia, and Australia. It includes about
fifty-seven species and subspecies, including some of the largest snakes in the
world. The only species included in this book is the Burmese python (*Python
bivittatus*), which was introduced to North America.

Burmese Python (*Python bivittatus*)

This tropical giant, one of the largest snake species on the
planet, was introduced to southern Florida, and it has now
established a population in the wild. The media has focused its
attention on this invasive species for its potentially dangerous
and damaging presence within native ecosystems. Their exact effect on the
Everglades ecosystem is still under investigation as of the publishing of this
book. Generally speaking, this species is not directly harmful to humans with
the exception of those who seek them out and catch them. Their strong jaws
and large teeth can cause painful lacerations. Despite their imposing size,
they are known to be mild mannered in captivity.

Wild adults in Florida grow to between six and seventeen feet in total
length, though some captive snakes have grown to over twenty feet. Males
are generally smaller than females, and have a slimmer build. The typical
color is yellowish to tan or gray, with large series of dark brown blotches bordered in black. A great assortment of colors and patterns exists in captive
populations.

These pythons eat a wide variety of prey, and stomach contents of wild captured Burmese pythons have included deer mice, cotton rats, rabbits, raccoons, opossums, squirrels, waterfowl, deer, and alligators. The last two are
likely only preyed upon by the largest pythons, given the physical constraints
of consuming even relatively small specimens of deer or alligators.

Left: The rectilinear trail of a very large Burmese python. *Right:* The lateral undulation trail of a slow-moving Burmese python.

This heavily built snake tends to favor rectilinear locomotion whenever possible. Snakes rapidly escaping danger, especially smaller pythons, will also use lateral undulation.

You'll most likely see pythons' rectilinear trails crossing open, muddy wetland areas, or moving across sandy trails between wetland areas.

Trail:　Rectilinear Locomotion　Overall Trail Width: 3–6 in (7.62–15.24 cm)
　　　　　　　　　　　　　　　　Inner Trail Width: 2–5 in (5–12.7 cm)

Notes:
- Finding the trail of a large adult Burmese python can be really awe inspiring. Despite being nonnative, they are an exciting animal to track and to see in the wild.
- This species may or may not leave a tail drag. It often carries the tail tip straight out as it moves forward.
- The habitat of the large eastern diamond-backed rattlesnake (*Crotalus adamanteus*) overlaps with the Burmese python, and this snake has similar measurements to smaller python specimens. It is more likely, however, that the diamond-backed rattlesnake will use slow lateral undulation rather than rectilinear locomotion.
- There are annual hunts organized for this python in southern Florida, with the intent to remove as many of these nonnative snakes from the wild as possible.

The close-up of a rectilinear trail from a large Burmese python.

Despite being large snakes, they are cryptic and difficult to find in the wild. The greatest threat to this species, however, is extended cold periods with temperatures at or below freezing, though further studies might reveal other significant sources of mortality.

Coachwhip (*Masticophis flagellum*) [previously *Coluber flagellum*]

This very wide-ranging snake lives largely in the warmer southern half of the United States. It has a very long, lean to moderately thick build with large eyes and an alert disposition to match them. This is arguably the fastest snake in North America, and is able to speed up to a maximum of eight to ten miles per hour. This might not seem all that fast when compared to the land speeds of fleet-footed mammals or the extreme speeds of a stooping falcon. This snake, however, is incredibly rapid and easily outpaces other snake species. It also often outpaces and outmaneuvers would-be human captors.

The largest coachwhip specimens are found in the southeastern part of the United States, and some have measured well over eight feet long. Most adults fall in the three- to five-foot range. They are active hunters, and sneak up on prey as close as they can before capturing them with a final burst of speed. They use both their sense of smell and their keen eyesight for hunting.

Scientific studies by Kevin van Doorn, PhD, and Professor Jacob Sivak from the Faculty of Science have shown that coachwhips can improve their vision by changing the blood flow to their spectacle, the clear scales over their eyes that act like a protective contact lens. In a controlled environment,

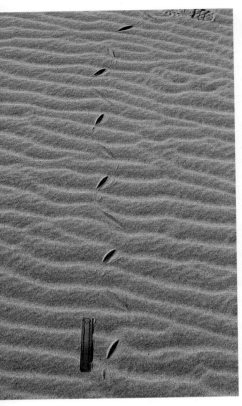

The typical lateral undulation trail of a rapidly moving coachwhip snake coming toward the camera.

the blood flow to the spectacle was minimized to ensure the best visual clarity in the presence of a threat.

This snake is generally a ground predator, but will climb into trees to seek climbing lizards and nesting birds. It preys on a wide variety of animals including small mammals, lizards, frogs, snakes (including rattlesnakes), and birds. Prey is generally grabbed with the mouth and quickly swallowed. This species may hold down larger items with its body, but it does not constrict prey.

This species includes several recognized subspecies in the United States, including the Baja California coachwhip (*Coluber flagellum fuliginosus*), lined coachwhip (*Coluber flagellum lineatulus*), red racer (*Coluber flagellum piceus*), San Joaquin coachwhip (*Coluber flagellum ruddocki*), western coachwhip (*Coluber flagellum testaceus*), and eastern coachwhip (*Coluber flagellum flagellum*). Each subspecies has distinctly different markings and different ranges. Three of the subspecies, *C. f. flagellum*, *C. f. piceus*, and *C. f. fuliginosus*, frequently include partially or largely melanistic individuals.

Trail: Lateral Undulation   Overall Trail Width: 2.76–6.7 in (7–17 cm)
                              Inner Trail Width: 0.36–1.2 in (0.91–3 cm)
                              Wave Length: 8.97–15 in (22.78–38.1 cm)

This species travels using lateral undulation, and holds its head well off the ground. As it travels rapidly, the lateral undulation trail it leaves tends to break up into parts limited to the ridges of displaced soil that are angled in an alternating pattern of short lines set at about 45 degrees pointing in opposite directions. The inner trail width is relatively wide directly along and to the inside of these ridges. This implies the snake slides slightly as it pushes down and backward with only parts of its belly. At slower speeds, the lateral undulation trail may appear more continuous.

Left: The lateral undulation trail of a coachwhip, overlaid by its rectilinear trail. This may have been a male scent trailing a female. Both snakes were moving toward the camera. *Right:* The trail of a coachwhip struggling up a steep sandy slope and moving toward the camera. This is what a sidewinding trail looks like from a long-bodied snake.

Additional Locomotion Forms

This species generally travels and hunts using rapid lateral undulation. It may sometimes use rectilinear locomotion when moving into or through cover, or when stalking prey. It likely uses concertina locomotion when climbing into shrubs and small trees. This species rarely uses sidewinding locomotion.

Notes:

- The trail of this species might be confused with that of the smaller, skinnier patch-nosed snakes (*Salvadora* genus). Note that coachwhip trails show more sideways pushing force visible in their wider inner trail. The trails of patch-nosed snakes also tend to create more sharply angular waves of lateral undulation and proportionally narrower trail widths.

- This lively, rapidly moving animal is a joy to watch hunting in the field. It may sit in ambush if it spots potential prey approaching. Coachwhip snakes can lunge from perfectly still to a chasing gait with blinding speed.

Left: The lateral undulation trail of a coachwhip moving toward the camera. *Right:* The trail of the the Baja California coachwhip (*Coluber fuliginosus*). JEFF NORDLAND

- This species is known for its feisty demeanor, and will often strike when cornered. It will also thrash, expel musk and feces, and bite when picked up. Its small teeth hardly do any damage, but it can be rather intimidating to the would-be captor, especially since the snake often first strikes at the face. If biting and fleeing don't work, coachwhip snakes also sometimes play dead.
- This species prefers relatively high temperatures for activity, and can be surface active at temperatures of over 100 degrees Fahrenheit. I observed an individual crossing a road near midday at 103 degrees Fahrenheit. Robert C. Stebbins observed a coachwhip consuming a desert iguana (*Dipsosaurus dorsalis*) close to midday when air temperatures were well over 104 degrees Fahrenheit.
- Studies by Bogert and Cowles in 1947 on coachwhip snakes from Florida demonstrated that this species is highly resistant to desiccation. This no doubt is a significant aid in allowing this snake to be active during hot temperatures and to survive in some of the most arid parts of North America.

Common Kingsnake (*Lampropeltis getula*)

This medium to large, very widespread snake grows between 36 and 82 inches in total length. It is a popular species in the pet trade. There are seven commonly recognized species, including the eastern kingsnake (*Lampropeltis getula getula*), Florida kingsnake (*Lampropeltis getula floridana*), California kingsnake (*Lampropeltis getula californiae*), speckled kingsnake (*Lampropeltis getula holbrooki*), black kingsnake (*Lampropeltis getula nigra*), Outer Banks kingsnake (*Lampropeltis getula sticticeps*), and desert kingsnake (*Lampropeltis getula splendida*). Each subspecies has distinct color patterns, and shows a significant amount of variability in appearance as well.

The habitat preferences vary from subspecies to subspecies, but include dry pine forests, deciduous woodlands, marshes, prairies, farmlands, piñon-juniper woodlands, chaparral, and deserts. In particularly arid regions, you'll most commonly encounter them in riparian areas or near water.

Kingsnakes feed on a wide variety of prey, including small mammals, birds

and their eggs, lizards, frogs, and small turtles. Most famously, they also consume other snakes, including venomous species such as coral snakes (*Micrurus* ssp.) and pit vipers such as rattlesnakes (both *Crotalus* and *Sistrurus* ssp.). They have been shown to have venom-neutralizing elements in their blood serum, which helps them

The beautiful lateral undulation trail of a common kingsnake, rapidly moving from right to left. KEVIN STOHLGREN

The slow lateral undulation trail of a California kingsnake moving left to right.

deal with the defensive venomous bites of these snakes (Ernst and Ernst 2003). Kingsnakes generally kill prey using constriction, though they may simply grab and swallow small animals.

Rattlesnakes sometimes recognize the scent or sight of an approaching kingsnake and flee, or assume a unique defensive position in which they hide their heads under their coils and strike out with the forward part of their bodies to thump the attacking kingsnake and discourage it.

Kingsnakes are one of the top reptilian predators where they are present. They are not, however, without their own predators, which include large hawks, golden eagles, and coyotes. Kingsnakes are also known to be cannibalistic.

Watch for their trails on dirty roads, trails, and muddy areas on the edges of wetlands.

Trail: Lateral Undulation   Overall Trail Width: 2–5.3 in (5–13.46 cm)
Inner Trail Width: 0.3–1.7 in (0.76–4.32 cm)
Wave Length: 5–13.8 in (12.7–35 cm)

Notes:
• This moderately thick snake typically moves using slower forms of lateral undulation.

Desert Striped Whipsnake (*Coluber taeniatus*)

This long, thin, and fast-moving snake can grow between 40 and 72 inches in length. In the southern parts of its range, you'll typically find this species in semiarid mountains and open woodland habitats. Farther north, you can also find it in sagebrush steppe, juniper woodlands, and desert scrublands. It has a wide range in the Southwest, from central Texas to eastern California and as far north as central Washington State.

The dorsal surface of this snake is typically dark brown, black, or grayish. There are two yellowish lines on either side of the body that are divided by the dark base body color. The ventral surface is pale, often creamy yellow becoming pink toward the tail.

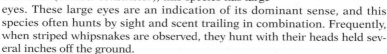

Like its relatives the coachwhip (*Coluber flagellum*) and racer (*Coluber constrictor*), this species has large eyes. These large eyes are an indication of its dominant sense, and this species often hunts by sight and scent trailing in combination. Frequently, when striped whipsnakes are observed, they hunt with their heads held several inches off the ground.

This alert species will quickly disappear into cover, even before you spot it. It will climb into shrubs if they are available, and may travel from shrub to shrub without touching the ground.

This species hunts during the day, feeding largely on lizards. It also catches and consumes smaller snakes, small mammals including bats, and some large insects. Typically, this snake approaches prey slowly, but will give chase if the prey animal moves as the snake approaches.

Watch for the trails of this species on dusty roads or paths, or in sandy spots in shrublands throughout its range.

Trail: Lateral Undulation Overall Trail Width: 5–7 in (12.7–17.78 cm)
Inner Trail Width: 0.5–1.3 in (1.27–3.3 cm)
Wave Length: 10–12 in (25.4–30.48 cm)

The lateral undulation trail of a desert striped whipsnake. Notice the relatively wide overall trail width, and the repeated marks of the body in the inner trail.

The rectilinear trail of a desert striped whipsnake. Notice the sand deposited on the stick where the snake climbed over it.

Another rectilinear trail made by a desert striped whipsnake. JOSHUA JONES

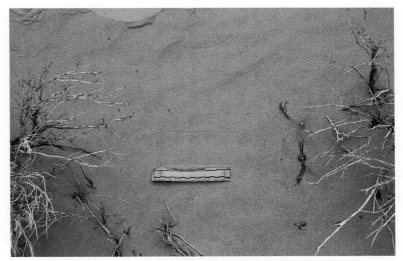

The very subtle mark made by only the tail tip touching down as a desert striped whipsnake climbed from one bush to another.

The lateral undulation trails of this species typically show large, repeating waves leaning backward away from the direction of travel. Due to this snake's very long, tapered body, repeated overlays of inner trail appear within each wave, each slightly narrower than the other.

Tail Drag: 0.1–0.2 in (0.25–0.5 cm)

Tail drags from this species are very thin, when present.

Notes:

- Observing this species in the field takes a considerable amount of luck and patience. Slowly walking and looking closely in prime habitats along the edges of riparian areas through arid and semiarid landscapes is a good approach.

- In areas that include many large shrubs, look for this species' sheds as an indication of its presence. This species often leaves its shed in the higher branches of shrubs.

- This species has declined significantly in Washington State, and may be threatened with extinction there without more study and protection. The reasons for the decline likely include habitat destruction, declines in the abundance of lizards (such as the sagebrush lizard), and changes in local ecosystems due to invasive species of plants such as cheatgrass.

Garter Snakes (*Thamnophis* ssp.)

These snakes are arguably the most widely recognized and commonly encountered in the United States. At least sixteen species of garter snakes are found in North America, north of Mexico.

Though harmless, garter snakes are technically venomous. They can deliver a mildly toxic saliva to their prey or would-be predators through enlarged rear teeth. This venom is effective on their prey, but at worst causes only minor albeit unpleasant reactions in people.

The typically messy lateral undulation trail of a common garter snake rapidly moving left to right.

This group of snakes has moderately large eyes, three dorsal stripes, and a slim to moderate build. All garter snakes have keeled scales and relatively wide ventral scales. Typically these snakes inhabit wetlands, moist meadows, fields, and open woodlands. Many species are most abundant near water.

Garter snakes have many predators including large fish, herons, birds of prey, raccoons, opossums, foxes, and coyotes. Their primary method for avoiding predation is camouflage and escape. When grabbed, they will typically expel musk and feces, and some may bite.

Their trails feature typical lateral undulation, and are often messy and inconsistent. Very large specimens may also use rectilinear locomotion when moving slowly. Occasionally, in areas of large, open sand, some garter snakes will also sidewind.

Common Garter Snake (*Thamnophis sirtalis*)

This is arguably the most widely distributed snake species in all of North America. Typically, adults have three stripes down the length of their body, one being middorsal with one stripe to each side. This species, however, can vary considerably in color and pattern. There are about twelve subspecies of the common garter snake, recognized by differences in range and physical appearance.

This is one of the best-studied reptiles in North America, and much is known about its behavior and ecology. In colder regions of the United States, this species uses hibernaculum (winter dens) which it shares with other garter snakes and sometimes other snake species. It can survive freezing temperatures, but will die when frozen for prolonged periods.

This small to moderately large snake grows from 18 to nearly 55 inches. When handled, it will typically expel feces and musk, and may sometimes

The wide, lateral undulation trail of a common garter snake struggling up a sandy slope.

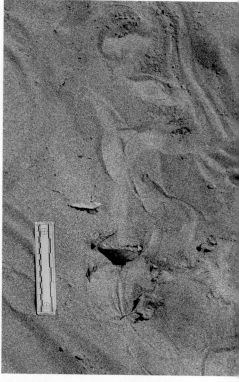

bite. This species feeds on a wide variety of prey, including frogs, toads, salamanders, large invertebrates, small snakes, young birds and their eggs, and small mammals. This is one of the only animals known to be able to eat the highly toxic rough-skinned newt (*Taricha granulosa*).

Watch for garter snake trails along the edges of wetlands, especially in mud along ponds or on dusty or sandy areas near wetlands. They will seek cover in a variety of natural and man-made debris such as logs, large rocks, bark, boards, tarps, and tin.

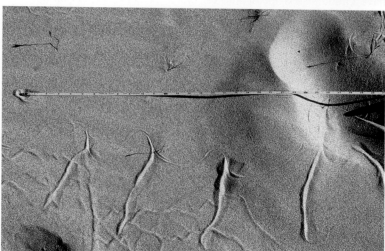

The sidewinding trail of a common garter snake moving left to right.

Very large specimens or gravid females may use rectilinear locomotion. In areas of large open sand, this species will also sidewind.

Trail: Lateral Undulation   Overall Trail Width: 3.5–7.2 in (8.89–18.29 cm)
                                 Inner Trail Width: 0.5–1.3 in (1.27–3.3 cm)
                                 Wave Length: 5.3–14.8 in (13.46–37.59 cm)
          Tail Drag: 0.1–0.3 in (0.25–0.76 cm)

Notes:
- This commonly encountered snake is an excellent animal to study in order to learn more about snake locomotion. Temporarily collecting these snakes, putting them in sandy areas, and following them to observe their trails can be very rewarding. Make sure to return the animals to where you found them, and respect your state and federal laws regarding handling or collecting these snakes.
- This snake is still a relatively common animal in the pet trade, despite the tendency for it to musk and bite. Captive snakes may acclimate to being handled.
- This species is more cold tolerant than most other snakes, and is also the snake species found farthest north in North America.

Northwestern Garter Snake (*Thamnophis ordinoides*)

This small to medium-sized snake has a proportionately small head, barely wider than its neck. Typically, this species has three longitudinal stripes down the length of the body: one middorsal stripe and a stripe on either side. This species is surprisingly variable in color and in pattern. The body color is typically dark green, brown, or gray, sometimes with dispersed reddish spots. The dorsal stripe can be yellow, green, tan, or orange. The lateral stripes are often yellow or tan. Some individuals only clearly show a single middorsal stripe. Others have a middorsal stripe that is very thin or disappears lower on the body. Occasionally, very dark (melanistic) individuals are found. Adults typically grow between 15 and 30 inches in total length, with most averaging toward the smaller end of the spectrum.

This species is restricted largely to the moist, sunny habitats such as forest openings, meadows, and fields along the edges of coniferous and mixed forests. Although you'll find this snake species in wet meadows, it doesn't typically inhabit extensive wetland areas such as ponds, sloughs, or lakes. It chooses to hunt in more upland and slightly drier sites than the common garter snake, but both species may be found together under the same cover object. When pursued near the water's edge, this species does not typically take to water to escape danger.

This species feeds on earthworms, slugs, and, to a lesser extent, small amphibians. Like all garter snake species, this one just grabs prey and swallows it live.

When handled, this snake typically does not bite but will often expel musk and feces.

Watch for the trail of this snake species in dust under bridges or crossing trails through fields during the summertime. Boards, large pieces of bark, and similar cover objects often harbor multiple northwestern garter snakes underneath them.

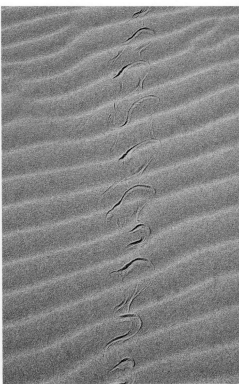

Left: The lateral undulation trail of a northwestern garter snake turning around.
Right: The sidewinding trail of a northwestern garter snake transitioning into a lateral undulation trail as it moves toward the camera.

   This species typically moves in a lateral undulation but will also use other gaits in certain situations, such as sidewinding in wide, open sandy areas.

Trail: Lateral Undulation   Overall Trail Width: 3–6 in (7.62–15.24 cm)
                               Inner Trail Width: 0.5–1 in (1.27–2.54 cm)
                               Wave Length: 4.8–9.1 in (12.19–23.11 cm)
         Tail Drag: 0.1–0.3 in (0.25–0.76 cm)

Notes:

- This small, gentle snake is a good animal to handle for those unfamiliar with snakes since it rarely bites.
- This snake is surprisingly cold tolerant, and its activity has been observed on sunny winter days in the lowland valleys of the Pacific Northwest. It is not typically very active on cool, rainy days.
- Along with the common garter snake, this species is an excellent one to keep around the garden as it helps control pests such as slugs. You can construct small rock piles to give them good basking and overwintering sites if you want them to stick around.

Glossy Snake (*Arizona elegans*)

This medium-sized snake—it grows from three to six feet long—is found from California to eastern Texas and lives in a wide variety of habitats featuring sandy or loose, loamy soils, including grasslands, chaparral, sagebrush steppe, rocky washes, and desert scrub. They are skilled burrowers, spending the heat of the day submerged in cooler substrate.

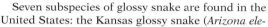

Seven subspecies of glossy snake are found in the United States: the Kansas glossy snake (*Arizona elegans elegans*), Texas glossy snake (*Arizona elegans arenicola*), Mojave glossy snake (*Arizona elegans candida*), desert glossy snake (*Arizona elegans eburnata*), Arizona glossy snake (*Arizona elegans noctivaga*), California glossy snake (*Arizona elegans occidentalis*), and Painted Desert glossy snake (*Arizona elegans philipi*).

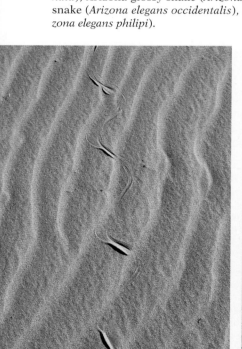

Left: The lateral undulation trail of a glossy snake moving away from the camera. *Right:* The lateral undulation trail of a glossy snake parallel to the smaller trail of a western shovel-nosed snake.

Glossy snakes are smooth scaled, with somewhat pointed heads and countersunk lower jaws. Their body color is generally tan or cream, with numerous rounded gray or brown blotches with black edges. The number of blotches, darkness of overall color, and presence of brown speckling varies between the seven subspecies.

This species is sometimes confused with visually similar gopher snakes (*Pituophis catenifer*), which have keeled scales and lack the countersunk lower jaw. Also, glossy snakes are generally crepuscular and nocturnal.

At night, these snakes will seek out buried prey such as lizards and small mammals. They dig through the soil and plug burrow entrances with their heads before grabbing the prey, or they constrict prey by pressing it with their bodies against the wall of the burrow. Glossy snakes also catch prey on the surface, including small birds.

Watch for their trails in open, sandy areas of deserts and grassland habitats of the western United States.

Trail: Lateral Undulation   Overall Trail Width: 2.6–9 in (6.6–22.86 cm)
Wave Length: 9.8–14 in (24.89–35.56 cm)
Inner Trail Width: 0.5–1.2 in (1.27–3 cm)

The lateral undulation trail of this species resembles, but is much larger than, the consistent sine waves of shovel-nosed snakes (*Chionactis occipitalis*). Like that species, the glossy snake's inner trail tends to remain narrow when the animal travels on open ground, on level substrates. These will change somewhat with directional or speed alterations.

Tail Drag: 0.1–0.2 in (0.25–0.5 cm)

Tail drags are not common in this species, as it often holds the tail tip straight out as it slithers.

Notes:

- Along with the much smaller shovel-nosed snake, this snake's trails are relatively common in open, sandy areas of the desert Southwest.
- Their trails can be distinguished from those of gopher snakes, because the trails of glossy snakes show much less body sliding when they travel via baseline lateral undulation. You can observe this in the narrower inner trail width, and smaller ridges thrown up by the glossy snake. Remember to take into account substrate, slope, and aging of the trail.

Gopher Snake (*Pituophis catenifer*)

This is a widespread and familiar species to many, especially those living in rural settings. It is often locally referred to as a "bull snake" or "blow snake." This is likely a reference to its defensive behavior, as it puffs itself up, holds its ground, and hisses loudly. This harmless species can often grow large, and includes six commonly recognized subspecies: the Pacific gopher snake (*Pituophis catenifer catenifer*), Great Basin
gopher snake (*Pituophis catenifer deserticola*), San Diego gopher snake (*Pituophis catenifer annectans*), Sonoran gopher snake (*Pituophis catenifer affinis*), Santa Cruz Island gopher snake (*Pituophis catenifer pumilis*), and bullsnake (*Pituophis catenifer sayi*).

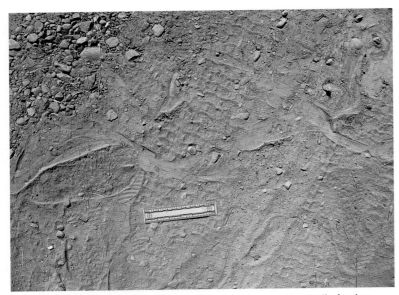

The lateral undulation trail of a slow-moving Pacific gopher snake moving right to left. KIM CABRERA

The subspecies vary somewhat in size, coloration, and pattern. The smallest subspecies is the Santa Cruz Island gopher snake, with adults growing to just over 36 inches in length. The largest subspecies are either the bullsnake or the Sonoran gopher snake, with records for both growing to 100 inches in length.

Gopher snakes are moderate to thickly built snakes, with long bodies and large heads. They have a series of large, repeating saddles down the middle of the dorsal surface and smaller repeating blotches along each side. The body color is yellowish to tan, while the saddles and blotches are brown to black. A small proportion of

The rectilinear trail of a Pacific gopher snake. KIM CABRERA

The lateral undulation trail of a rapidly moving Great Basin gopher snake moving left to right.

Left: The trail of a Sonoran gopher snake that slid down a steep sandy slope, keeping its body in a fixed position. This snake was moving toward the camera.
Right: The lateral undulation trail of a Sonoran gopher snake climbing a steep sandy slope, moving away from the camera.

The rectilinear trail of a gopher snake traveling down a steep slope.

some populations show a striped pattern, especially in populations of the Pacific gopher snake.

Gopher snakes are found in a wide variety of habitats across the United States, including: farms, grasslands, chaparral, sagebrush-steppe, oak savannas, open coniferous forests, wetlands, and even suburban yards. Generally active during the day, they become more nocturnal during the heat of summer.

Gopher snakes are constrictors and often favor small mammals as prey. They are known to kill prey by constricting either in coils or by pressing the animal against the side of a burrow. They also sometimes eat birds, bird and turtle eggs, and some lizards.

Gopher snakes vary considerably in temperament. One might immediately respond to an approaching person with puffing, hissing, and even striking, while another might turn tail and slither rapidly away. Finally, a third may act very placidly, merely looking up and flicking its tongue, even when picked up.

Watch for their trails crossing dirt roads or sandy or dusty trails, in washes, and along the edges of sand dunes. They frequently use lateral undulation as a baseline form of movement when traveling, but will also regularly use rectilinear locomotion. This tendency to use rectilinear locomotion is especially true for larger individuals.

Trail: Lateral Undulation   Overall Trail Width: 4.3–10 in (10.92–25.4 cm)
                                 Inner Trail Width: 0.4–3 in (1–7.62 cm)
                                 Wave Length: 6.6–18.5 in (16.76–47 cm)
This form of locomotion is used at both high and low speeds. At higher speeds, this species creates large waves that may or may not be consistent. At slower speeds, the waves are typically inconsistent and interspersed with straight portions of the trail. The inner trail width increases and the overall trail width decreases as the animal slows.
Rectilinear Locomotion   Overall Trail Width: 1.5–4 in (3.81–10.16 cm)
                          Inner Trail Width: 0.6–2.5 in (1.52–6.35 cm)
This species often uses this very slow style of locomotion when it moves in or toward an area of cover when trying to avoid detection. Large

individuals may also move this way when crossing trails or dirt roads. Gopher snakes also tend to use rectilinear locomotion when traveling down steep slopes on loose substrates.

Tail Drag: 0.1–0.2 in (0.25–0.5 cm)

Tail drags are uncommon for this species, but when present are usually very thin.

Notes:

- This is one of the most commonly encountered snake trails where this species is abundant. Its trails might be confused with those of other large colubrids, such as common kingsnakes (*Lampropeltis getula*), that share its range and habitat. When compared to similar sized kingsnakes, gopher snakes tend to create larger waves with greater wave heights when traveling at moderate to rapid speeds using lateral undulation.

- I have witnessed one particular individual—an adult Sonoran gopher snake—use a passive sliding form of locomotion on a steep sand dune slope. This unusual movement form created what resembled a rectilinear trail, except that the snake moved forward with parts of its body still bent, which created a wider trail than typical rectilinear trails. See the photo of the trail on white sand on page 261 for an example.

- Along with garter snakes, this is one of the best species to study to learn more about snake locomotion, since it can demonstrate a wide variety of movement styles.

Hog-nosed Snakes (Genus *Heterodon*)

Snakes in this small group have relatively short, thick-set bodies. They are mildly venomous, rear-fanged predators of small animals, especially toads. These snakes are named for the upturned rostral scales, with which they dig in the soil.

Eastern Hog-nosed Snake (*Heterodon platirhinos*)

This medium-sized snake of the Midwest and eastern United States is well known for its defensive behaviors, which include puffing up, hissing, flattening out the neck like a cobra, and rolling over and playing dead. Adults measure between 20 and 45 inches in total length.

This species' ground color is yellow, olive, brown, gray, or black, with transverse dorsal blotches and often two rows of smaller dark-brown blotches on each side. A dark, V-shaped mark reaches from the neck onto the parietal scales. A transverse dark bar runs between the eyes and extends downward to the corner of the mouth. The rostral scale is turned up, and is used for digging.

This species preys on a variety of small frogs, salamanders, mammals, and large invertebrates. They predominately feed on toads.

Eastern hog-nosed snakes are found in habitats including woodlands, grasslands, sand dunes, and cultivated fields. Typical habitats include areas with relatively dry, loose, and especially sandy or loamy soils. Certain species of toads, such as the Fowler's toad (*Anaxyrus fowleri*), also favor these soils.

The typical lateral undulation trail of an adult eastern hog-nosed snake. Notice the wide inner trail width. This snake was moving left to right. JOHN VANEK

The lateral undulation trail of an eastern hog-nosed snake moving right to left across compacted sand. JOHN VANEK

This lateral undulation trail moves away from the camera. The snake was followed as part of a scientific study using radio telemetry and wildlife tracking skills. JOHN VANEK

This species will use its enlarged rear fangs to puncture toads and inject them with venom when they inflate defensively. The venom of this species is not dangerous to humans.

The eastern hog-nosed snake is known to dig nesting chambers in sandy soil for its eggs.

Trail: Lateral Undulation   Overall Trail Width: 4–6.5 in (10.16–16.51 cm)
Inner Trail Width: 1–2 in (2.54–5.08 cm)
Wave Length: 8–14 in (20.32–35.56 cm)

The lateral undulation trail of this species commonly has a relatively wide inner trail width in proportion to the overall trail width. Although the waves of undulation are relatively wide, they are also relatively short in height (as demonstrated by the proportionately short overall trail width).

Tail Drag: 0.1–0.2 in (0.25–0.5 cm)

Notes:

- This snake shows its thickness in its lateral undulation trail patterns.
- You may confuse the trails of this species with those of rattlesnakes, except that eastern hog-nosed snake trails frequently show a ridge of soil crossing the area of the inner trail width at abrupt angles. The trails of rattlesnakes also show a thick tail drag.
- This snake is sometimes seen in the pet trade. They are challenging to keep, however, as they require amphibians to eat, and will suffer poor health when fed rodents.
- Despite all the dramatic defensive behavior of wild snakes, hog-nosed snakes are mild mannered when handled and infrequently bite.

Patch-nosed Snakes (Genus *Salvadora*)

The snakes in this small North American group are slim and move fast, and feature a raised, wraparound rostral scale. Some evidence points to these snakes favoring the eggs of reptiles such as lizards and snakes as a major food source. The specialized rostral scale is not used to dig burrows; rather, it's likely these snakes use it to dig out reptile eggs.

Western Patch-nosed Snake (*Salvadora hexalepis*)

This beautiful, slim western snake dwells in the Mojave Desert, Colorado Desert, Sonoran Desert, and a small part of the Great Basin Desert. One subspecies is also found in coastal chapparal in southern California. This species is typically present in desert flats, washes, and canyon bottoms with gravel or sandy soils.

This is a small to medium-length snake with adults measuring between 12 and 46 inches in total length. It is alert and elusive,

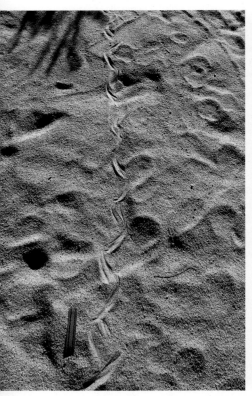

Left: The lateral undulation trail of a rapidly moving desert patch-nosed snake.
Right: Another lateral undulation trail of a rapidly moving desert patch-nosed snake.

and can be difficult to observe in the wild. This species is most active in the morning or late afternoon into the evening.

There are three subspecies of the western patch-nosed snake: the desert patch-nosed snake (*Salvadora hexalepis hexalepis*), Mojave patch-nosed snake (*Salvadora hexalepis mojavensis*), and coast patch-nosed snake (*Salvadora hexalepis virgultea*). The body appears pale tan to yellowish in color, with dark stripes down the length of the back subdivided by a single pale middorsal stripe running the length of the body between the dark stripes. The scales are smooth and the eyes are relatively large.

This species feeds largely on reptiles, especially lizards and their eggs, small snakes, and frogs. They hunt both by sight and scent. Patch-nosed snakes are fed on by a variety of species, including larger snakes such as coachwhip snakes and kingsnakes.

Watch for this snake's distinctly narrow, rapidly lateral undulation trails crossing sandy patches of open ground, such as washes. When startled, these snakes will find cover on the ground under dense shrubs.

Trail: Lateral Undulation   Overall Trail Width: 1.7–5.1 in (4.32–12.94 cm)
Inner Trail Width: 0.5–1 in (1.27–2.54 cm)
Wave Length: 7–11.6 in (17.78–29.46 cm)

The trails of this species show a relatively consistent pattern of shallow, repeating waves when moving rapidly. The apex of the waves may be rounded or come to a point. The overall trail width is narrow when compared to the trails of other snake species of a similar size.

Tail Drag: 0.1–0.2 in (0.25–0.5 cm)

Tail drags of this species are inconsistent and very thin.

Notes:

• This species has been observed moving back and forth while in tall grass gently blowing in a breeze. It is possible such behaviors are meant to further disguise this snake's shape.

• This snake's trail is most distinct when it moves rapidly across open areas. Watch for the narrow overall trail width and shallow waves of undulation.

• This elusive snake can be difficult to see in the field, but following the fresh trails of this species greatly increases the chances of encountering them.

Rattlesnakes

This group belongs to the pit viper family, Viperidae, that includes species in the genera *Crotalus* and *Sistrurus*. All have keeled scales, vertical pupils, and loreal pits used to sense heat. They also all possess a segmented rattle composed of dead skin, which they use to warn would-be predators or respond to the threat of potential trampling by large animals.

Rattlesnakes are venomous, and some species have the potential to seriously or fatally injure humans. Fatalities are rare, however, and most bites occur when someone is trying to pick up, handle, or harm a rattlesnake. Rattlesnakes that puff up and stand their ground feel cornered, and will rattle and hiss loudly to broadcast that they will defend themselves if pushed. They will only strike as a last resort. Generally though, rattlesnakes are likely to turn tail and slither away as soon as they can. Give them their space to avoid injury, and observe these fascinating animals at a distance.

A prairie rattlesnake moving slowly away from the camera. Notice the tail base leaving a wide tail drag in the trail. JOE LETSCHE

Female rattlesnakes give live birth, and do not lay eggs. In many species, females remain with the young for several days or more, or until the baby snakes have shed their first skin. This likely gives these young snakes a head start by protecting them from predation.

All rattlesnake species have very wide ventral scales in proportion to other snake species. These wide scales help rattlesnakes move efficiently over a variety of substrates.

Rattlesnake species, especially those belonging to the genus *Crotalus*, tend to create similar trail patterns to each other when moving using a baseline lateral undulation style of locomotion. They travel relatively slowly, resulting in wide inner trail widths with a thick tail drag down the middle of the trail. Frequently, the ridges formed by the backward pressure of the snake's body will directly cross the middle of the trail, traveling nearly or completely across the width of the inner trail. Rattlesnakes also use a type of locomotion that combines elements of both lateral undulation and some rectilinear locomotion.

Where several rattlesnake species of a similar size are found together, telling the species apart from their trails may be very challenging.

Sidewinder Rattlesnake (*Crotalus cerastes*)

This small, desert-adapted rattlesnake species is also sometimes referred to more casually as the "horned rattlesnake" for its enlarged supraocular scales found directly above its eyes, which point upward. Its range is limited largely to the Mojave, Colorado, and Sonoran Deserts in California, Nevada, extreme southwestern Utah, and southwestern Arizona. Its prime habitat generally includes loose sand dunes or sandy washes, but you'll also sometimes find it in more loose, gravelly sites, in rocky areas, or out on hardpan.

This species includes three recognized subspecies that are distinguished by their location: Mojave Desert sidewinder rattlesnake (*Crotalus cerastes cerastes*), Colorado Desert sidewinder rattlesnake (*Crotalus cerastes laterorepens*), and the Sonoran sidewinder rattlesnake (*Crotalus cerastes cercobombus*). Its background body color matches the dominant color of

The trail of a sidewinder rattlesnake moving toward the camera.
CAMERON ROGNAN

soils where it lives. Subspecies from extreme southwestern Utah and southern Nevada, areas that include red sandstone, will be more orange or pinkish in color. Meanwhile, individuals from other parts of this species' range will generally show much more subtle variations of yellows and tans.

Sidewinders are small, stout rattlesnakes that can move very efficiently on loose, fine sediments such as sand. They do this by using a sidewinding locomotion, hence their name. This movement entails lifting and depositing the body in a diagonal fashion, which often leaves the impressions of the scutes (enlarged belly scales). This snake species tends to hold its tail well off the ground as it

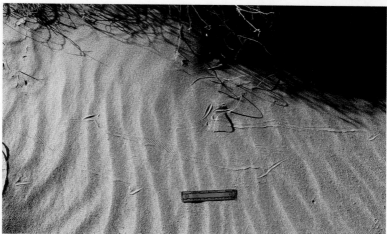

The beautiful trail of a sidewinder rattlesnake. Look close for the marks made by the sliding ventral scales.

A close-up of the curved end of the sidewinding mark. This is made by the neck of the snake and can be used to determine the direction of travel.

travels. This leaves diagonal lines that are cleanly separated by negative space. Using this gait, sidewinders are able not only to travel on loose substrates effectively, but are also better adapted than other snakes at traveling over very hot substrates. This is because they have reduced body contact with the ground, which continuously changes (Stebbins 2012).

Like other rattlesnakes, sidewinders are ambush predators. They often position themselves at the base of a shrub or rock or near similar cover along a rodent runway. This species may partially bury itself while lying in ambush to reduce exposure to rapid temperature change or perhaps for greater concealment from prey. Juve-

The rectilinear trail of a slow-moving sidewinder rattlesnake. Notice the thick tail drag typical of rattlesnakes.

The subtle rectilinear trail of a sidewinder leading to its lay. The snake later exited the lay heading toward the right using sidewinding locomotion.

nile sidewinders feed largely on lizards, while adults favor small rodents, though they will also eat larger lizards.

Trail: Sidewinding   Distance between Sidewinding Marks: 3.4–6 in (8.64–15.24 cm)
Inner Trail Width: 0.3–1.3 in (0.76–3.3 cm)
Length of Individual Sidewinding Marks: 12–26.4 in (30.5–67 cm)

The sidewinding trails are neat, with each separate line being relatively similar in length and angle to the previous one. The space between these lines also stays fairly consistent until the animal changes direction, adjusts for steepness of landscape, or significantly changes speed. Overall, each line appears as an elongated, backward "J" shape. The curve of the J is made by the part of the snake very near the neck. The tip of the curve of this J points in the direction of travel.

Lateral Undulation   Overall Trail Width: 2–4 in (5–10.16 cm)
Inner Trail Width: 0.5–1.5 in (1.27–3.81 cm)
Wave Length: 5–7.2 in (12.7–18.29 cm)

Tail Drag: 0.2–0.5 in (0.5–1.27 cm)

This species does not commonly use lateral undulation, but when it does, it is likely transitioning from rectilinear locomotion to sidewinding, or vice versa.

Rectilinear Locomotion   Overall Trail Width: 1–1.7 in (2.54–4.32 cm)
Inner Trail Width: 0.6–1.5 in (1.52–3.81 cm)

Tail Drag: 0.2–0.6 in (0.5–1.52 cm)

This snake also often uses a rectilinear locomotion for short distances, most typically when entering or exiting a burrow, lay, or some form of cover. It rarely uses any other form of locomotion.

Notes:

- This unique serpent is mesmerizing to watch as it travels across loose sand with ease. It leaves some of the most beautiful tracks in the desert. This species is largely nocturnal, though you may sometimes observe it during the day in the morning or late afternoon.
- Being a pit viper, this animal does possess potent venom that is potentially dangerous to humans. Though not aggressive, it has a fast, twitchy disposition and should never be handled. It deserves respect, so carefully approach the snake to observe it, and maintain a safe distance.
- The lays of this species are commonly found in open, sandy areas at the bases of rocks or shrubs, or against embankments in washes.

Western Rattlesnake (*Crotalus oreganus*)

This species is widespread in the western United States, and includes six subspecies north of Mexico: the midget faded rattlesnake (*Crotalus oreganus concolor*), Grand Canyon rattlesnake (*Crotalus oreganus abyssus*), Arizona black rattlesnake (*Crotalus oreganus cerberus*), southern Pacific rattlesnake (*Crotalus oreganus helleri*), northern Pacific rattlesnake (*Crotalus oreganus oreganus*), and Great Basin rattlesnake (*Crotalus oreganus lutosus*).

This species varies in size from subspecies to subspecies. The smallest is the midget faded rattlesnake, which grows between 18 and 24 inches in total length. Larger subspecies such as the Great Basin rattlesnake and the northern Pacific rattlesnake grow from 15 to 55 inches.

This species preys on a wide variety of small animals including small birds, mice, voles, gophers, and ground squirrels. Juvenile western rattlesnakes also eat large insects and lizards. This is one of the most familiar rattlesnakes, as

The lateral undulation trail of a northern Pacific rattlesnake. Notice the ridges of sand sharply crossing the span of the inner trail width.

The lateral undulation trail of a southern Pacific rattlesnake.

it is found across much of the western United States, including in suburbs of cities such as Sacramento, San Francisco, and Los Angeles.

The venom of this species varies considerably from one population to the next, with some having largely hemotoxic venom and others having both hemotoxic and neurotoxic venom. Hemotoxic venom damages blood and tissues, while neurotoxic venom affects the nervous system. The venom of all populations is considered potent, so give this species a respectful distance to avoid getting bitten.

Northern Pacific Rattlesnake (*Crotalus oreganus helleri*) and Southern Pacific Rattlesnake (*Crotalus oreganus oreganus*)

Trail: Lateral Undulation Overall Trail Width: 1.8–5.4 in (4.57–13.72 cm)

Inner Trail Width: 0.9–2.2 in (2.29–5.59 cm)

Wave Length: 5–13.9 in (12.7–35.3 cm)

Rattlesnakes create this type of trail at a variety of speeds. When moving rapidly, they will leave rolling, sine-wave patterns of undulation with a wider overall trail width. At increasingly slower speeds, the overall trail width decreases as these waves becomes shallower and the inner trail width tends to increase.

Tail Drag: 0.3–0.6 in (0.76–1.52 cm)

The tail drag of this species is often present and proportionately wide relative to the inner trail width. The marks are made by the mid-portion of the tail, as the lower end of the tail (including the rattle) is generally held up and well clear of the substrate. Sometimes, while most of the snake's body moves relatively straight, the tail creates a smaller, wavier undulating trail of its own.

Midget Faded Rattlesnake (*Crotalus oreganus concolor*)

I've presented the smallest subspecies of western rattlesnake separately here to emphasize the smaller measurements of its trails.

Trail: Lateral Undulation   Overall Trail Width: 1.8–3.2 in (4.57–8.13 cm)

                                     Inner Trail Width: 0.5–1.2 in (1.27–3 cm)

                                     Wave Length: 3.8–6.3 in (9.65–16 cm)

Rattlesnakes create this type of trail at a variety of speeds. When moving rapidly, they will leave rolling, sine-wave patterns of undulation with a wider overall trail width. At increasingly slower speeds, the overall trail width decreases as these waves becomes shallower and the inner trail width tends to increase.

Rectilinear Locomotion   Overall Trail Width: 1–1.7 in (2.54–4.32 cm)

                                     Inner Trail Width: 0.8–1.2 in (2–3 cm)

This type of trail is relatively straight, as the animal exerts very little pressure backward as it moves forward. Very small ridges of substrate can be seen directly to the side of the trail and in line with the length of the trail. The difference between the inner trail width and overall trail width is small and due only to minor changes in direction, rather than waves of side-to-side undulating motion.

Tail Drag: 0.2–0.4 in (0.5–1 cm)

The tail drag of this species is often present and is proportionately wide relative to the inner trail width. The marks are made by the mid-portion of the tail, as the lower end of the tail (including the rattle) is generally held up and well clear of the substrate. Sometimes, while most of the snake's body is moving relatively straight, the tail will create a smaller, wavier undulating trail of its own.

The rectilinear trail of a midget faded rattlesnake.

The lateral undulation trail of a midget faded rattlesnake.

Notes:
- Other commonly encountered signs of this species are sheds and lays. Multiple sheds found close together in one location, especially in a boulder pile or near a rocky crevice, may indicate a hibernacula or a nursery.
- If you repeatedly find trails or lays of this species in the same areas, this most likely means that the location combines good cover and a consistent food source. Generally such spots are important to this snake species and should be left undisturbed.
- Please do not harm or collect this species, as they play an important role in their local ecosystems as predators of small mammals.

Western Diamond-backed Rattlesnake (*Crotalus atrox*)

This is the largest western rattlesnake species, and can grow to a truly impressive size and girth. Adults can reach between 30 and 90 inches in total length. Average adults often grow between forty-eight and sixty inches in total length.

This species ranges from the extreme southeastern parts of the Colorado Desert in California, and across the Southwest throughout much of the Sonoran and Chihuahuan desert regions. It prefers the canyons, lower mountain slopes, and rocky foothills in arid and semiarid areas including relatively open, sparsely vegetated areas and well-vegetated ranchlands.

This species feeds on a variety of prey, including small birds, rats, mice, ground squirrels, and even small rabbits. Juveniles will also feed on large insects, lizards, and frogs.

The rectilinear trail of an adult western diamond-backed rattlesnake.

The venom of this species is potent, and given the size of this snake, a large amount of venom can be delivered in a single bite. Give this species a respectful distance.

Trail: Rectilinear Locomotion   Overall Trail Width: 2–3.7 in (5–9.4 cm)
Inner Trail Width: 1.5–3 in (3.81–7.62 cm)
Tail Drag: 0.5–1 in (1.27–2.54 cm)
The tail drag of this species is often present and is proportionately wide relative to the inner trail width. The marks are made by the mid-portion of the tail, as the lower end of the tail (including the rattle) is generally held up and well clear of the substrate. Sometimes, while most of the snake's body is moving relatively straight, the tail will create a smaller, wavier undulating trail of its own.

Notes:

• This large, heavy-bodied snake often travels using a rectilinear locomotion. Its wide, slow-moving trails are a common sight in washes and arroyos throughout the more southerly desert regions of the southwestern United States.

• Being large and heavily built, this species will even leave conspicuous lays in relatively coarse gravel soils. These are relatively common near the bases of large shrubs or saguaro cacti in southern Arizona.

• This species is more likely than other rattlesnakes to puff up and hold its ground. Given space, though, they will turn tail and escape. Please do not harm or collect this species, as they play an important role in their local ecosystems as predators of small mammals.

Red Diamond Rattlesnake (*Crotalus ruber*)

This large species of rattlesnake is very similar to the western diamond-backed rattlesnake (*Crotalus atrox*), but inhabits a separate range. In the United States, this species is restricted completely to southwestern California. It is found largely along the coast, down through the mountains of the Peninsular Ranges and as far east as the western edge of the Colorado Desert. It is orange-red in color.

Adults grow between 30 and 65 inches in total length, and average between 24 and 48 inches. They feed on woodrats, ground squirrels, and even small rabbits. Adults and especially juveniles will also eat lizards, birds, and sometimes also carrion.

The venom of this species is potent, and it can deliver a large amount in a single bite. Give this species a respectful distance.

Trail: Lateral Undulation   Overall Trail Width: 4–8.1 in (10.16–20.57 cm)
Inner Trail Width: 2–2.8 in (5–7.1 cm)
Wave Length: 9–13.1 in (22.86–33.27 cm)

Left: The rectilinear trail of a large, captive red diamond rattlesnake. *Right:* A close-up of the rectilinear trail of the large, captive red diamond rattlesnake.

Tail Drag: 0.4–0.9 in (1–2.29 cm)

The tail drag of this species is often present and is proportionately wide relative to the inner trail width. The marks are made by the mid-portion of the tail, as the lower end of the tail (including the rattle) is generally held up and well clear of the substrate. Sometimes, while most of the snake's body is moving relatively straight, the tail will create a smaller, wavier undulating trail of its own.

Notes:

- Like its relative the western diamond-backed rattlesnake, this species is large and heavily built. Therefore, it favors rectilinear locomotion to travel slowly through its home range. It also frequently uses a slow lateral undulation form that includes some features of rectilinear locomotion, by moving less side to side and creating small ridges of substrate mostly to the sides of the trail.

- It is not uncommon to find a large male in the same location as a female at any time of year, though especially between February and June during their breeding season.

- These snakes are generally mild mannered and nonaggressive. Please do not harm or collect this species, as they play an important role in their local ecosystems as predators of small mammals.

Speckled Rattlesnake (*Crotalus mitchellii*)

This is arguably the most cryptically colored of all the rattlesnake species in the United States. Several subspecies exist, and there is considerable variation in pattern and color within each. Generally, this species has colors and patterns that match those found on the dominant rocks in their ranges. This species is a skilled climber, and may be found high on rock platforms or in rock crevices.

Speckled rattlesnakes mainly live in rocky hills in coastal, semiarid, and arid landscapes. You can also find them in areas of loose soil or sand vegetated with creosote and other desert shrubs. Adults grow between 23 and 52 inches in total length. They feed largely on small mammals, as well as on lizards and some small birds.

The venom of this species is considered potent. Be alert in the rocky hillsides where this species lives, as it has incredible camouflage. Give this species a respectful distance.

The slow lateral undulation trail of a speckled rattlesnake.

Trail: Lateral Undulation   Overall Trail Width: 3–5.5 in (7.62–13.97 cm)
Inner Trail Width: 0.9–1.5 in (2.29–3.81 cm)
Wave Length: 5–11.6 in (12.7–29.46 cm)
Tail Drag: 0.3–0.6 in (0.76–1.52 cm)

The tail drag of this species is often present and is proportionately wide relative to the inner trail width. The marks are made by the mid-portion of the tail, as the lower end of the tail (including the rattle) is generally held up and well clear of the substrate. Sometimes, while most of the snake's body is moving relatively straight, the tail will create a smaller, wavier undulating trail of its own.

Notes:

- Finding the trail of this snake is uncommon, due to its tendency to spend much of its time in and around rocky areas. You'll most likely encounter these trails crossing rocky areas.

- When climbing or scrambling in rocky areas in this snake's habitat, watch for spots on or between boulders that collect sand or dust. These are likely locations to find the lays of this species.

- Please do not harm or collect this species, as they play an important role in their local ecosystems as predators of small mammals.

Shovel-nosed Snakes

This group of snakes includes two species found north of Mexico. The Sonoran (organ pipe) shovel-nosed snake (*Chionactis palarostris*) only ranges north into the United States in the Organ Pipe Cactus National Monument in Arizona. These snakes prefer to live in habitats such as washes, dunes, and flats that include sand or loose gravel.

These small snakes feed on invertebrates such as scorpions, centipedes, spiders, solpugids, and a variety of insects. These are caught and swallowed without any constricting.

This group of snakes is generally nocturnal, and may be one of the most abundant snake species in the habitat you find them in. Their beautiful, perfectly curving trails are a common sight in sandy areas within their range. The best time to see their tracks is on warm, still mornings between May and June (Ernst and Ernst 2003). Individuals have been observed from as early as mid-March until as late as November.

Western Shovel-nosed Snake (*Chinoactis occipitalis*)

This is the more wide-ranging of the two shovel-nosed snake species, and you can find it in deserts of southeastern California, the southern tip of Nevada and western Arizona down to south-central Arizona. It tends to prefer dry desert habitats, especially in flats with scant vegetation and on gently sloping alluvial fans that include loose sandy areas.

This small snake only grows to about 16 or 17 inches in length. Despite their diminutive size, they can put on quite a show of bluffing if they feel cornered. This performance might include coiling, striking, hissing, and even biting. They may also expel foul-smelling musk and feces when handled.

This species travels largely on the surface, though it might also travel short distances just under the surface in loose substrates such as sand. It may travel this way to be less obvious to potential predators, as well as to seek out buried prey. When buried, it can't see the prey, and likely can't smell it. It can, however, feel the vibrations made by the movement of prey on the sand, and the snake might hone in on these vibrations.

Trail: Lateral Undulation Overall Trail Width: 1.19–1.89 in (3.02–4.8 cm)
Inner Trail Width: 0.164–0.35 in (0.416–0.889 cm)
Wave Length: 2.97–4.91 in (7.54–12.74 cm)

A Colorado Desert shovel-nosed snake leaving a beautiful trail. JEFF NORDLAND

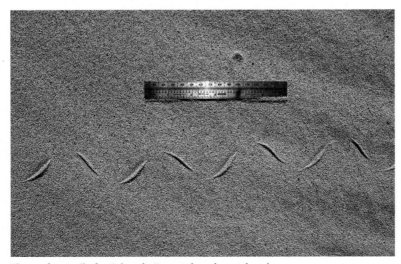

The perfect trail of a Colorado Desert shovel-nosed snake.

The raised ridge of sand in the center of the photo hides the body of a Mojave shovel-nosed snake moving just under the surface.

The waves in the trail left by this species are often nearly perfect, consistent sine waves. The snake creates the even, repeating pattern of waves while traveling, but the pattern might vary if the snake slows down, suddenly speeds up, or changes direction. The waves of lateral undulation are considerably shorter than they are wide. Also, the inner trail width tends to remain fairly narrow under most conditions.

Additional Locomotion Types

This species may use sidewinding locomotion for short distances on extremely hot substrates (Cowles 1977). It is capable of other forms of locomotion such as rectilinear and concertina, though it appears to prefer using lateral undulation whenever possible.

Notes:

- This is one of the classic characters of the sandy portions of the Mojave, Colorado, and Sonoran Deserts. Though these snakes are common, you'll likely never see them unless you venture out into the dunes or sandy flats on a warm evening with a flashlight. You may also spot them crossing roads on warm desert nights.

- This species is able to disappear under the surface of the sand within a few seconds. Along with the defensive behaviors mentioned above, this ability to burrow rapidly also plays a role in helping this species avoid potential predators.

- This species is largely crepuscular, though is sometimes seen out in the open during early morning or late afternoon.

Leaf-nosed Snakes (Genus *Phyllorhynchus*)

This small group of desert-dwelling snakes includes two species: the saddled leaf-nosed snake (*Phyllorhynchus browni*) and spotted leaf-nosed snake (*Phyllorhynchus decurtatus*). They are small, short, and stoutly built. Both possess an enlarged, triangular, leaf-shaped rostral scale that curls back over the blunt snout. They use their specialized snouts for burrowing and for finding prey.

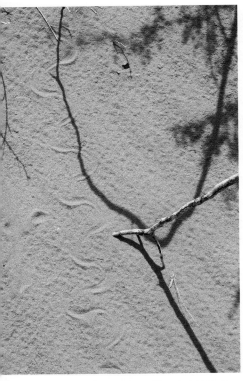

Spotted Leaf-nosed Snake (*Phyllorhynchus decurtatus*)

This is a small snake, as adults only grow between 12 and 20 inches in total length. The body color is generally pale tan, gray, or cream. A dark band crosses the head between the eyes and continues below the eyes to the corner of the mouth. Rectangular brown blotches appear on the length of the dorsal surface. Below and in between each blotch is a smaller brown spot. The scales of this snake are smooth and the pupils are vertical.

This species is surface active strictly at night. You'll often find it in habitats such as open flats that include some

The lateral undulation trail of a spotted leaf-nosed snake moving up a slight slope.

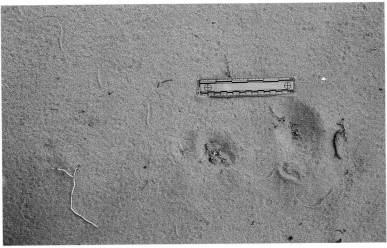

The lateral undulation trail of a spotted leaf-nosed snake moving left to right. Notice the several indentations where the snake attempted to burrow.

areas of gravel or sandy-soil deserts. Often they inhabit areas that include plants such as catclaw acacia, creosote bushes, and cacti such as saguaro.

This species prefers reptile eggs whenever it can find them, but it also eats small lizards and some insects. I witnessed a juvenile snake burrowing into an ant nest and feeding on the larvae.

This snake is an effective burrower, but not as efficient at moving through sand as the shovel-nosed snakes. It is not known to travel under the surface of the sand; rather, it travels on the surface and digs down to locate prey. It will also hunt lizards, especially nocturnal species such as banded geckos (*Coleonyx* ssp.), on the surface.

Watch for the trail of this species in low, sandy spots in open desert areas or in sandy patches throughout rocky desert areas. In parts of the southwestern United States, it is one of the most frequently encountered snake species crossing roads at night.

Trail: Lateral Undulation   Overall Trail Width: 1.5–3.5 in (3.81–8.89 cm)
Inner Trail Width: 0.3–0.7 in (0.76–1.78 cm)
Wave Length: 3–5.5 in (7.62–13.97 cm)
Does not frequently register an obvious tail drag.

Notes:

• This species' trails are similar to those of the shovel-nosed snakes, but there are subtle differences. The waves in the lateral undulation trails of the spotted leaf-nosed snake tend to be somewhat taller and the overall trail width is greater, as is the inner trail width.

• Spotted leaf-nosed snakes are not typically found out in large areas of sand dunes, but are more commonly observed along the edges of large sandy areas or in washes.

Variable Sand Snake (*Chilomeniscus stramineus*)

This species of snake finds the northern limits of its range in southern Arizona. It prefers to dwell in uplands of Sonoran Desert scrub, with loose gravel and sand soils. These snakes are found in riparian areas, arroyos, and washes. Preferred habitats generally include vegetation such as saguaro and palo verde, thorn scrub, or, at lower elevations, mesquite and creosote.

This small, nocturnal snake has a short, stout build. It is somewhat similar to the western shovel-nosed snake (*Chionactis occipitalis*) and Sonoran shovel-nosed snake (*Chionactis palarostris*), though it is thicker and shorter tailed than either species. The variable sand snake is generally orange or buff with narrow or broad dark-brown to black bands crossing the tail and body. The orange body color fades to cream on the lower sides, and the belly is yellow or white. This species has a flattened snout and countersunk lower jaw.

You can find this snake traveling on the surface at night, but it will also travel just under the surface of loose soils. Its surface trails are even, sine-wave lateral undulations similar to those made by shovel-nosed snakes. It hunts invertebrates, and is known to consume sand cockroaches, centipedes, and grasshoppers.

The variable sand snake is active for much of the year, from April to November (Ernst and Ernst 2003). There is still a great deal about this small snake's ecology that we don't know, and more field studies are necessary to help us better understand this fascinating desert dweller.

Trail: Lateral Undulation Overall Trail Width: 1.4–1.9 in (3.56–4.83 cm)
Wave Length: 2.8–3.4 in (7.1–8.64 cm)
Inner Trail Width: 0.3–0.5 in (0.76–1.27 cm)

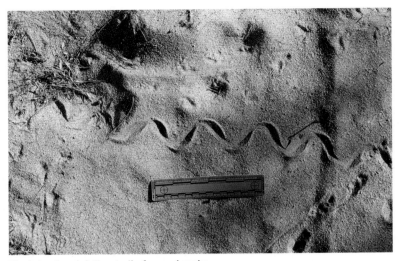

The lateral undulation trail of a sand snake.

A close-up of a lateral undulation trail of a sand snake.

The trail made by a sand snake traveling just under the surface.

The trail left by this species often features nearly perfectly consistent sine waves. The snake makes the evenly repeating pattern of waves while traveling, but these might vary if the snake slows down, suddenly speeds up, or changes direction. The waves of lateral undulation are generally shorter than they are wide. Also, the inner trail width tends to be moderately wide proportionate to the overall trail.

Additional Locomotion Types

This species can travel under the surface using the sand-swimming form of modified lateral undulation shared by shovel-nosed snakes, as well as California legless lizards (*Anniella* ssp.) and the Florida sand skink (*Neoseps reynoldsi*). This movement creates consistent sine wave trails that are caused by the collapsing of tunnels as the snake's body exits them.

Notes:

- The shorter, stouter build of this snake helps it move even more efficiently through loose sand than the shovel-nosed snakes. It can disappear under the surface of sand or loose gravel in two to three seconds.

- The fresh surface lateral undulation trails of shovel-nosed snakes might be confused with those of the sand snake. Look closely at the inner trail width, and note that those of the sand snake tend to be proportionately wider. Older trails will be more difficult to distinguish, as when weathering happens the trail will appear wider.

- It is possible that, like the shovel-nosed snake, the sand snake can also locate prey through vibrations transmitted through the loose sand.

Western Threadsnake (*Rena humilis*)

This is a very small snake; adults only grow between 7 and 16 inches long. The diameter of an adult's body is smaller than that of a number 2 pencil. This species is purple-brown or pink, and to the untrained eye might look more like an earthworm than a snake.

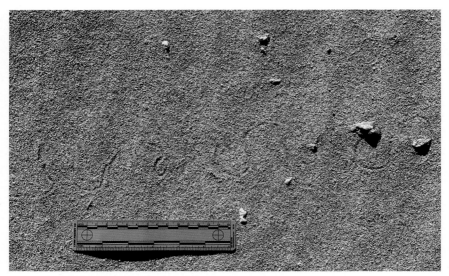

The subtle lateral undulation trail of a western threadsnake.

The body of this serpent is covered in tightly fitting, very smooth and reflective cycloid scales that are uniform in appearance around the entire body. Indeed, no enlarged scutes appear on the ventral surface. It has a blunt head and countersunk lower jaw. The eyes are reduced to mere dark spots, giving rise to its other name: the western slender blind snake. All of these are adaptations for a largely subterranean life.

This species and its related threadsnakes are almost entirely insectivores. They particularly prefer ants and termites.

You'll find this species in grasslands, deserts, and in foothills or mountain slopes with loose soils. It may be most abundant or most visible in areas with running water. You can usually find it under cover objects such as logs, rocks, or piles of plant matter. Western threadsnakes may also be surface active at night and might be found on roads during periods of high humidity or rain.

Although not uncommon across its range, this species can be difficult to see due to its nocturnal tendencies and the fact that it spends the majority of its life underground.

Trail: Lateral Undulation   Overall Trail Width: 1.5–2.64 in (3.81–6.7 cm)
Wave Length: 1.5–2 in (3.81–5.08 cm)
Inner Trail Width: 0.13–0.23 in (0.33–0.58 cm)

The waves in the trail left by this species are often wildly erratic. Though in general the waves are tall and narrow in shape, they may show sudden and inconsistent variation, such as the appearance of very small waves. These occur due to the erratic action of the snake's short, stubby tail whipping around. This snake moves mostly using a form of lateral undulation. At slower speeds, this species' trail may resemble that of other snakes as it takes on a more regular pattern of wide waves of lateral undulation.

Notes:

- Arguably one of the most unusual snakes in the United Snakes, both in appearance and life history, because of its largely subterranean life and specialized diet. Its body is adapted to dig for invertebrates and follow them through their narrow tunnels.
- The erratic patterns and incredibly narrow trail width help distinguish the trails of this species from those of similarly sized colubrid snakes.

Sharp-tailed Snake (*Contia tenuis*)

This is a small snake, pink-brown to gray-brown in color. Adults grow between 8 and 12 inches in total length. This species has a sharply pointed scale at the tail tip. A narrow line of reddish-copper color typically appears on either side of the back. The top of the head and dorsal surface is somewhat flattened, likely helping this snake fit into tight crevices under bark and inside logs.

You'll largely find this species in oak woodlands, as well as in dense coniferous forests, open pine forests, and grasslands. It prefers moist microhabitats under cover objects.

This species was divided into two species in 2010: the common sharp-tailed snake (*Contia tenuis*) and forest sharp-tailed snake (*Contia longicaudae*). The forest sharp-tailed snake has a significantly longer tail, different scale count, and can grow to larger size, between 14 and 18 inches in total length. Even a very large sharp-tailed snake is still likely to be barely the diameter of a number 2 pencil.

These little snakes largely eat slugs, and they can be active in relatively cool temperatures down to 50 degrees Fahrenheit (Stebbins 2012). This species

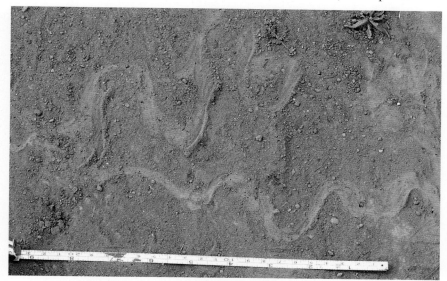

The trail of a sharp-tailed snake moving through shallow dust in a rapid lateral undulation at the top, then turning around and moving more slowly below.

The trail of a sharp-tailed snake moving at a moderately rapid rate using lateral undulation through deep dust.

has long teeth relative to its overall size, especially the teeth located in its lower jaw.

This snake's trails are not found very far from cover. It spends a majority of its life under cover objects, where it finds shelter and prey.

Trail: Lateral Undulation   Overall Trail Width: 1–2.6 in (2.54–6.6 cm)
                                   Wave Length: 2.2–3.2 in (5.59–8.13 cm)
                                   Inner Trail Width: 0.2–0.3 in (0.5–0.76 cm)
           The lateral undulation trails of this species are small, and often can be
           somewhat messy.
Notes:
• You may mistake the trail of this small, reclusive snake with those of other
  very small colubrid snakes in its range, like the ring-necked snake (*Diadophis
  punctatus*).

Turtles and Tortoises   7

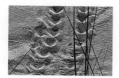

From the *Teenage Mutant Ninja Turtles* to *The Tortoise and the Hare*, this group of shelled reptiles is well known and celebrated by many people. Turtles such as the red-eared slider (*Trachemys scripta elegans*) or the eastern box turtle (*Terrapene carolina carolina*) are very popular in the reptile pet trade. Some species possess beautiful coloration, interesting behaviors, or fascinating physiology. With all this in mind, what makes a turtle a turtle?

First, all turtles have a two-part shell, made up of the upper portion (carapace) and the lower portion (plastron). These shells are made of fused bone, and, in many species, are covered in special scales called "scutes." They also possess sharp-edged, beak-like mouths that lack any teeth.

Turtles in North America can be organized into four main categories according to their habitats: marine, fully aquatic, semiaquatic, and terrestrial. Marine turtles are more commonly called sea turtles (including Chelonidae and Dermochelyidae). Turtles considered fully aquatic are species such as snapping turtles (Chelydridae) and softshell turtles (Trionychidae). Semiaquatic turtles include mud and musk turtles (belonging to Kinosternidae), and pond turtles (belonging to Emydidae). Terrestrial turtles include species such as box turtles (also belonging to Emydidae) and tortoises (belonging to Testudinidae).

Marine turtles are truly aquatic, living in oceans and only coming on to land to lay eggs. These turtles have fully formed, paddle-like flippers. They are mostly found in tropical waters, though a few species do venture into temperate waters. Many sea turtles travel amazing distances throughout their lives. The leatherback (*Dermochelys coriacea*) travels up to ten thousand miles annually across the Pacific Ocean in search of its favorite food: jellyfish.

The fully aquatic turtles include two groups that are physically built very differently. Snapping turtles are very large, with heavy carapaces, and reduced, cross-shaped plastrons. They have large, heavy legs, and large webbed feet. They swim or plod along on the bottom of waterways. In contrast, softshell turtles have flat, pancake-like carapaces covered with thick, soft, and leathery skin. Their feet are fully webbed, and paddle-like with well-developed claws. They are capable of swimming rapidly and are generally wary of close approach. Both softshell turtles and snapping turtles are well known to give powerful defensive bites if handled or harassed.

Semiaquatic turtles all have well-developed, hard-shelled carapaces and plastrons. All have legs and feet developed for walking, with some degree of webbing and obvious claws. This group of turtles is arguably the most visible to the casual observer, as they spend hours every day basking on the rocks, logs, and banks of various waterways.

Terrestrial turtles have thick carapaces and plastrons. They tend to have legs that can only be called "elephantine." In other words, they are robust, and columnar, with thick, often relatively blunt claws. These turtles tend to be less dependent on, or completely independent of, nearby aquatic habitats. The most terrestrially adapted group of turtles is tortoises. For example, desert tortoises (*Gopherus agassizii*), which have adapted incredibly to extreme desert life, can go without drinking water for years.

Turtle Locomotion

All turtles basically move in a similar fashion, due to their shared physiology. Turtles essentially move one foot at a time. This provides more stability as they make forward progress, by using the support of three stationary limbs. The sequence of footfalls might look something like this:

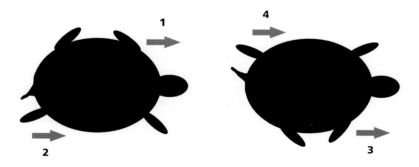

Notice that the turtle begins with the left front foot. This is followed by the right hind, then the right front, and finally the left hind. JOE LETSCHE

If the turtle were to lift both legs on the same side of the body simultaneously, it would fall over onto that side. It must first plant the front leg before it can bring the hind forward. Turtles generally only do understep walks. Mud and musk turtles moving at top speed may be the exception. They leave front and hind tracks that land nearly parallel, with the front tracks landing just slightly further back and to the inside of the hind tracks. Determining the speed of a turtle can be a bit confusing if you are used to looking at the walking gaits of mammals.

The further behind the front track is to the hind track, the slower the animal was moving. You might be deceived, however, as a hind track may appear to be overstepping the front track. This, however, is not what it appears at first glance. Rather, the hind track is landing ahead of the previous front track. In other words, the turtle is going so slow, it is basically taking a half step. Strange as it might sound, when the steps are very short, the hind track can be two steps behind the front track.

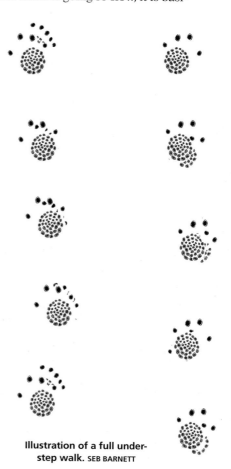

There are essentially three types of understep gaits that turtles can use. Two of these gaits are a form of extreme understep. The first is what can be called the "full understep walk." In this case, the hind track lands on top of the front track—as in a direct register gait—except that it is actually landing on the front track made two steps back. In other words, the footfall of the hind is so short it is always one step behind the front foot.

The second of these extreme understep gaits can be called the "half understep walk." This gait appears to be an overstep gait, where the hind track lands just ahead of the front track. This is also deceptive, as the hind track is not actually overstepping the most recent front track; rather, the hind track is overstepping the front track that was made previously (one step behind the most recent front track).

Lastly, the third turtle gait, which is also the most commonly observed,

Illustration of a full understep walk. SEB BARNETT

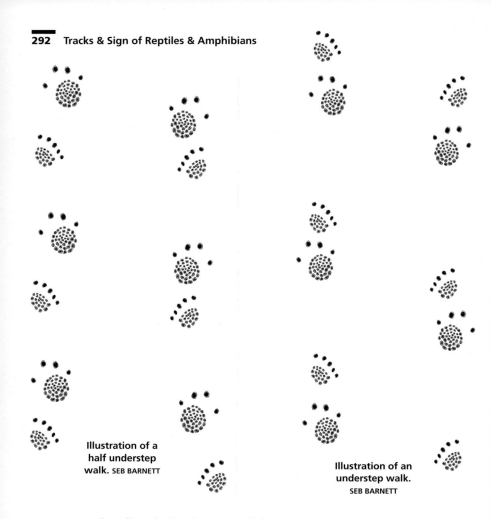

Illustration of a
half understep
walk. SEB BARNETT

Illustration of an
understep walk.
SEB BARNETT

is simply called the "understep walk." In this gait, the turtle essentially places its hind foot directly behind where the front foot is or was a moment ago. This is a gait many vertebrates use at slow speeds, including many mammals, lizards, salamanders, and crocodilians.

Gaits and Patterns as Indicators of Behavior

The speed of a turtle, as expressed in the tracks, can indicate its state of mind. A turtle walking very slowly in an extreme understep walk might be very relaxed or cautious. Meanwhile, one moving in a rapid understep—with the hind tracks landing tightly behind the front tracks—

Notice the pattern of the tracks this desert tortoise created while it fed. It approached the small grass clump from the background of the photo, faced it, and turned in a slow circle as it fed. Finally, it turned away from the food and walked away toward the right.

might mean it is making a dash for water or the cover of a nearby burrow or other form of shelter.

The pattern left by the general movement of the turtle can also be an indication of certain kinds of behavior. Take a look at the photos on this page.

You can easily observe directional changes in turtle trails by simply looking at the footfalls. The presence of the nails indicates the orientation of the feet, with nails pointing generally in the direction of travel.

Here is another example where the same desert tortoise approached another small clump of grass. It approached from the lower left and circled clockwise toward the right. Then, it stopped feeding, stepped over the clump of grass, and resumed walking off toward the upper right.

This animal was traveling from left to right in the frame. Notice how it takes a step out of line at first with its right front foot; then, a few steps after, it does the same with its left front foot. The smoothed area in the middle of the trail was made by the temporary dragging of the plastron. JOE LETSCHE

As a turtle turns, one of its front feet must step out of line with the previous footfalls. If the animal continues to turn, then the trail will show an overall shift in direction and continue to one side or the other. If, however, the animal does not continue turning in the direction initiated by that first step, then you will only notice one front track out of line and the trail will continue as it did before.

Tracks forming a "mating circle" from a pair of desert tortoises. TREVOR OSE

Avoiding Confusion

Because they have large, shelled bodies, turtles leave very distinct trails, as their feet must move in parallel lines. Notice that the foot falls do not land very far in toward the center of the trail. This is an excellent way to distinguish turtle trails from those of other animals. Here are some photos to demonstrate this.

Right: These tracks, left by a species of waterfowl, are large relative to the overall trail width. They also fall well into the middle of the trail and leave very little negative space down the trail's midline. JOE LETSCHE

Left: These two sets of raccoon trails are also easily distinguished from turtle trails. Notice they are also very large in relation to the overall trail width. The raccoon tracks also demonstrate a very long stride and leave almost no negative space along the midline of the trail. *Right:* Now here is a trail from a turtle. Notice two parallel lines of tracks that are spaced well apart. The tracks appear relatively small compared to the overall trail width. Therefore, it is easy to determine turtle trails from those of mammals or birds. You may occasionally confuse very obscure turtle trails with the trails of other reptiles. In the case of the trails of very small turtles, you might confuse them with especially large invertebrates. JOE LETSCHE

Tail Drags

Tail drags commonly appear in turtle trails. Basically, all turtle tails are thickest at or near the base, and more tapered toward the point. Observing how quickly the tail tapers, or remains thick along its length, can help determine a particular group, species, or sex of turtle. Tail drags can also help with overall interpretation.

The very short tail of the eastern box turtle. JOE LETSCHE

The very long tail of an eastern snapping turtle. JOE LETSCHE

A very thin line registering in the middle of the trail likely indicates the tail tip, while a much wider and deeper line likely indicates it is being made by the portion of the tail closer to the base.

Some turtles have very short, stubby tails, while a few have very long tails. Terrestrial turtles and tortoises tend to have short tails. Softshell turtles have short, thick tails that can create wide tail drags when they are pressed downward as the turtle moves forward. The turtles with the longest tails are snapping turtles.

Softshell turtles have short, thick tails. JOE LETSCHE

Here is the trail left by a spiny softshell turtle as it entered shallow water, showing the wide tail drag. JOE LETSCHE

Naturally, the longer the tail, the more likely it is to register as part of the overall trail left by the turtle. Turtles with longer tails are more likely to leave more continuous tail drag marks as they travel. Those with shorter tails may only register occasionally, and for short distances. This might change if the shorter-tailed species is moving through very deep substrate, or if the animal is moving in a manner where it is sliding forward on its plastron.

All turtle tail drags that register consistently will show some degree of undulation as the animal makes forward progress. Even in a seemingly straight tail drag, subtle waves of undulation will show up as very small raised ridges on either side of the drag line. A noticeable change in the pattern made by the tail drag can indicate a change in speed. Take for example the following sequence from the trail of a young western pond turtle (*C. marmorata*).

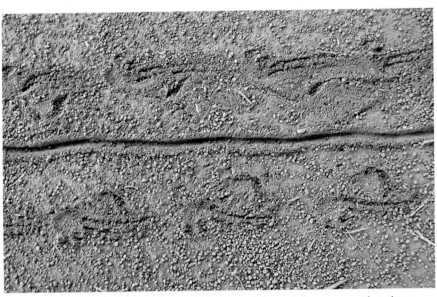

Here the animal is moving slowly from left to right, and shows a smooth and continuous tail drag. JOE LETSCHE

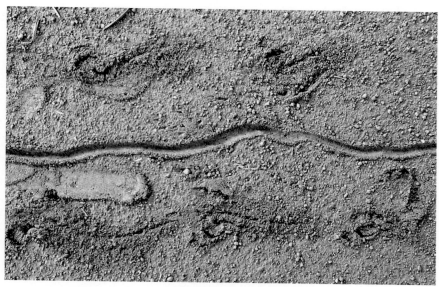

Here the animal is starting to speed up. Notice how the line left by the tail drag curves more. JOE LETSCHE

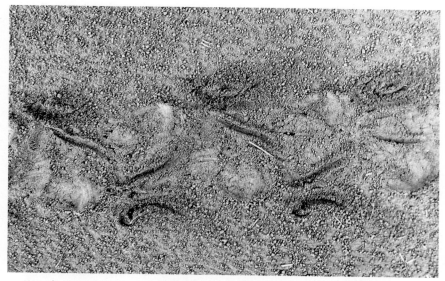

Here the animal is throwing itself forward so rapidly that it is destabilizing its entire body. With each lunging step of the hind feet, it lifts its body so high its tail loses contact with the ground. Then each time it comes down, the rear of its plastron bumps into the ground. This kind of transition in speed is well demonstrated by both the tail drag and the marks made by the plastron. In this particular situation, the turtle was fleeing as rapidly as possible. This is not an efficient way of moving for any turtle, and this one likely moved this way in response to a perceived threat. JOE LETSCHE

Plastron Drags

As seen in the previous series of photos, a turtle's plastron can also register as part of the trail. This might happen for several possible reasons, such as when the turtle changes direction. As the turtle steps out with one of its front feet, the plastron will dip downward toward that foot, registering if the substrate is deep enough. Deep substrate might also show a continuous plastron drag because the turtle is essentially walking with the lower portion of its shell bottoming out. This is more common with species that are fully or semiaquatic, and tend to carry their plastron closer to the ground.

Turtles may also leave plastron marks when they intentionally present their carapace surface to an approaching predator. The likely purpose of this is to protect the legs on the side facing the predator by lowering the shell over them like a shield. This presents less of an obvious target and also guards the more vulnerable regions that would otherwise be exposed. It may also reduce the likelihood of a turtle being flipped over by the approaching predator. A turtle on its back exposes its legs, tail, head, and neck as it tries to right itself again. Unless it is a species that can close up its shell, such as a mud turtle, the turtle will be vulnerable to attacks to any of its exposed limbs, its tail, or its head.

Below is a good example of plastron tilting behavior.

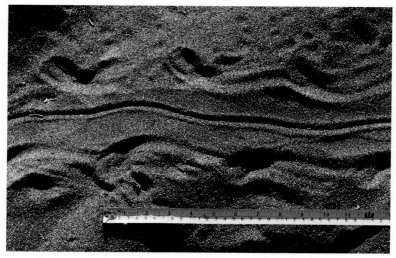

Here is an example of a painted turtle trail showing a continuous plastron drag. Notice the unbroken line made by the dragging tail, and the continuous smoothed-out area made by contact with the plastron. This animal was traveling from left to right.

This common map turtle (*G. geographica*) was traveling from the lower left to upper right. Note the tears in the substrate in the lower center of the image, along the turtle's right side. These marks are made by the edge of the plastron. JOE LETSCHE

Some turtles, especially those with close ties to water, will slide along on their plastrons intentionally for short distances. This is common for turtles entering or exiting the water at a basking site. This can leave a characteristic form of sign, where a trail leads out of the water and stops at a spot where the turtle turned around.

Plastron drag marks left by a passing turtle on especially smooth mud may be very subtle. You may only see small rips and U-shaped tears in the substrate to hint at its passing. The U-shaped tears will be open in the direction of travel.

The contact of the plastron with the substrate creates friction, and is therefore not a very energy-efficient means of travel. Turtles generally raise their shells clear of the ground surface when walking, especially over extended distances. Due to the added friction, turtles will have a narrower trail width in deeper substrates as they lift their plastrons as high as possible. The legs, which are normally in a more sprawling position, will be made more vertical to provide this added lift.

Plastron drag may also be used as a means of controlled braking, for instance when a turtle travels down a steep slope. Many turtles also frequently drop their plastrons onto the ground just as they enter water. The water's edge is a great place to look for both plastron drags and tail drags, which may help in the identification process.

Pauses and full stops in the trail will also show some degree of plastron contact. When a turtle pauses, a small area of the plastron—usually

Here is an example of tears in the mud made by the sliding contact of a passing turtle's plastron. This animal was moving from right to left. Notice the difference in texture of mud where it has been disturbed compared to the outside of the trail where it has not been disturbed. JOE LETSCHE

the rear portion—will make contact with the ground. When it comes to a full stop, it may drop its plastron fully to lie on the ground. If the substrate is appropriate for capturing tracks, then relatively clear impressions of the plastron surface might be visible.

Lays are made by turtles that stop for extended periods to rest or thermoregulate.

Burying: Dig-ins and Dig-outs

Many turtles will, at certain times, take shelter underneath a substrate. This might be to hide from potential predators, to avoid extreme heat, or to hibernate during cold periods of the year. Burying techniques can be divided into two categories, based on directional movement through a substrate: dig-ins and dig-outs. When turtles burrow into mud, sand, or leaves, their entries are dig-ins. Dig-outs, on the other hand, are created when turtles emerge. It is useful to distinguish these from long-term burrows, which the same animal uses repeatedly over time. Both dig-ins and dig-outs are usually used only once. Dig-ins and dig-outs are more often made by water-associated turtles, Emydids and Kinosternids.

Here is an example of a dig-in made by a striped mud turtle. Part of the turtle's shell is visible on the right. In this case the turtle dug itself in under the leaf litter.

Determining Species via Tracks and Sign

Identifying a turtle down to its species level can be challenging. First, use a local field guide to determine what species you may encounter in the area you are exploring. Some regions of the United States—such as the Pacific Northwest—have very few native turtle species. In contrast, the southeastern portion of the country has a wide diversity of native turtle species. Use information on range and habitat preferences to narrow down your choices. Next, look at the overall trail characteristics and the specific track morphology.

The next section includes detailed information on how to identify a variety of turtle species using their tracks. There are a variety of details you should consider. First, determine the overall trail width. Then, look at the size of the tracks relative to the trail width. Look also at the specific track morphology such as track orientation, claw length and position, foot scalation, and overall shape.

The understep walking trail of an ornate box turtle. JONAH EVANS

Box Turtles (Genus *Terrapene*)

This group of turtle species has high-domed carapaces and hinged plastrons. Two species are found in the United States: the eastern box turtle (*Terrapene carolina*) and ornate box turtle (*Terrapene ornata*). These terrestrial turtles may travel far from water, though they will also use water's edge habitats, and may sit and soak in shallow water.

Box turtles were named for the high-domed, hinged shell that allows them to close up and put much of their body out of danger. The middle of the plastron features a hinge, which the turtle can open or close at will. This protects the vulnerable soft parts of the turtle from would-be predators.

These turtles have thick, long legs covered in large scales. Their legs look somewhat like those of a tortoise, though more slim and with sharper claws. Male box turtles have a concave plastron, or a bowl-shaped depression on the lower plastron that allows them to fit snugly onto the carapace of females during mating.

Eastern Box Turtle (*Terrapene carolina*)

This commonly encountered turtle in some parts of the United States has four subspecies: the Gulf Coast box turtle (*Terrapene carolina major*), three-toed box turtle (*Terrapene carolina triunguis*), eastern box turtle (*Terrapene carolina carolina*), and Florida box turtle (*Terrapene carolina bauri*).

This is the most wide-ranging turtle species, found from eastern Texas and Oklahoma, north to Massachusetts, and south to the Florida Keys. You'll generally find this species in open woodlands, although it's also found in pastures, wet meadows, and riparian zones. Although terrestrial, some subspecies require wet microhabitats, and may be found in vernal pools, farm ponds, shallow streams, and marshes.

This species is an omnivore, with a diet that includes mushrooms, fruit, seeds, leaves, slugs and snails, earthworms, a wide variety of invertebrates, fish, frogs, and carrion.

The front and hind tracks of a three-toed box turtle.

Watch for its tracks on sandy roads, sandy or muddy spots near creeks, and on the edges of wetlands.

Track: Front: 0.3–0.5 in (0.76–1.27 cm) x 0.4–0.8 in (1–2 cm)

Tracks are small. Front feet possess five toes, and all register in the tracks. The track is symmetrical. The claws are moderate in length and found on all toes. Claws are similar in length, with the claw on toe 3 projecting the furthest forward. No webbing is present. The metacarpal area is covered in medium-sized, rounded scales with the largest toward the back of the foot. Fronts face directly forward or pitch toward the middle of the trail up to about 45 degrees.

Hind: 0.3–1.1 in (0.76–2.79 cm) x 0.4–0.7 in (1–1.78 cm)

Tracks are small. Hind feet possess four toes, and all register in the tracks. One subspecies, the three-toed box turtle (*Terrapene carolina triunguis*) only has three toes on the hind foot. The track is asymmetrical. Toes 1–3 have large claws. Toe 4 has a small claw. Claws form a crescent arcing on the outer side of the foot. Males have claws that are thick, longer, and curve downward. No webbing is present. The metatarsal area is covered in medium-sized, rounded scales, with the largest toward the back of the foot. In deeper substrate, part of the lower leg may also register. Hind tracks face directly forward or pitch slightly to the outside of the trail. The hind tracks are usually in line with or slightly more to the outside of the trail than the front tracks.

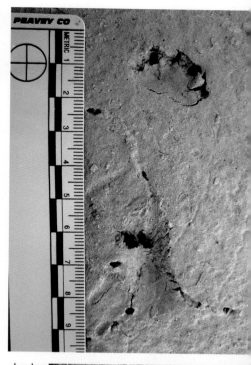

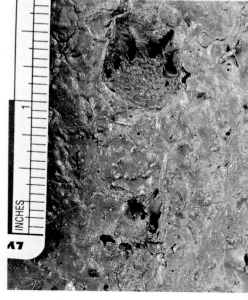

The front and hind tracks of a three-toed box turtle showing the scalation pattern.
STEVE FORTIN

 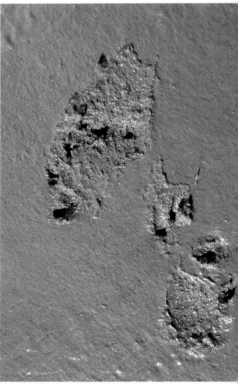

Left: The understep walking trail of an eastern box turtle. JOE LETSCHE *Right:* The front and hind tracks of an eastern box turtle. This was most likely a female turtle, as the claw marks in the hind track imply thinner, straighter claws. JOE LETSCHE

Trail: Understep Walk   Trail Width: 2.5–3.8 in (6.35–9.65 cm)
                           Stride: 2.6–4.6 in (6.6–11.68 cm)

Tail drags are infrequent with this species since the tail is moderate in length and the plastron is carried relatively high. When present, the tail drag usually forms a broken, zigzagging line.

Notes:

- One of the signs of this species is a "form" created when the turtle wallows in the shallows of a muddy waterway. Look on page 353 to see an example.
- This species will often seek log or brush piles in woodland habitats as shelter.

Ornate Box Turtle (*Terrapene ornata*)

The ornate box turtle has two commonly recognized subspecies: ornate box turtle (*Terrapene ornata ornata*) and desert box turtle (*Terrapene ornata luteola*). The two subspecies appear somewhat different in pattern; the desert

The understep walking trail of an ornate box turtle.

box turtle has ten to sixteen radiating lines on the costal scale over the second pleural bone, whereas the ornate box turtle only has five to nine lines.

This species of box turtle is found in arid and semiarid treeless, sandy plains and grassy open country such as prairies. It sometimes enters woodlands along riparian areas. In the Southwest, you may also find it on the edges of deserts.

Unlike the eastern box turtle, the ornate box turtle is largely carnivorous. It feeds on earthworms, snails, grasshoppers, beetles, and a variety of vertebrates, mostly as carrion. It will also occasionally eat cactus fruit and stems, dandelions, and other plants.

Watch for its tracks in sandy washes, prairie dog towns, and sand dunes.

Box turtle trail.
SEB BARNETT

Track: Front: 0.3–0.9 in (0.76–2.29 cm) x 0.4–1.1 in (1–2.54 cm)

Tracks are small. Front feet possess five toes, and all register in the tracks. The track is symmetrical. The claws are moderate in length and found on all five toes. Claws are similar in length, with the claw on toe 3 projecting the furthest forward. No webbing is present. The metacarpal area is covered in medium-sized, rounded scales with the largest toward the back of the foot. Fronts face directly forward or pitch toward the middle of the trail up to about 45 degrees.

Hind: 0.3–1.4 in (0.76–3.56 cm) x 0.4–0.9 in (1–2.29 cm)

Tracks are small. Hind feet possess four toes, and all register in the tracks. The track is asymmetrical. Toes 1–3 have large claws. Toe 4 has a small claw. Males have claws that are thick, longer, and curve downward. No webbing is present. The metatarsal area is covered in medium-sized, rounded scales, with the largest toward the back of the foot. In deeper substrate, part of the lower leg may also register. Hind tracks face directly forward or pitch slightly to the outside of the trail. The hind tracks are usually in line with or slightly more to the outside of the trail than the front tracks.

Trail: Understep Walk   Trail Width: 2.5–4.3 in (6.35–10.92 cm)

Stride: 2.6–4.6 in (6.6–11.68 cm)

Tail drags are infrequent with this species since the tail is moderate in length and the plastron is carried relatively high. When present, the tail drag usually forms a broken, zigzagging line.

Notes:

• This species is dependent on burrows for its survival, especially in the more arid parts of its range. The turtles likely adopt these burrows from other animals, such as kangaroo rats. They may also share burrows with other species, such as prairie dogs or ground squirrels.

• Trails of this species in loose sand tend to show repeating crescents formed by the dragging claws of the hind feet.

Desert Tortoise (*Gopherus agassizii*)

This is the largest terrestrial turtle species in the United States, and resides in the Mojave Desert ecoregion. Records exist of individuals with a carapace length of at least 15 inches (Stebbins 2012).

Desert tortoises are brown to tan terrestrial turtles that have physically adapted to the arid and semiarid habitats where they live. Their shell is high domed and is covered in large scutes that show growth rings. The rear of the carapace appears toothed. The plastron is wide at the center, with pointed gular horns at the anterior end in males. The limbs, head, neck, and tail are brown to grayish brown. Large, pointed scales cover their limbs. Males have longer claws and longer, thicker tails.

This desert creature has made amazing physiological and behavioral adaptations to the extreme conditions of the Mojave Desert. These tortoises can hold water in their bladders for months at a time as a way to conserve water. The concentrated urine is often held until the animal can drink water, at which time it is expelled and the tortoise fills up again. They are able to drink large amounts of water at one time (Lovich and Ernst 2009).

Desert tortoises sometimes dig out small catchment basins during desert rain events, which capture and hold water for

Left: **The half understep walk of a desert tortoise.** CAMERON ROGNAN *Right:* **The full understep walk of a desert tortoise walking up a sand dune.** JONAH EVANS

extended periods, allowing the tortoises to drink more. Please see more on this behavior in the Signs section on page 350, which includes a photographic example of this kind of tortoise sign.

This species has a rather long period of activity that starts between February and April and lasts until September or even as late as November. This active phase varies according to latitude, elevation, and patterns of precipitation.

Desert tortoises rely heavily on burrows for safety, thermoregulation, avoiding desiccation, and hibernation. They dig these burrows themselves, scooping with their forelegs. When a burrow is deep enough, the tortoise will turn around and push the dirt out with the sides of its shell like a bulldozer. These burrows may be straight, curved, or even forked on the inside. Burrows also often have an enlarged chamber within.

Fresh tracks going in or out of a burrow, or an accumulation of fresh nearby scats, may indicate the shelter is currently being used. During periods of inactivity—such as during hibernation—tortoises may remain in a burrow for several months without emerging.

The front and hind tracks of a desert tortoise.

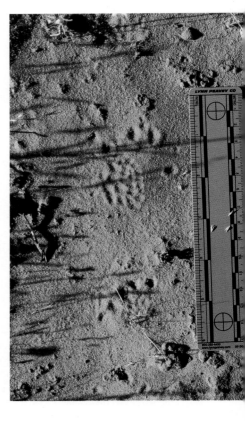

Despite their seemingly ponderous build, desert tortoises are surprisingly good climbers and often dig their wintertime dens at the tops of steep slopes or banks.

Track:

Front: 0.8–1.3 in (2–3.3 cm) x 1.3–1.8 in (3.3–4.57 cm)

Tracks are medium in size. Front feet possess five toes and front tracks register five claws that are spaced relatively close together. Claws are short. There is no webbing on the foot. The metacarpal region is covered in large scales, with the largest at the back of the metacarpal region. The foot is angled so that much of the contact is made by the claws, and only a small area of the metacarpal touches down. More of the foot surface will register when the tortoise is walking with legs held more upright. A relatively small area of scalation is visible in the track, shaped like a narrow oval. Front tracks pitch at an angle of about 45 degrees toward the center line of the trail.

Hind: 1.5–2.5 in (3.81–6.35 cm) x 1.3–2 in (3.3–5.08 cm)

Tracks are medium in size. Hind feet possess four toes and hind tracks register four claws that are spaced well apart. The claws reach well ahead of the metatarsal region. There is no webbing present. The metatarsal region is covered in large scales, toward the back of the foot. The scales furthest back on the metatarsal region are sometimes long, and can make marks similar to claw marks in deeper substrates. The contact point of the foot is significantly larger than on the front feet. A large, roughly circular area of scalation is visible in the track. Tracks tend to be longer than wide. Hind tracks pitch at an angle toward the outside of the trail, up to about 45 degrees.

Trail: Extreme Understep Walk   Trail Width: 7–9.2 in (17.78–23.36 cm)

Stride: 4–6 in (10.16–15.24 cm)

Tail drags from this species are rare, as the tail is relatively short and the animal carries its carapace high on the legs.

Notes:

• Tracks on firmer substrates may be limited to claw marks.

• Where these tortoises find shelter in a cave, burrow, or at the base of a shrub, you may be able to find marks from their plastron contacting the ground.

This is an example of a trail left by a tortoise as it walked across harder packed sand in a desert flat. Notice the lines left on either side of the trail—those were made by the claws of the front feet. The tip of the tape measure is next to one of the claw marks.

The feeding sign of a desert tortoise on common stork's-bill.

- Finding the tracks of this threatened tortoise is a real treat, and following them can teach you a great deal about their lives. If you are lucky enough to see a desert tortoise in the wild, please leave the animal alone. It is illegal to harass or collect them, and picking them up often stresses them to the point where they will void their bladder. This can be devastating to the tortoise.

The feeding sign of a desert tortoise on desert dandelion.

- Once thought to be a single species, it is now divided into two distinct species: the desert tortoise (*Gopherus agassizii*) and Sonoran Desert or Morafka's Desert tortoise (*Gopherus morafkai*). The second species is found in Arizona, east of the Colorado River and south to about Tucson.

Gopher Tortoise (*Gopherus polyphemus*)

This large, burrowing tortoise is found in well-drained sandy soils from eastern Louisiana to the southern tip of South Carolina and down through Florida. Typically it is found in beach scrub, pine-oak woodlands, and pine flats.

Gopher tortoises are terrestrial, and have adapted to the dry southern woodlands. They are brown to tan in color. This species grows to a maximum carapace length of about 15 inches. Its shell is high domed and covered in large scutes that show growth rings. The carapace appears flattened on top. The plastron is wide at the center, with pointed gular horns at the anterior end in males. The limbs, head, neck, and tail are brown to grayish brown. The limbs are covered in large, pointed scales. Males have longer claws and longer, thicker tails.

This species—similar to its relative, the desert tortoise—makes extensive burrows, which vary in depth from about eight to twenty feet. The gopher tortoise overwinters in them, and also uses them during periods of extreme heat. Burrows can serve as refuges during forest fires, which naturally occur with moderate frequency in pinewood and scrub habitat. They also serve as refuges for nesting, thermoregulation, and other needs for over one hundred other species of animal, including a variety of small and medium-sized mammals, snakes, lizards, frogs, and invertebrates. Studies have shown that these burrows are especially important to the eastern diamond-backed rattlesnake (*Crotalus adamanteus*) and allow this snake to have larger popula-

The understep walking trail of a gopher tortoise.

The full understep trail of a gopher tortoise entering its burrow. Notice the smoothed area created by the plastron dragging.

tions in habitats that are otherwise not ideal for its needs (Lovich and Ernst 2009).

This species feeds on a variety of plants including wildflowers, succulents, grasses, and fruits.

Watch for the trails of this species in sandy soils, especially around burrow entrances. Following a trail may lead to feeding sign on grass, cactus, or other plants. Its scats resemble those of desert tortoises.

Track: Front: 1–1.4 in (2.54–3.55 cm) x 1.8–2.5 in (4.57–6.35 cm)

Tracks are medium in size. Front feet possess five toes and front tracks register five claws that are spaced relatively close together. Claws are short. There is no webbing on the foot. The metacarpal region is covered in large scales, with the largest at the back of the metacarpal region. The foot is angled so that much of the contact is made by the claws, and only a small area of the metacarpal touches down. More of the foot surface will register when the tortoise is walking with legs held more upright. A relatively small area of scalation is visible in the track, shaped like a narrow oval. Front tracks pitch at an angle of about 45 degrees toward the center line of the trail.

Hind: 1.8–2.5 in (4.57–6.35 cm) x 1.2–2.3 in (3–5.84 cm)

Tracks are medium in size. Hind feet possess four toes and hind tracks register four claws that are spaced well apart. The claws reach well ahead of the metatarsal region. There is no webbing present. The metatarsal region is covered in large scales, toward the back of the foot. The scales furthest back on the metatarsal region are sometimes long, and can make marks similar to claw marks in deeper substrates. The contact point of the foot is significantly larger than on the front feet. A large, roughly circular area of scalation is visible in the track. Tracks tend to be longer than wide.

A gopher tortoise burrow, with the trails of hatchling tortoises dispersing from it.
DIRK STEVENSON

Hind tracks pitch at an angle toward the outside of the trail, up to about 45 degrees.

Trail: Understep Walk   Trail Width: 7.2–10.3 in (18.28–26.16 cm)
Stride: 5.4–10.6 in (13.72–26.92 cm)

Tail Drag: Tail drags from this species are rare, as the tail is relatively short and the animal carries its carapace high on the legs.

The pale, dry grass is a very old scat of an adult gopher tortoise.

The shell of a dead adult gopher tortoise, showing scutes on the carapace and the exposed bone underneath.

Notes:
- Where this tortoise is still found in abundance, its trails are a common sight in sandy substrates.
- Inhabited burrows show the marks of plastron drags and tracks coming in or out of the burrow entrance. The throw mounds in front of the burrow entrance can be great spots to look for the tracks of many animals.
- This is a threatened species that should be protected wherever it is found. Its burrows are vital habitat features for many species. Gopher tortoises should not be harmed, harassed, or collected.

Mud and Musk Turtles (Genus *Kinosternidae*)

This is a group of small, semiaquatic turtles that have broad plastrons and often relatively high-dome carapaces. All four of their feet are webbed. Mud turtles have two hinges on their plastrons, while musk turtles have only one. The plastrons of musk turtles are reduced in width, while those of mud turtles are broad and well developed. The pectoral scutes are rectangular in musk turtles and triangular in mud turtles. Both mud and musk turtles have foul-smelling scent glands on the sides of their bodies; those in musk turtles are arguably more potent.

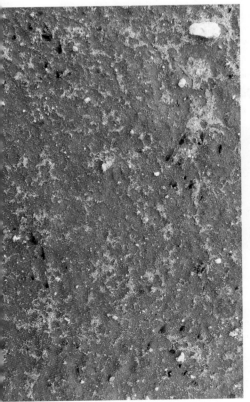

The front tracks of mud turtles are somewhat similar to the tracks of moles, as these turtles tend to turn their front feet on their sides so that they walk mostly on their claws. They share this trait with musk turtles (genus *Sternotherus*), but it distinguishes them from all other turtles. Also, the front tracks land just to the inside and slightly behind the hinds. Often the front and hind tracks are separated by very little space. In loose sand or deep mud, they appear as a single track.

Tail drags are unusual for this species except in very deep substrates. When present it will be very thin, 0.1–0.2 inches wide. Male mud and musk turtles have tail tips that end in a thick spine. Male mud and musk turtles also have longer, thicker tails than females.

The claws in an understep walking trail of a yellow mud turtle. JOE LETSCHE

Yellow Mud Turtle (*Kinosternon flavescens*)

This species of turtle lives in the south central United States, and can be found in some very arid landscapes. Its habitats include swamps, sloughs, sinkholes, rivers, creeks, ponds, reservoirs, cattle tanks, and lakes in grasslands, deserts, and open woodland areas. They prefer water with sand or mud at the bottom with stands of emergent aquatic vegetation. They require loose soil to burrow into to a depth below the frost line during the winter.

This turtle is small, but is large for a mud turtle, growing between 3.5 to 6.44 inches in carapace length.

This species has two hinges on the plastron, allowing it to close both the anterior and posterior portion to protect the limbs, tail, and head. The plastron is often yellowish or light brown. The carapace is brown to olive-green, and each scute has a dark border. The lower edge of the carapace may show some degree of yellow and the turtle may also have yellow on the chin, cheeks, and neck.

This species feeds on a wide variety of animals including worms, leeches, beetles, spiders, dragonflies, fish, clams, and amphibian larvae. It consumes some aquatic plant materials such as algae.

You'll most likely be able to observe their trails when the body of water these turtles use is at a low point, forcing them to move. You may find associated dig-ins and dig-outs with trails emerging from or leading to them.

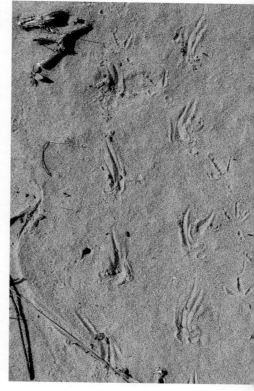

Track: Front: 0.4–0.5 in (1–1.27 cm) x 0.6–1 in (1.52–2.54 cm)

Tracks are small. Front feet possess five toes, and the claws of four or five toes usually register in the track. The claws are well developed, moderately long, and similar in length on all the toes. The two middle claws reach equally far forward and are ahead of the other toes. Webbing is present between all the toes. The metacarpal area is covered in small, relatively flat scales, and this region is shaped

The understep walking trail of a yellow mud turtle in sand. JOE LETSCHE

like a narrow oval oriented sideways. Generally only a small part of this region registers in the tracks, as much of the contact made by the front foot is limited to the nails. The front tracks face directly forward or pitch slightly to the inside. They also tend to land more to the outside of the trail, and appear noticeably wider than the hind tracks.

Hind: 0.5–0.9 in (1.27–2.28 cm) x 0.4–0.7 in (1–1.78 cm)

Tracks are small. Hind feet possess five toes, and only four possess claws. The claws are well developed, moderately long, and similar in length on all toes. Webbing is present between all toes. The metatarsal region is covered in small, relatively flat scales and is oval in shape. Hind tracks tend to register closer to the inside of the trail and orient directly forward or pitch slightly to the outside. They frequently land very closely behind and nearly on top of the front tracks.

Trail: Understep Walk   Trail Width: 3.6–5.2 in (9.14–13.2 cm)
Stride: 2.5–3.8 in (6.35–9.65 cm)

Notes:

• This small, water-dwelling turtle can be a real surprise to find in the widely scattered and sometimes well-separated waters throughout the semiarid and arid landscapes where it dwells.

• Watch for mole-like tracks around the edges of drying ponds or shallows of intermittent muddy or sandy river pools.

• This species tends to walk using the full length of its long legs, with the shell well clear of the substrate. Most commonly, only the nails register clearly in firm or very shallow substrates, while substrates such as deep mud may show most of the foot, especially in the hind tracks.

Striped Mud Turtle (*Kinosternon baurii*)

This small mud turtle can be found in the southeastern portion of the United States, from Florida to Virginia. Adults grow between 3 and 4.75 inches in carapace length. This species is named for the three pale lines that run down the middle of the full length of the carapace. The carapace may be tan to black, and in some specimens the stripes may be subtle or even occasionally absent. The plastron has two well-developed hinges. The skin appears tan to black, and the neck and head region may show some dark mottling. There are typically two light broken or continuous stripes extending from the eyes and back at least partially onto the neck.

This species prefers temporary freshwater wetland areas with soft bottoms, including ponds, canals, hardwood or cypress swamps, and streams. You'll occasionally find this species in brackish waters of relatively low salinity. Sometimes, this species inhabits the lodges of round-tailed muskrats (*Neofiber alleni*) in southern Florida.

This species is an omnivore, and will feed on algae, cabbage palmetto seeds, insects, and small snails.

Striped mud turtles will move overland during periods of heavy rainfall. Look for their trails between bodies of water.

Track: Front: 0.2–0.3 in (0.5–0.76 cm) x 0.3–0.5 in (0.76–1.27 cm)

Tracks are small. Front feet possess five toes, and the claws of four or five toes usually register in the track. The claws are well developed, moderately long, and similar in length on all the toes. The two middle claws reach equally far forward and are ahead of the other toes. Webbing is present between all the toes. The metacarpal area is covered in small,

The understep walking trail of a striped mud turtle in loose sand.

relatively flat scales, and the region is shaped like a narrow oval oriented sideways. Generally only a small part of this region registers in the tracks, as much of the contact made by the front foot is limited to the nails. The front tracks face directly forward or pitch slightly to the inside. They also tend to land more to the outside of the trail, and appear noticeably wider than the hind tracks.

Hind: 0.3–0.5 in (0.76–1.27 cm) x 0.2–0.3 in (0.5–0.76 cm)

Tracks are small. Hind feet possess five toes, and only four possess claws. The claws are well developed, moderately long, and

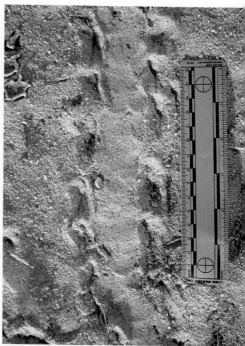

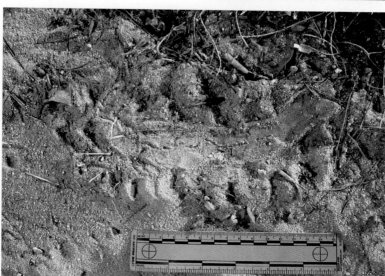

The trail left by a striped mud turtle that was attempting to dig down into the sand, moving from right to left.

similar in length on all the toes. Webbing is present between all the toes. The metatarsal region is covered in small, relative flat scales and is oval in shape. Hind tracks tend to register closer to the inside of the trail and orient directly forward or pitch slightly to the outside. They frequently land very close behind and nearly on top of the front tracks.

Trail   Understep Walk   Trail Width: 2–2.5 in (5–6.35 cm)
                             Stride: 1–2 in (2.54–5 cm)

Notes:
- This species tends to walk using the full length of its long legs, with the shell well clear of the substrate. Only the nails commonly register clearly in firm or very shallow substrates, while substrates such as deep mud may show most of the foot, especially the hind tracks.
- You can sometimes encounter this species crossing roads between wetlands, especially on rainy nights.

Eastern Mud Turtle (*Kinosternon subrubrum*)

This commonly encountered mud turtle ranges widely, from eastern Texas across the southeastern United States and up the Atlantic coast to as far north as Long Island. This species includes several subspecies, such as the eastern mud turtle (*Kinosternon subrubrum subrubrum*), Mississippi mud turtle (*Kinosternon subrubrum hippocrepis*), and the Florida mud turtle (*Kinosternon subrubrum steindachneri*). Adults measure between 3 and 4.75 inches in carapace length.

The understep walking trail of a Florida mud turtle. Notice the plastron drag and the thin tail drag.

This turtle prefers slow-moving shallow waters with abundant vegetation and soft bottoms, including ditches, canals, streams, rivers, wet meadows, farm ponds, marshes, and tidal flats. The coastal populations show a high tolerance for brackish water.

The eastern mud turtle feeds on both plants and animals, including crayfish, small crabs, beetles, tadpoles, seeds, and leaves of aquatic plants, even sometimes cereal grains (Lovich and Ernst 2009).

This species may move out of temporary wetlands as they dry, or it may burrow down into the soil and stay buried until the water returns. Of all the eastern mud turtles, this one will leave trails in surprising places like flooded agricultural or athletic fields, roadside ditches, and similarly temporary bodies of water.

Track: Front: 0.3–0.4 in (0.76–1 cm) x 0.6–0.8 in (1.52–2 cm)

Tracks are small. Front feet possess five toes, and the claws of four or five toes usually register in the track. The claws are well developed, moderately long, and similar in length on all toes. The two middle claws reach equally far forward and are ahead of the other toes. Webbing is present between all the toes. The metacarpal area is covered in small, relatively flat scales, and is shaped like a narrow oval oriented sideways. Generally only a small part of this region registers in the tracks, as much of the contact made by the front foot is limited to the nails. Front tracks face directly forward or pitch slightly to the inside, and tend to land more to the outside of the trail, appearing noticeably wider than the hind tracks.

Hind: 0.5–0.8 in (1.27–2 cm) x 0.3–0.5 (0.76–1.27 cm)

Tracks are small. Hind feet possess five toes, and only four possess claws. The claws are well developed, moderately long, and similar in length on all toes. Webbing is present between all toes. The metatarsal region is covered in small, relatively flat scales and is oval in shape. Hind tracks register closer to the inside of the trail and orient directly forward or pitch slightly to the outside. They frequently land very close behind and nearly on top of the front tracks.

Trail: Understep Walk   Trail Width: 2.8–3.7 in (7.11–9.4 cm)

Stride: 2.2–3.8 in (5.6–9.65 cm)

Notes:

• This species tends to walk using the full length of its long legs, with the shell well clear of the substrate. Most commonly, only the nails register clearly in firm or very shallow substrates, while substrates such as deep mud may show most of the foot, especially in the hind tracks.

Pacific Pond Turtle (*Actinemys marmorata*)

This species used to be split into two subspecies: the northwestern race (*Actinemys marmorata marmorata*) and southwestern race (*Actinemys marmorata pallida*). These have since been named as two distinct species: the northwestern pond turtle (*Actinemys marmorata*) and southwestern pond turtle (*Actinemys pallida*).

This turtle is found strictly along the Pacific coast of the United States, and populations are found in Washington, Oregon, and California. This species strongly prefers a variety of wetlands habitats, including lakes, ponds, reservoirs, streams, and rivers. You can sometimes find it in brackish estuarine waterways.

The understep walking trail of a Pacific pond turtle that has been preserved in drying mud. MARCUS REYNERSON

This pond turtle generally has a tan to brown carapace, with black spots or lines radiating from the center of each scute. The plastron is usually yellowish tan, with variable degrees of dark uneven markings. The head, neck, limbs, and tail appear tan to brown with fine black speckles. This species has a long, tapered tail. Males have a concave plastron, flatter carapace, and thicker, longer tail. They also usually have a paler throat with fewer markings, whereas females have more speckling on the throat. Females have flat plastrons, more highly domed carapaces, and longer claws on their hind feet. Adult males also have two raised areas visible at the back of the head.

This species prefers habaitats that include basking sites, emergent vegetation, and safe refuges, such as undercut banks, submerged vegetation, mud, rocks, and logs. This species of turtle often uses terrestrial habitats, such as rock crevices and secluded patches of dense leaf litter that are above the flood line, near to their aquatic habitats. This is especially true of those turtles that live near streams, as they often overwinter in terrestrial refuges (Lovich and Ernst 2009). Those who inhabit ponds and lakes tend to overwinter in the aquatic habitats themselves.

Females may travel more than a quarter mile from water to a nesting site during the breeding season. Watch for the tracks of these turtles traveling up and down stream corridors along the edges of reservoirs and lakes.

Track: Front: 0.7–1 in (1.78–2.54 cm) x 0.8–1.2 in (2–3.05 cm)

Tracks are small to medium. Front feet possess five toes, and all register in the tracks. The claws are well developed, moderate in females, and slightly longer in males. The claws of toes 1–4 are of similar length, while the claw on toe 5 is smallest. Toes 1 and 5 are positioned on the foot on the same plane. Webbing is present between all of the toes and reaches to the toe tips. The metacarpal region is covered evenly in small scales of mostly uniform size. This part of the foot tends to register as a sideways oval, being obviously wider than long. Front tracks register strongly pitched toward the center line of the trail, up to about 45 degrees. The front tracks fall closer to the inside of the trail than the hind tracks.

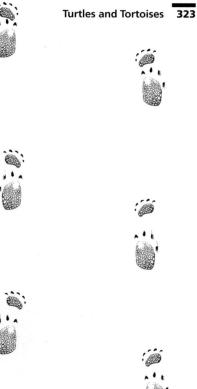

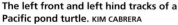

The left front and left hind tracks of a
Pacific pond turtle. KIM CABRERA

Hind: 0.6–0.95 in (1.52–2.41 cm)
x 0.8–1.2 in (2–3.05 cm)

Tracks are small to medium. Hind
feet possess five toes, but usually
only four are clearly visible. The
claws are well developed and
moderate in length. Claws are
present on toes 1–4. Toe 5 is
positioned lowest on the foot.
The claws of females are almost
twice as long on the hind feet as

**Pacific pond turtle
understep walk.**

those of males. Webbing is present between all toes and reaches to the
toe tips. The metatarsal region is covered evenly in small scales of a mostly
uniform size. Hind tracks face forward in the direction of travel or may
pitch slightly toward the outside of the trail. The hind tracks fall in line
with or slightly more to the outside of the trail relative to the position
of the front tracks.

Trail: Understep Walk   Trail Width: 5–6.5 in (12.7–16.51 cm)
                         Stride: 5–6 in (12.7–15.24 cm)

Tail Drag: 0.2–0.4 in (0.5–1 cm)

This species has a relatively long tail, so tail drags are common. The tail
creates a gently undulating continuous line with a consistent width. The

tail drags of males are slightly wider than those of females. At higher speeds, especially if the turtle is scrambling to escape a potential predator, the tail drag will become a broken, inconsistent line.

Notes:

- This species has experienced dramatic declines in significant portions of its range. It has been overharvested for the pet trade, as well as in the past for the meat market. It has also been affected by growing urban sprawl and by a variety of pollutants contaminating the waterways where it lives.
- This turtle is considered a species of special concern in the states of California and Oregon. It is endangered in the state of Washington.
- If you are lucky enough to observe this species in the wild, please do not disturb or collect it. Help protect the wetland and stream habitats this animal is dependent on.

Painted Turtle (*Chrysemys picta*)

Arguably, this is one of the most beautiful turtles in North America. There are several painted turtle subspecies, most of which show some degree of difference in coloration. Generally, this turtle has red markings on the marginal scutes and a notched-up jaw. The carapace is relatively flattened and smooth. It is generally olive-green or black, with yellow or reddish-colored borders.

Some populations have a red line running the length of the carapace down its center. The plastron is relatively flat, and may be solid yellowish or bright red with dark green or black scrawling that is mirrored nearly perfectly on either side.

The degree of darkness of this species is a reflection of the dominant substrate where it is found. The head, neck, limbs, and tail are dark green to black and striped in yellow or red. A yellow stripe extends behind the eye and down the neck. A pair of yellow lines extends up the lower neck and onto the chin. Males have longer tails and longer claws on forelimbs.

The understep walking trail of a painted turtle. JOE LETSCHE

Drought conditions often force this species to bury itself and wait (aestivate) in the muddy bottom of a drying water body, or can send it out to seek a more persistent body of water. You'll most likely find the trails of painted turtles under these conditions.

Painted turtles are known to move long distances, sometimes overland, in search of new bodies of water or to look for mates. Males generally seek females, and are therefore more likely to immigrate to new bodies of water. Some populations of painted turtles are known to migrate to and from such sites regularly. This species has a well-developed homing ability: in one study, researchers released painted turtles back into the wild, and many of the animals traveled several kilometers back to their initial point of capture.

Perhaps the most widely known research done on this species centered on the natural freeze tolerance of hatchlings. The young of painted turtles overwinter in the nesting cavities where their eggs were laid. In these chambers, hatchlings survive temperatures as low as 12 degrees Fahrenheit. There is some indication that "cryoprotectants" such as glucose and lactate—which build up in the muscles, heart, brain, and internal organs—might be used as a form of antifreeze (Lovich and Ernst 2009).

Painted turtles are opportunistic omnivores, willing to consume a wide variety of animals and plants, including carrion. Their diverse diet and cold-tolerant hatchlings no doubt contribute to their widespread range. They are found from the Pacific coast in southern British Columbia to the Atlantic coast in Nova Scotia and down to the Gulf Coast.

Watch for the tracks of this species in mud or sandy banks where this species is basking, when it crosses overland from one wetland area to another, or when females are moving into upland areas to find nesting sites.

Painted turtle trail.
SEB BARNETT

Track: Front: 0.6–0.8 in (1.5–2 cm) x 0.7–0.85 in (1.78–2.16 cm)
Tracks are small. Front tracks have five distinct claws, all of which generally register. The claws are well developed

The tracks of a painted turtle that was missing its front right foot.

and moderate in length on females, while those on males are several times longer. The longest claws are on toes 3 and 4, and toe 2 is a little shorter. Toes 1 and 5 are very similar in length and are the shortest. Toes 1 and 5 are positioned on the foot on the same plane. The claws of adult males can double the length of the track. Webbing is present between all of the toes and reaches to the toe tips. The metacarpal region is covered evenly in small scales of mostly uniform size. This part of the foot tends to register as a sideways oval, being obviously wider than long. Front tracks face the direction of travel or pitch inward toward the center line of the trail, but often at less than 45 degrees. The front tracks fall closer to the inside of the trail than the hind tracks.

Hind: 0.8–1.3 in (2–3.3 cm) x 0.6–0.75 in (1.53–1.9 cm)

Tracks are small to medium in size. Hind tracks have four distinct claws, all of which generally register. Hind tracks are longer than they are wide, and oval in shape. Toes 2 and 3 are similar in length and are the longest. Toes 1 and 4 are similar in length and shorter. Toe 5 is the shortest, but does not possess a claw and may not be obvious in the track. Webbing is present between all toes and reaches to the toe tips. The metatarsal region is covered evenly in small scales of mostly uniform size. The hind track angles forward or may pitch slightly toward the outside of the trail. Hind tracks fall in line with or slightly more to the outside of the trail relative to the position of the front tracks.

Trail: Understep Walk   Trail Width 4–6.6 in (10.16–16.76 cm)
                        Stride: 2.3–5.5 in (5.84–13.97 cm)

Tail Drag: 0.3–0.4 in (0.76–1 cm)

The tail mostly registers in deeper substrates, otherwise it's infrequently observed. You'll see it as a continuous, slightly undulating line of consistent width.

Notes:
• When compared to the tracks of pond sliders (*Trachemys scripta*), the front tracks of painted turtles are often less dramatically pitched inward.

- This species carries its plastron fairly low, and may drag it in deeper substrates. It has a very large, wide plastron. The plastron drag can reach out nearly to the edge of the tracks.
- This species is shy and will quickly dive into the water when approached closely in the wild. If you wait patiently, you may see it emerge and return to nearly the same spot where it was initially basking.
- Painted turtles have been introduced to many suburban areas surrounding major metropolises along the Pacific coast.

Common Musk Turtle (*Sternotherus odoratus*)

This is the most common and widespread musk turtle species, and is also commonly referred to as the "stinkpot" for its strong-smelling secretions. These are small, mainly aquatic turtles, with adults growing between 3 and 5 inches in carapace length.

The carapace of this species is highly domed, and may be smooth or have three keels. It is typically olive, brown, or dark gray, but may be covered in a layer of algae. All musk turtles have a relatively small plastron. They have two pale stripes on their heads, as well as two fleshy barbels on their chins or sometimes their throats.

Though small, this is a feisty species and males will bite in defense if handled. They do not typically wander very far from water, and prefer muddy-bottomed, freshwater environments with still or slow-moving water. They are generalists in diet, eating plant and animal matter, including insects, tadpoles, small fish, mollusks, and sometimes carrion.

Watch for their trails in the mud of shallow wetlands or wetland edges during low-water periods. This species can climb nearly vertical trees in order to bask, and is most often found basking on small-diameter logs and branches hanging down into water.

Track: Front: 0.2–0.3 in (0.5–0.76 cm) x 0.3–0.5 in (0.76–1.27 cm)

Tracks are very small. Front feet possess five toes, and the claws of four or five toes usually register in the track. The claws are well developed, moderately long, and similar in length on all the toes. The three middle claws reach equally far forward and are ahead of the outer two toes. Webbing is present between all toes. The metacarpal area is covered in small, relatively flat scales, and

A small musk turtle in the middle of the trail of a large snapping turtle.
STEVE FORTIN

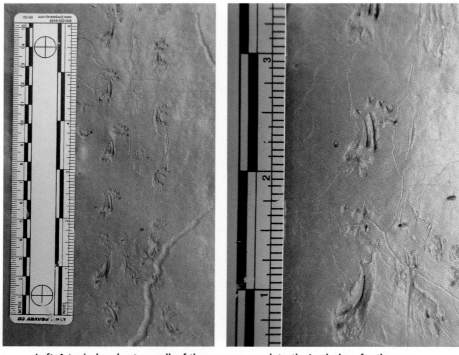

Left: A typical understep walk of the common musk turtle. Look close for the very thin tail drag. JASON KNIGHT *Right:* Close-up of tracks from the same common musk turtle trail. Here you'll see the mole-like hind tracks and the front track that appears similar to the hind track of a rodent. JASON KNIGHT

this region is shaped like a narrow oval oriented sideways. Only a small part of this region generally registers in the tracks, as much of the contact made by the front foot is limited to the nails. Front tracks face directly forward or pitch slightly to the inside. The front tracks tend to land more to the outside of the trail, and appear noticeably wider than the hind tracks.

Hind: 0.3–0.5 in (0.76–1.27 cm) x 0.2–0.3 in (0.5–0.76 cm)

Tracks are very small. Hind feet possess five toes, and only four possess claws. The claws are well developed, moderately long, and similar in length on all the toes. The claws are arranged in a strong arc, similar to that made by the claws of a mole. Webbing is present between all toes. The metatarsal region is covered in small, relatively flat scales, and is oval in shape. Hind tracks tend to register closer to the inside of the trail and orient directly forward or pitch slightly to the inside. They frequently land very close behind and nearly on top of the front tracks.

Trail: Understep Walk   Trail Width: 1.2–2 in (3.05–5.08 cm)
Stride: 0.9–2 in (2.28–5.08 cm)

Notes:

• Given this species' reduced plastron and relatively long legs, it can take long steps. The hind tracks often land immediately behind the front tracks, or sometimes even parallel with them.

An example of a dig-out of the common musk turtle. The turtle traveled out of the mud and headed to the left. JASON KNIGHT

- The front tracks of this species are typically wider than long. The hind tracks are typically longer than wide.
- The tail drag of this species is very thin when present, and is usually made by the hardened spine at the tail tip.

Sliders (Genus *Trachemys*)

This group of turtles includes two species north of Mexico: the Big Bend slider (*Trachemys gaigeae*) and pond slider (*Trachemys scripta*). The first species is very poorly known in the wild and has a range limited largely to the Rio Grande drainage. The second species is common across much of United States.

Both are highly aquatic turtles, and commonly bask socially on logs, rocks, or banks. Both begin life as largely carnivorous and become more omnivorous as they mature. They eat invertebrates, fish, and several types of aquatic plants.

Pond Slider (*Trachemys scripta*)

These are some of the most widely introduced turtles on the planet. This is largely due to the popularity of the red-eared slider (*Trachemys scripta elegans*) in the pet trade. Their native ranges include the central United States, from southern Michigan and northern Illinois down to the Gulf Coast, then west from New Mexico to the Atlantic coast in Virginia, and down the coast to Florida.

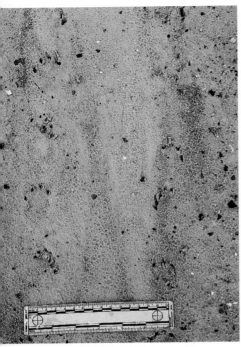

The half understep trail of a red-eared slider. STEVE FORTIN

Adults have a low, smoothly domed shell. The rear edge of the carapace is jagged, with an increasingly saw-toothed edge in older individuals. Juvenile turtles have a single low keel down the central line of the carapace. The rear marginal scutes on the cara-pace come to a point, giving the back edge a toothy appearance. Young tur-tles have greenish shells with greenish lines on each scute, often with a thicker central yellow line. The head, neck, tail, and limbs are marked in alternating green and yellowish lines. Behind each ear is an oblong red, orange, or yellow-colored mark. The plastron is yellow overall, with green ovals or blotches that become darker with age, sometimes to the point of turning black. Adults often show more faded colors, especially on the cara-pace, which may become brown with age. Also, adults show some degree of increased melanism with age, with a few becoming completely black. This species is sexually dimorphic, with males having longer tails (about twice that of females) and much longer claws on the forelimbs. Females average larger in size than males at maturity, and also have a higher domed carapace. Adult females' hind claws are almost twice as long as those of adult males, which they likely adapted to assist in nest excavation.

This species occupies a variety of freshwater habitats, including lakes, swamps, farm ponds, canals, sloughs, slow-moving rivers, and reservoirs. It may be active in all months of the year in warmer regions in the South. Meanwhile, in northern parts of its range it will hibernate in burrows in muddy bottoms, hollow stumps, muskrat lodges, and other shelters. This species tends to become inactive when water temperatures reach below 50 degrees Fahrenheit. Individuals have sometimes been observed moving underwater below a layer of surface ice.

This species often spends considerable time underwater. This is especially true during winter hibernation. Juveniles can survive over 180 days submerged at normal levels of dissolved oxygen in the water. Outside of hibernation, adults are sometimes known to dive for several hours at a time. This turtle species is

This is an adult red-eared slider making a dig-in in deep mud. JOE LETSCHE

almost entirely dependent on breathing air by surfacing.

Pond sliders may have extensive home ranges that include several bodies of water between which they make frequent overland journeys. This means that you'll likely observe this species' tracks more than others between such sites. Generally, this species travels overland to other bodies of water to feed, bask, reproduce, and to seek shelter for periods of dormancy.

In the United States, this species tends to nest between the months of April and July. Females tend to select open, sun-exposed areas with non-muddy soil that are relatively close to water. Pond slider females may return to the same site over the course of their reproductive lives. Just like with other turtle species, pond slider females often dig several test holes.

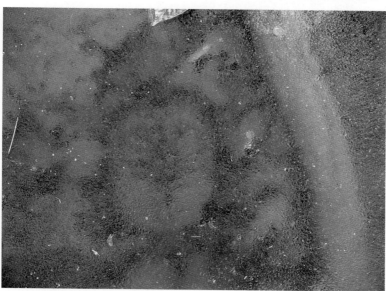

This is an adult red-eared slider moving very slowly under ice. JOE LETSCHE

These holes often contain roots, rocks, and other barriers to aid in digging. Nests are oval in shape and are made in soft soils, with the entrance being about a quarter of the diameter of the nest cavity itself. If the ground is hard, the female will release fluid from her cloacal bladder to help soften it for digging (Lovich and Ernst 2009). Once egg laying is completed, the female pushes loose soil into the cavity and kneads the soil with her hind legs and the rear part of her plastron. Then, she voids fluids again, resulting in a soil plug that usually stays separate from the eggs.

This species is known to be omnivorous. Juveniles are, however, highly carnivorous at first, but focus progressively more on plant foods as they age. This species' broad dietary preferences include a wide variety of plants and animals, and its varied diet has likely contributed to its ability to thrive in many areas where it has been introduced.

Track: Front: 0.5–1 in (1.27–2.54 cm) x 1–1.5 in (2.54–3.81 cm)

Tracks are small to medium. Front feet possess five toes, and all register in the tracks. The claws are well developed and moderate in length on females, while males can have claws on their front feet that are several times longer. The longest claws are on toes 3 and 4 and toe 2 is a little shorter. The claws on toes 1 and 5 are similar in length and are the shortest. The claws of adult males can double or even nearly triple the overall length of the track. Toes 1 and 5 are positioned on the foot on the same plane. Webbing is present between all of the toes and reaches to the toe tips. The metacarpal region is covered evenly in small scales of mostly uniform size. This part of the foot tends to register as a sideways oval, being obviously wider than long. Front tracks strongly pitch toward the center line of the trail, often at between 45 and 90 degrees. The front tracks fall closer to the inside of the trail than the hind tracks.

Hind: 0.8–1.4 in (2–3.56 cm) x 0.8–1.4 in (2–3.56 cm)

Tracks are small to medium in size. Hind feet possess five toes, and all register in the tracks. Claws are present on toes 1–4. Toe 5 is positioned lowest on the foot. The claws are well developed and moderate in length. The claws of females are twice as long on the hind feet as those of males. Webbing is present between all toes and reaches to the toe tips. The metatarsal region is covered evenly in small scales of a mostly uniform size, although slightly larger scales are located at the bottom of the metatarsal region near the base of toe 5. Hind tracks face forward in the direction of travel or may pitch slightly toward the outside of the trail. The hind tracks fall in line with or slightly more to the outside of the trail relative to the position of the front tracks.

Trail: Understep Walk   Trail Width: 5–7.5 in (12.7–19.05 cm)
                          Stride: 5–7.5 in (12.7–19.05 cm)

Tail Drag: 0.3–0.5 in (0.76–1.27 cm)

Tail drags are common for this species mainly in deep substrates. In shallow substrates they register infrequently. They'll usually appear as a continuous, slightly undulating line of consistent width.

Notes:

• This species carries its plastron fairly low and may drag it in deeper substrates. It has a very large and wide plastron. The plastron drag can reach out to the edge of the tracks.

• You'll most frequently encounter the tracks of this turtle on muddy or sandy banks where it comes out of the water to bask, or when it moves from one wetland area to another. Also, you may see tracks made by females seeking upland areas in which to build a nest.

Snapping Turtles

This is a group of turtles with large bodies, large heads, and powerful jaws. They have a rough, well-developed, and keeled carapace, with a strongly serrated posterior region. The greatly reduced plastrons form a cross of bone across the midsection of the abdomen.

There are only two species found in the United States: the alligator snapping turtle (*Macrochelys temminckii*) and snapping turtle (*Chelydra serpentine*). The former species is one of the largest freshwater turtles on earth. The smaller, more common species is still an impressive animal to see at its full adult size.

Snapping Turtle (*Chelydra serpentina*)

The most common and widespread snapping turtle species in the United States ranges over the eastern two-thirds of the country. Small populations of snapping turtles have been introduced across the western states. This species can be found in nearly every kind of freshwater habitat, but are most likely to live in slow-moving water with sandy or muddy bottoms and an abundance of submerged woody debris or aquatic vegetation. A few populations of snapping turtles are found in coastal estuaries.

This species has long, powerful legs and a large carapace. The carapace is serrated at the rear edge, and has three low keels formed by knobs near the center of the scutes running its length. The plastron is small, very narrow, and cross-shaped overall. It is attached to the carapace by a narrow bridge of bone on either side. Large adults can have a carapace length of up to 19 inches.

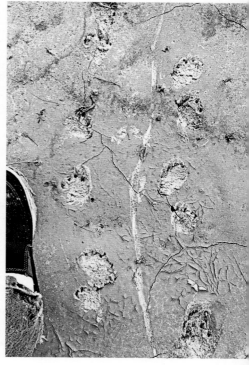

The understep walking trail of a snapping turtle in drying mud. STEVE FORTIN

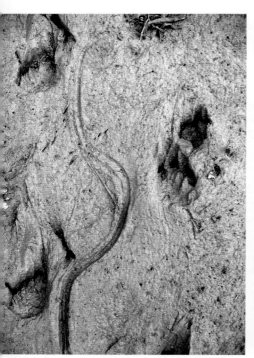 

Left: **The understep walking trail of a snapping turtle. Notice the relatively wide tail drag.** JOE LETSCHE *Right:* **A close-up of some snapping turtle tracks, showing the large holes created by the claws.** CHRIS HYDE

This species has a very large and long tail, and often leaves significant tail drags in association with its tracks. Also, the feet and legs are large and powerful. When this species walks, it usually carries its shell well above the ground on fully extended legs.

The tracks of large adult snapping turtles are impressive, with deep, large claw marks and large tracks. Watch for them in sandy areas or muddy spots at the edges of or between wetland areas. They may also plow wide, undulating trails through deep mud in ditches. Nesting females can travel far over land—distances have been recorded to over half a mile (Lovich and Ernst 2009).

Track: Front: 1.3–1.7 in (3.3–4.32 cm) x 1.5–2.6 in (3.81–6.6 cm)

Tracks are medium in size. The track has five toes and all tend to register. Toe bone outlines are often visible in the track in deeper substrates. Claws are very well developed, long, and thick. They can make large gouges at the tips of the fingers. In loose substrates, claws can create significant claw drag marks as well. Webbing is present between all the toes and reaches to the toe tips. The metacarpal area of the foot is large, and covered in scattered scales of variable size. Scales are often largest at the back of the metacarpal region, near the base of toe 1 and progressively smaller and more dispersed across the foot toward toe 5. The foot scalation is likely distinct for each individual. Front tracks pitch at about 45 degrees toward the center line of the trail. The front tracks frequently fall slightly more to the inside of the trail than the hind tracks.

Hind: 1.5–2.8 in (3.81–7.1 cm) x 1.5–2 in (3.81–5.08 cm)

Tracks are medium in size. The foot has five toes, but generally only four register clearly. Only four toes have well-developed claws. The claws on the hind feet are even longer than those on the front feet. Toe 5 is smallest and possesses only a short, rounded claw. Toe bone outlines may be visible in the track. Webbing is present between all toes and reaches to the toe tips. The metatarsal region is large, long, and covered in scattered scales of variable size. Scales are often largest at the back of the metatarsal region, and tend to be smallest near the toe tips. The foot scalation is likely distinct for each individual. Hind tracks face directly forward in the direction of travel and tend to fall slightly more to the outside of the trail than the front tracks.

Trail: Understep Walk   Trail Width: 8–10.5 in (20.32–26.67 cm)

Stride: 7–8.5 in (17.78–21.59 cm)

Tail Drag: 0.34–1 in (0.86–2.54 cm)

Of all the turtle species included in this book, the snapping turtle is most likely to leave a tail drag. The tail drags of this species are often relatively wide and continuous.

Notes:

• This is a big, powerful turtle that looks prehistoric and demands respect. It has a very long neck and powerful bite. It can act rather pugnacious when it feels threatened.

• Unlike many other turtles, snapping turtles often walk with their shells raised high off of the ground. They will sometimes push their tails down into the ground to act almost as a fifth limb. This high clearance means that the plastron is unlikely to register in

Snapping turtle understep trail.

This is a wide trough-like trail of a snapping turtle through deep mud at the edge of a wetland. JOE LETSCHE

the trail, except when the turtle is in very deep substrate or when it comes to a full stop.

- This turtle is frequently killed when crossing roads. Most often, these are females moving inland to a nesting area. This species is also hunted in some parts of the United States for food. Though widespread and common, it is a long-lived species and needs to be conserved locally.

Softshell Turtles

This small group is made up of three species (five if you include the two introduced species in Hawaii). Softshell turtles are distinct in appearance, because they do not possess an obvious external bony shell. The carapace and plastron of these turtles lack the horny scutes that make other turtles so easily recognizable. And, they are covered in thick, leathery skin.

The neck of a softshell turtle is very long and retractile. It has compressed and paddle-like limbs, with three claws on each foot. The surface area of the forelimbs is nearly twice that of a pond slider (*Trachemys scripta*), and with extensive interdigital webbing, these turtles have considerable thrusting power for swimming.

Their snouts extend in tube-like proboscides that they essentially use as snorkels, allowing them to maintain a minimal profile above the water's surface. They are nearly invisible from potential predators while in the water.

These turtles are fast and agile swimmers, and most are omnivorous. They are highly aquatic, spending most of their time water foraging or floating at the surface, and they bury themselves in the soft bottom with only their heads and necks protruding. This kind of submergence—called a dig-in—is one of the major signs of softshells, especially spiny softshells.

The trail of a spiny softshell turtle entering water. Notice the way the tracks become wider as the turtle moves into deeper water. JOE LETSCHE

They can also stay submerged for extended periods of time. Spiny softshells (*Apalone spinifera*) have been recorded voluntarily staying submerged for up to fifty minutes. Generally while diving, softshell turtles are under for less than twenty minutes. This same species is known to have an almost fish-like ability to extract oxygen from the water they are in. It has been recorded that up to 38 percent of their oxygen exchange and 85 percent of their carbon dioxide exchange occurs through their skin when they are submerged in water.

Softshell turtles can also grow quite large. The largest species is the Florida softshell (*Apalone ferox*), whose maximum recorded carapace length measured just over 26 inches.

Because of all of their permeable, exposed skin—including the skin that covers their expansive carapace and plastrons—softshell turtle species are more prone to dehydration than other turtles with hard shells. If forced to stay out of water for several days, they may die of dehydration (Lovich and Ernst 2009).

This group of turtles is also known to have an extremely keen sense of smell. They may locate food in large part through olfactory cues, especially while in the water.

Spiny Softshell (*Apalone spinifera*)

This turtle can grow up to a carapace length of about 21 inches. It also has the widest range of any softshell turtle species in North America. It gets its name from the knobs or spines on the front end of its carapace.

This species is sexually dimorphic, with adult males being less than half the size of adult females. Mature males have relatively long, thick tails with cloacal openings near the tips. Females have short tails that are mostly hidden beneath the carapace when viewed from above. Males frequently have black dots or circles spread relatively evenly around the olive-green carapace. They may also have a thin, yellow line running along the edge of the carapace. Female shells are more uniform in color, or their carapaces may be covered in large, darker green patches with ragged edges. Both sexes have a

The understep walking trail of a large spiny softshell leaving a drying wetland.

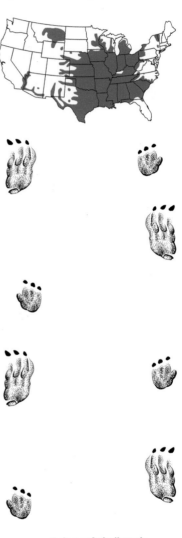

black-bordered yellow line running from the nose, through each eye, and down the neck. There is often also a second black-bordered yellow line running from the nose, across the length of the mouth and at least partially down to the neck. Older individuals—especially older females—may become darker with less distinct markings as they age.

This turtle species generally prefers rivers, as well as small creeks, channels through marshes, irrigation ditches, ponds, river oxbows, and some reservoirs. They often prefer soft mud or gravel bottoms with at least some aquatic vegetation (Lovich and Ernst 2009).

Spiny softshell turtle understep walking trail.
SEB BARNETT

Left: The front and hind track of a large female spiny softshell in drying mud.
Right: This is the front foot of a spiny softshell turtle, showing three clawed toes, two clawless toes, and extensive webbing. JOE LETSCHE

This species is mostly carnivorous, feeding on a wide variety of fish, frogs, and invertebrates, including snails, dragonfly larvae, and crayfish. It has a long neck, sharp beak, and quick strike, making it a surprisingly effective predator. It will also lunge its head to the full length of its neck and bite anyone foolish enough to casually pick this species up.

The spiny softshell is potentially one of the fastest-moving turtles on land in North America, especially when it is fleeing from danger. It is also capable of swimming rapidly, aided by its expansive, flipper-like feet.

Track: Front: 1–2 in (2.54–5.08 cm) x 1–2 in (2.54–5.08 cm)
Tracks are medium sized. Front feet possess five toes, but register only three claws. The claws are moderately long, and register as the deepest part of the track. The front foot has extensive webbing between all of the toes and extends partially up the outside of the foot from toe 5. The metacarpal area is covered in smooth, scaleless skin. Front tracks tend to register facing in the direction of travel. The line made by the three claws angles down toward the center line of the trail.

Hind: 2–3.3 in (5.08–8.38 cm) x 1–1.5 in (2.54–3.81 cm)

Tracks are medium to large in size. Hind feet possess five toes, but register only three claws. The claws are moderately long and register as the deepest part of the track. The hind foot has extensive webbing between all toes and extends partially up the outside of the foot from toe 5. When extended, this webbing makes the foot more paddle shaped. The metatarsal region is covered in smooth, scaleless skin. There is a hardened, crescent-shaped scale at the base of the metacarpal region. Hind tracks register facing in the direction of travel. The line made by the three claws is less sharply angled down toward the center line of the trail than the line made by the claws in the front tracks.

Trail: Understep Walk   Trail Width: 8–11.5 in (20.32–29.31 cm)
                         Stride: 7–10.4 in (17.78–26.42 cm)

Tail Drag: 0.3–1 in (0.76–2.54 cm)

Tail drags are generally infrequent, but when observed are most often seen near the water's edge or even as part of trails in the water in deep substrates. The tail of this species is thick and may be pushed down into the ground as the turtle travels, creating a relatively thick tail drag. The tail drags of males are more obvious, as males possess longer and thicker tails.

Notes:

• This species is graceful in the water, and due to its wary behavior and tendency to flee when approached, can go unobserved. When frightened, it may observe with just its nose and eyes breaking the surface.

• This species has been introduced widely outside of its native range, in several locations in western Washington State, and even on several Hawaiian islands. Unlike other turtle species that were likely spread in part through the pet trade, this species was spread via its use as a food animal.

Digs, Lays, and Burrows

Many reptiles and amphibians make digs, lays, nests, or burrows. Digs are depressions or cavities, typically excavated in search of food or as a test site for a potential shelter, such as a burrow or nesting cavity. Lays are made by the animals resting their bodies on a substrate as part of thermoregulation, camouflage, or during interactions with other animals. Nests are excavations created specifically to contain and shelter eggs. Burrows are cavities excavated for the purpose of creating shelter for that animal. These can be shallow or deep, and typically are at least several times the animal's body length in depth.

Some species have specially adapted features on their feet to help facilitate more efficient digging, such as the enlarged tubercles on the feet of many toad species or the enlarged claws on the front feet of desert iguanas. Here are some commonly encountered examples of these kinds of sign on the landscape.

Digs

Some species of reptiles will make digs in pursuit of prey or to escape potential danger. Some species will make swift digs by diving headlong into loose substrate and disappearing from sight while pursued by a predator. Others make very deliberate digs with their forelimbs to seek buried prey or to escape hot temperatures.

Lizards

Whiptails

The various species of whiptails make frequent digs while foraging for prey. Their digs are shallow, often U-shaped depressions with a throw mound width similar to the widest portion of the dig itself. Where whiptails choose to make a dig may be dependent on where they pick up the scent of prey.

Fringe-toed Lizards

This group of lizards lives in areas of loose sand, such as open dunes and large, sandy washes. They make digs for several reasons: to escape predators, to cool down, and for nighttime shelter.

Lizards make predator-escape digs by diving headfirst into the sand. They do this so swiftly that they can seem to disappear in an instant when observed by a human pursuer. They tuck in their front limbs,

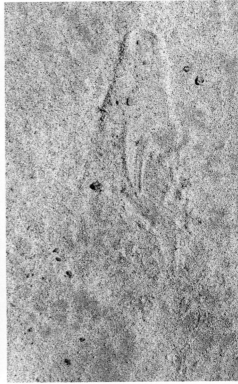

Left: Tiger whiptail dig. *Right:* The dig-in of a Coachella Valley fringe-toed lizard.

The morning dig-out of a Colorado Desert fringe-toed lizard.

vibrate their heads so rapidly that the sand appears to have liquid properties, and flick their tails rapidly for complete submersion.

When following the tracks of this species, such digs are most obvious in places where the trail of a fringe-toed lizard suddenly ends in an area of open sand. The actual spot of the dig may be easy to overlook as they are quite subtle.

You can distinguish this kind of dig from that of a nighttime shelter by the direction of the trail. The digs created for nighttime shelters have trails that emerge from them and move away. You'll often find a scat nearby that the lizard deposited when it first emerged in the morning.

Digs created for the purpose of thermoregulation during extreme heat are usually created in the shade of nearby plants and often have walking trails leading up to them.

Burrows

Desert Tortoise

These tortoises create a variety of burrows. Shallow ones barely larger than the tortoise itself to about four feet deep are known as "pallets." These are short-term shelters used to avoid extreme heat or to rest out of sight. "Dens" are deeper burrows that the tortoises dig horizontally in banks or washes, and are between six and thirty inches deep. Summer burrows are found in flats and benches. Tortoises dig these at angles between twenty and forty degrees and depths of six to eight feet (Lovich and Ernst 2009).

The entrances to the burrows of adult desert tortoises are crescent or half-moon shaped. In loose substrate or in front of recently excavated burrows, you may find throw mounds. The burrow entrances measure between 3.5 and 9 inches tall and 7 and 14 inches wide.

The typical burrow and throw mound of an adult gopher tortoise.

A desert tortoise burrow.

The pallet of a desert tortoise under a dense shrub.

Gopher tortoises have similar-sized burrow entrance measurements to those of desert tortoises, between 4 and 10 inches tall and 7 and 14.5 inches wide. Gopher tortoises prefer softer, sandy soils for burrowing. The burrows of this tortoise species are vital to over one hundred species of vertebrates as shelter, including lizards, snakes, frogs, toads, mice, and even burrowing owls. They are especially important refuges during hot summer temperatures, as well as during frequent low-intensity forest fires in southern pine woodlands and sand scrub.

Box Turtles

Box turtles generally dig burrows in areas of sandy soils, such as sand dunes. They are smaller than tortoise burrows, generally measuring between three and five inches tall and five and seven inches wide.

The burrow entrance and trail of an ornate box turtle. JONAH EVANS

The shallow retreat burrow of a Great Basin collared lizard.

Lizards

Some lizard species dig their own long-term burrows. Some species such as side-blotched lizards or collared lizards dig shallow burrows that are only a few inches long. They tend to dig these directly under large cover objects such as boulders or boards. Such shallow burrows are used for shelters during the night and cold periods, or as a spot to quickly escape from pursuing predators.

Desert Iguanas

These large lizards dig their own long-term burrows. The deep burrows are often found at the bases of creosote bushes or along the edges of washes. Desert iguanas are territorial during the breeding season, but for much of the year will tolerate each other. Several of their burrows may be found within a small area. Burrows give these lizards a good place to escape the extreme heat or approaching predators, as well as relatively safe nighttime retreats. Their long forelimb claws help them dig through even relatively compacted desert pavement. They may dig their burrows among those of kangaroo rats, or even share existing burrows with them.

The burrow entrance of a desert iguana.

The burrows of desert iguanas are round or slightly wider than tall. They measure between 1.5 and 3 inches wide. They may be between 10 and over 36 inches in depth. Frequently used burrows may have associated scats, tracks, or pieces of shed skin.

Whiptail Lizards

Whiptails will often use burrows as nighttime retreats. They may use burrows made by other animals—such as those of large scorpions, which they modify—or they may dig their own. Whiptail burrow entrances are

The low, crescent-shaped burrow of a little white whiptail.

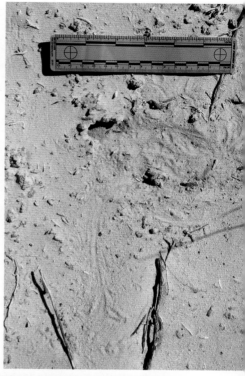

small, ranging from 0.5 to about 2 inches wide. They are usually wider than tall, with a low ceiling and crescent-shaped entrance.

Frogs and Toads

Frogs and toads will use burrows made by other species, especially those of ground squirrels. They also commonly seek shelter in natural hollows in logs, trees, rock crevices, and even cracks in drying mud. Many true toads and spadefoot toads dig their own burrows, and some even create new burrows daily. These animals dig using the enlarged tubercles found on the underside of the hind feet.

The burrows of toads have small entrances the same width or slightly wider than the animal itself, and are generally oval in shape. When an

The dig-in site of a Great Basin spadefoot.

The dig-in site of a Great Plains toad.

obvious burrow opening is visible, it's actually most likely to be empty. Toads usually only use dig-outs one time.

Toads—both true toads and spadefoots—dig by alternating with their hind feet and sinking backward into the substrate. They tend to follow a corkscrew or spiral pattern as they make deeper progress. Burrows that are still inhabited will be plugged with soil. If the substrate surrounding the digs is conducive to leaving tracks, you can often identify the species using it.

These burrows are especially important for toads in semiarid and arid environments. They can be quite deep, especially in hot desert habitats. Under these extreme conditions, the burrows can reach a depth of twelve to over twenty inches. These burrows allow the moisture-dependant amphibians to remain in significantly cooler and more humid environments. In regions that lack rain for longer periods, these toads may remain in their burrows for months at a time, conserving energy and water as they await the return of the rains. The sudden emergence of hundreds of toads during a monsoon rain is one of the natural wonders of the desert.

Alligators

The American alligator makes its burrows in banks or dense piles of vegetation. It uses these shelters during colder months, and these burrows are especially important in parts of the southeastern United States that have cold periods in the winter. In overall appearance, these burrows are similar to those of tortoises: they are crescent or half-moon shaped. The

The vegetation-covered entrance to an American alligator burrow.

burrows of adult alligators are eleven to eighteen inches high and fifteen to over thirty inches wide. The entrances of these burrows generally face the water, usually only a short distance from the water's edge. Burrows that have been recently used will show a flattened area in front of the entrance, and often an established trail leading to the water.

Other Digs

Tortoises

Desert tortoises live in dry to extremely dry environments in the desert Southwest. They can survive without rainfall for long periods of time by storing water in their bladders. When rain falls, however, they will gladly drink their fill. These tortoises dig shallow basins in muddy spots to collect water following periodic rains. They will use the water they collect for drinking and soaking. These basins can also provide water for other animals.

Alligators

American alligators act as ecological engineers, as they create very large, bowl-shaped digs that stretch between six to more than twenty feet across. These large depressions are called "gator holes" and are vital

The catchment basin dug out by a desert tortoise.

Two shallow ponds excavated by an alligator. DIRK STEVENSON

water-holding habitats in swamps and wetlands throughout the southeastern United States. Alligators dig these depressions during the dry season, and when holding water they become habitats for many insects, frogs, small fish, and wading birds. They also provide drinking water for other species such as deer, raccoons, and bobcats. In turn, all of these feed the alligators.

Large areas of seasonally flooded grassland and interspersed cypress hammocks exist in the Everglades. During the extremes of the dry season, some of the only water to be found is at the center of these tree hammocks, held either in solution holes in limestone sinks or in alligator-created water catchment basins.

Lays

Lays are marks left by sitting or lying animals. A variety of reptiles create lays, often for the purpose of thermoregulation. Some snakes, especially rattlesnakes, are also known to make lays while waiting in ambush for potential prey.

Tortoises

Tortoises can make several different kinds of lays, depending on location, that all serve to help them thermoregulate. They make them in hollows under large boulders, in small, shallow caves, or at the base of dense shrubs. Occasionally, they also make lays in more open areas when they have stopped to rest or warm up.

The lay of a desert tortoise in a shallow cave.

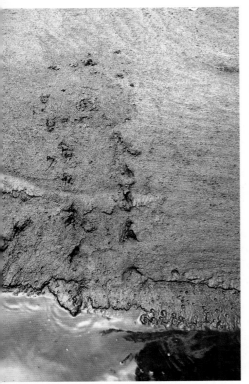

The basking site of a turtle. JOE LETSCHE

Look at the overall shape of the lay, and for the marks left by the scales of the plastron when it is set down or dragged across the soil, as well as the nearby associated sign such as tracks and scat to distinguish tortoise lays from those of other animals.

Turtles

Being more associated with water, turtle lays are often found at the edges of wetlands or waterways. Many turtle species will create short trails that lead into lays within a few feet of water. These are often circular spots where the turtle has basked, then turned around and returned to the water.

Box Turtles

Box turtles create lays, which are called "forms," along the edges of wetlands. These forms are made by soaking box turtles, and tend to be circular to oval in shape, measuring close to the length of the turtles' shells. They can be seen in the mud on the edges of ponds and other shallow wetlands.

Rattlesnakes

Some of the most common snake lays encountered in the field are those made by pit vipers, especially rattlesnakes. Other snakes sometimes create lays, but no other group of snakes consistently makes lays in such perfectly circular shapes. Rattlesnakes tend to curl up into tight spirals with one coil against the next, and leave neat circles pressed into the substrate. Often these are located at the base of a boulder or shrub or against a log or large cactus. The snakes create these for the purpose of thermoregulation or to wait in ambush for potential prey. Rattlesnakes often position themselves along active rodent runways, with the intention of catching the animals as they follow their routine pathways through the landscape. The snakes locate these runways through their sense of smell and may stay in position waiting for prey for days at a time.

The size of these lays reflects the size of the snake that made them. Generally, lays made by adult rattlesnakes are between three and twelve inches in diameter. They might be confused with the marks made by windblown grass, except that rattlesnakes indent the substrate more deeply and show significantly more displacement of substrate around the rim of the circular area.

The form created by two soaking eastern box turtles. JOHN VANEK

A sidewinder in its lay. TODD A. HOGGAN

The lay of a Great Basin rattlesnake. Look closely for the impression of the rattle near the center of the lay. CAMERON ROGNAN

The lay of a timber rattlesnake with a hat for scale. Notice the darker leaves within the lay. DENIS CASE

The lays of timber rattlesnakes—which frequently live in areas of dense forest floor debris—can be difficult to spot. Look for circular depressions in leaf litter near the bases of stumps or alongside logs. Leaves in these lays may appear flattened and hold moisture for longer periods, and so appear to be a slightly different color than the surrounding leaf litter.

Generally, rattlesnakes enter and exit these lays using rectilinear locomotion. Watch for signs of rectilinear trails near these lays. Multiple lays found in the same location indicate an area that is being heavily used by one or multiple snakes. This is a good area to search for sheds and scats. You'll often find the snake nearby.

Lizards

Lizards often lie down when thermoregulating, though many species do so on solid surfaces such as logs, walls, or rocks. Those that live and spend much of their lives on the ground and on fine substrates will leave distinct body impressions.

Horned Lizards

This group of lizards is often found in areas of loose, sandy soils. They behave similarly to fringe-toed lizards, emerging from shallow burrows in the morning, foraging and exploring, then digging down into another shallow burrow at night. As they forage, they will sometimes stop and lie in one place to thermoregulate their body temperature.

Given their squat, wide-bodied builds, horned lizards' lays are relatively distinct. Watch for impressions of short legs, wide bellies, and

The lay of a Texas horned lizard showing the fine belly scales.

The circular area of tracks on the right is the loafing spot of a tiger whiptail.

short tails with thick bases. Look closely at the associated tracks leading into and out of lays as supporting evidence.

Occasionally, when the substrate is very fine, a close inspection of the belly region of the lay will show the impression of the ventral scales. In horned lizards, these are small, round, granular scales that form a net-like pattern.

Whiptails

Whiptail lizards also live on loose substrates and create lays. They create these most frequently in the morning or late afternoon for thermoregulating. The lays of whiptails are often rather messy in appearance, as these lizards frequently move repeatedly when they create them. Whiptails will often turn in a circle and lie down multiple times. Though you can find them in the open, you'll more often find them near small clumps of vegetation or other cover.

Nests and Eggs

Amphibians that are found in the United States are not generally known to construct nests. Most simply deposit eggs in water on the surface, on the bottom, or attached to vegetation. Many salamanders exchange sperm in the form of a packet deposited on the ground or on the surface of the water so that the female can walk over it and pick it up with her cloaca. These packets are called spermatophores and can be a relatively common sign in vernal pools or ponds during the breeding season.

Some species—especially those belonging to the genus *Ambystoma*—deposit multiple spermatophores. Spotted salamanders (*Ambystoma maculatum*) in particular produce many spermatophores, producing between ten and eighty-plus during several nights of courtship. These little white globs are ephemeral signs of salamander breeding activity. The longer-lasting signs are the egg masses.

Generally speaking, amphibian eggs are surrounded by one or more layer of clear jelly. The egg masses of most toad species are laid in jelly strings. Those of true frogs are spherical or loose masses between the size of a tennis ball and a softball, which float on the surface or are attached to vegetation in shallow water. Tree frogs often lay small egg masses attached to thin stalks of grass. The size and shape of the eggs, jelly layers, and overall egg masses, as well as the location of where they are deposited, are all clues to identifying what species made them. Egg mass identification is beyond the scope of this book, but you can find this information in other books, such as the excellent *Salamanders of the United States and Canada* by James W. Petranka and *Frogs of the United States and Canada* by C. Kenneth Dodd Jr. Both of these volumes are also excellent resources for more information on amphibian behavior.

Species of amphibians that incubate their eggs on land find crevices in caves, inside logs, or under moisture-holding cover objects. Some salamander species will remain with their eggs until they hatch. Often,

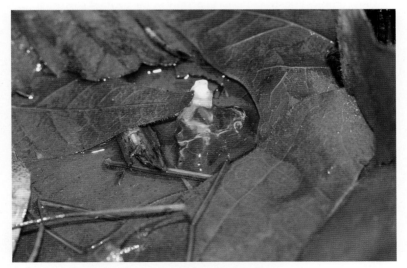

A close-up of a northwestern salamander spermatophore.

A female ensatina found attending her eggs. KIM CABRERA

they will lie with their bodies pressed against or surrounding the egg mass. This kind of attentive care may help prevent the eggs from drying out as well as reduce the risk of them being attacked by fungi or being consumed by small animals such as invertebrates or other amphibians, including other members of their own species.

Many reptiles construct nests, generally in the form of cavities in the soil. Some lizards simply use preexisting cavities not unlike some salamander species, but many others construct burrows into which they deposit eggs. Turtles create nests on land, even the fully aquatic alligator snapping turtle and sea turtles. Most turtle nests are bowl or flask shaped, with the widest portion being the deepest part of the cavity. Crocodilians also construct large nesting mounds out of soil and woody materials. Many snakes deposit eggs in hollows in logs, trees, brush piles, or deep in rock crevices. Others actively construct burrows into which they deposit their eggs.

Turtles

Turtles of all species in North America create nesting chambers for their eggs. Turtles generally dig their nests using alternating scooping movements with their hind feet. Many females will dig their nests as deeply as their legs allow them to stretch. Once the females deposit the eggs, they gently scrape the displaced soil back over the nest to close it up.

The most common way to encounter turtle nests is after they have been raided by predators. Common predators of turtle nests are raccoons, skunks, foxes, and, to a lesser extent, mink, coyotes, and snakes. Raided nests tend to show part of the nest chamber exposed, disturbed soil, and remains of eggshells scattered near the nest entrance. Nests of semiaquatic and aquatic turtles may be within a few yards of the water's edge or hundreds of yards away.

Some groups of turtles show differences in their egg shape, texture, and shell thickness. Snapping turtle eggs are generally spherical, with tough and leathery shells. Softshell turtle eggs are similar to bird eggs, with thick, brittle, rigid shells, and are spherical or ovoid in shape. Emydids, including pond sliders, cooters, painted turtles, map turtles, and box turtles, generally have elliptical eggs with thin, flexible shells that have a smooth or slightly pitted texture. The eggs of mud and musk turtles are brittle, with porcelain-like shells with a bluish-white or pinkish-white color. They sometimes deposit their eggs in piles of vegetation,

Left: A female painted turtle digging a nest on the edge of a gravel trail. Notice her hind foot digging down in the cavity. *Right:* A typical example of a turtle nest raided by a raccoon, in this case the nest of a snapping turtle. DAN GARDOQUI

The scat of a raccoon that has fed on the nest of a ornate diamond-backed terrapin. GEORGE L. HEINRICH

including into the nests of American alligators. Eggs of tortoises are often spherical or elliptical, with white, hard, brittle shells with a granular surface.

Alligators

The nests of crocodilians are proportionately large and generally found near water. American alligators (*Alligator mississippiensis*) make nests of vegetation and mud that are between seventeen and thirty-six inches tall and between five and eight feet wide. The nests of the American crocodiles (*Crocodylus acutus*) are usually made mostly of soil, especially sand and mangrove peat. These nests are

The nest of an American alligator in an urban park in St. Petersburg, Florida. Notice the raccoon mandible on the mound.

essentially huge incubating chambers that help maintain a relatively constant temperature and humidity, and help the embryos to develop.

The active nests are most commonly encountered during the summer, especially in June and July. Nests may remain long after the young have dispersed. Approach nests you believe to be active with caution, or do not approach them at all, since females are often nearby and may bluff charge or bite defensively.

Snakes

Some snakes actively construct nesting chambers in the soil, including the eastern hog-nosed snake (*Heterodon platirhinos*). Females of this species may use preexisting cavities or dig chambers in the soil using their snouts. These snakes are named for the large, curved rostral scale that they more commonly use to dig for toads.

You can recognize fresh nests by the small throw mound and flattened trail, by the smoothed floor of the chamber made by compression from the ventral scales, and by the presence of eggs within. The photo of the nesting hog-nosed snake below was taken as part of a long-term study involving radio telemetry (tracking using a radio signal).

Lizards

Lizards of various species also dig nesting chambers. This is especially true of many spiny lizards (*Sceloporus* ssp.) and whiptails (*Aspidoscelis* ssp.). These lizards are sometimes encountered digging nesting chambers in early to midsummer (usually from May to July). These nesting chambers are essentially shallow burrows 4 to 10 inches deep. They are 1 to 2.5 inches tall and between 1.5 and 3 inches wide.

The nesting burrow of an eastern hog-nosed snake. JOHN VANEK

The nesting burrow of a Texas spotted whiptail. When completed, this burrow typically has a plug of loose soil. CHRIS HYDE

The nest of a five-lined skink. CARL BRUNE

These tiny eggs belong to an unknown species of gecko and were found under a rock in the Florida Keys.

Nests may have a small, shallow throw mound in front of the entrance and a plug of loose soil closing up the entrance. You'll generally find them in areas of soft, loose, slightly moist soils. Lizard eggs are generally quite small and ovoid and have thin, leathery shells.

Sometimes only the eggshells are encountered. This likely occurs following predation on the nests, or when substrates such as sand dunes shift, exposing the eggs.

Sheds 9

A commonly encountered form of reptile sign is an entire or fragmented shed skin. These are commonly called "sheds." They are delicate, ghostly impressions of the scales and color patterns of the animals, and can tell us what species are found in a given location. They can also allow us to see the diversity of scales on the bodies of reptiles, and even show us the presence of scars or parasites.

All reptiles and amphibians shed their skins to some degree. Most amphibians, however, tend to consume their skins shortly after they are shed, and therefore rarely leave behind such evidence of their passing. When they do leave them behind, it's usually in water and and you won't commonly encounter them. Turtles and crocodilians may leave small fragments behind, generally only one or a few scales at a time. The scales of reptiles are part of the epidermal layer of the skin itself. Unlike humans, who generally shed a few cells at a time, reptiles and amphibians shed them in large areas.

Shed skins and skin fragments age quickly when exposed to ultraviolet light. Only relatively fresh sheds show clear, bold patterns, while those exposed to sunlight will bleach to pure white over time. Earthworms and other invertebrates consume sheds. Some birds and even small mammals will use them as nesting material. The California ground squirrel is known to seek out the sheds of rattlesnakes. It will chew them into a pulp and apply the pulp to the fur on its back and tail. It likely does this as a form of scent camouflage to reduce the risk of predation from the rattlesnakes themselves. Research has shown that rattlesnakes are less attracted to squirrels with a combination of rattlesnake and squirrel scent (Clucas, Owings, Rowe et al. 2008).

The most common groups of herps that leave behind identifiable shed skins are snakes and lizards. Snakes shed their skins in one single piece. This shed skin is usually turned inside out as it begins to peel at

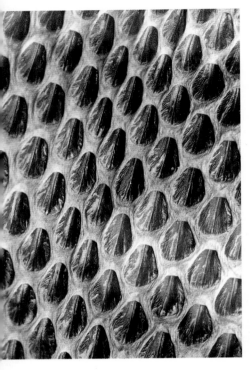

An extreme close-up of the scales of a northern Pacific rattlesnake.

the head and rolls back. Snakes often leave their shed skins under rocks or logs, at the bases of shrubs and trees, or in narrow crevices. Generally speaking, it will be in a spot where the snake can rub its body to help with the process of shedding, or "ecdysis."

In contrast, most lizards shed their skins in smaller, uneven segments, from only a portion of the body at a time. One exception to this rule is that alligator lizards (*Elgaria* and *Gerrhonotus* species) shed their skins in one single piece. Like snakes, lizards tend to leave the sheds, or shed fragments, under or near cover objects such as rocks, shrubs, or other similar physical structures.

Understanding Scales

In order to identify shed skins or skin fragments, it will help you to better understand the different types of scales that exist on the body of a reptile.

Lizards can commonly have one or several of the following types of scales: keeled, mucronate, cycloid, and beaded or granular. Keeled scales have a raised line traveling part or all the way down the middle. Mucronate scales end in an abrupt point; many keeled scales are also mucronate. Cycloid scales are semicircular and overlapping. They are similar in appearance to some tightly fitting fish scales. Granular scales appear as raised circular or oval bumps. Lizards that have particularly large granular scales are in the genus *Heloderma*, and these lizards are sometimes referred to as "beaded lizards."

Some lizards have granular scales interspersed with enlarged granular scales, called "tubercles." These should not be confused with the tubercles found on the feet of amphibians described earlier in this book.

Some lizards also have additional structures under their scales, called osteoderms. These small bones act to reinforce the scales, not unlike armor. This is a feature shared by crocodilians and turtles, but it is not found in snakes.

As a general rule, reptiles have larger and thicker scales in areas of the body that tend to be most exposed to physical stresses. Conversely, they will have smaller scales in parts of the body that need to move more freely. A good example of this is that alligator lizards have granular scales along their entire sides, which allows them to easily expand their abdomens during breathing, after a large meal, or when females are gravid.

Many lizards also have specialized scales along the insides of their rear thighs. These are called femoral scales. Femoral pores are openings of follicular glands that are used to excrete and deposit pheromones via a pasty substance often made of keratin and lipids. In many species that possess femoral pores, the males will have larger pores than females. Also, mature males will tend to have larger pores than juvenile males. Members of the following families of lizards found in the United States possess femoral pores: Anguidae, Helodermatidae, Scincidae, Crotaphytidae, Iguanidae, Phrynosomatidae, Xantusiidae, and Teiidae.

Snake body scales are generally smooth, keeled, or cycloid. One particular group of snakes—the blind snakes (Leptotyphlopidae and Typhlopidae)—are well known to have very tightly overlapping cycloid scales, even on their ventral regions. In appearance, the cycloid scales of snakes look very similar to the cycloid scales of lizards, appearing semicircular with rounded edges. Meanwhile, smooth scales are generally elongated, leaf-shaped, and lack a keel. Keeled scales are generally very similar in shape except that they possess a keel running part of the way down the middle, or all the way down the middle of the length of each scale.

The interlocking rattles of rattlesnakes are made of modified scales. Another segment is added to the rattle each time the snake sheds. Rattles cannot be used to accurately age a rattlesnake, and will also often become worn, or partially broken off, in the wild. Newborn rattlesnakes only possess a single button, and do not gain their first rattle segment until after their first shed.

When looking at the scales found on the sheds of snakes, look very closely for additional features. One of these is the presence or absence of an apical notch. This is a notch found at the tip of each scale. Also, many species show the presence of apical pits on each scale, which generally appear in pairs. Apical pits are very tiny holes that can be seen on each scale using a hand lens or microscope. The precise location and size of these apical pits is a useful diagnostic tool for identifying or differentiating among some species. These features will be mentioned in the following text whenever they are important to identification. It is useful to carry a hand lens to look at some of the very small features of sheds. Apical pits are best seen under a low-power microscope.

Further Scale Nomenclature

Often the most useful parts of a snake shed for identification purposes is the head portion. The heads possess a variety of specific scales arranged in particular numbers, shapes, and patterns that are diagnostic for many species or genera. Let's start here by taking a close look at these scales, their names, and typical locations.

The rostral scale is found at the tip of the snout on the dorsal surface. Behind the rostral scale are the nasal and internasal scales. The nasal scale or scales are those immediately surrounding and including the nostril opening. The internasals are found between the nasal scales on either side of the face. Behind the internasals are the prefrontals. These are framed on the other side by the loreal scale(s) in most species. The small scale or scales just ahead of the eye are known as the preoculars (pre = before). The loreal scale falls between the nasals and the preoculars. The scale directly covering the eye is called a spectacle (also called the brille). This is a transparent scale that acts similarly to a contact lens. Sitting just above, and in contact with, the spectacle is the supraocular (supra = above). There is one above each spectacle. Between these, in most species of snakes, is the frontal scale. Just behind the spectacle are the small scales known as the postoculars

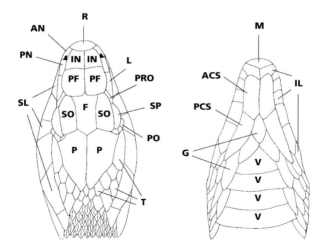

These are the major scales and scale regions as seen on a typical colubrid snake, in this case a garter snake species. R = rostral; IN = internasal; AN = anterior nasal; PN = posterior nasal; PF = prefrontal; F = frontal; SO = supraocular; P = parietal; SL = supralabial; T = temporal; PO = postoculars; SP = spectacle; PRO = preocular; L = loreal; M = mental; IL = infralabials; ACS = anterior chin shields; PCS = posterior chin shields; G = gulars; and V = ventrals.

(post = behind). Behind the supraoculars and the frontal scale are the large parietal scales. To either side of these are the anterior temporal scales. Along the edges of the mouth are the labial scales. Those found on the dorsal surface are the supralabials, while those found on the ventral surface are the infralabials.

The ventral surface of the face also has several other distinct scales. The scale at the tip of the mouth on the ventral surface is called the mental scale. The large, paired scales running down the middle of the ventral surface of the head just behind the mental scale are the anterior chin shields. Another pair of scales, the posterior chin shields, is found just behind these. Behind, and often between, the rear of these posterior chin shields are the small throat scales known as gulars. The broad ventral scales that run the length of the snake starting just behind the gulars are called the ventral scutes, or just the ventrals.

The rattlesnakes of the genus *Crotalus* have a different arrangement of scales on the dorsal surface of their heads. The largest scales on their heads are the supraoculars, which partially project over the spectacle. There are numerous small scales on top of the head, and in many species are found between the supraoculars, which are called inter-supraoculars. The exception is the Mojave rattlesnake (*Crotalus scutulatus*) which has two (and sometimes, though rarely, three) large scales between the supraoculars that span this space. The supraoculars are pointed and project upward and outward over the spectacle in sidewinder rattlesnakes (*Crotalus cerastes*). Many also have numerous scales before and after the nasals, which are referred to as postnasals and prenasals, respectively.

Counting middorsal scales on a snake can be an additional detail that can help you sort out similar species. Scale counts can be done in different ways; two of the most common are described here. Counting the scales diagonally across the back from one side, beginning at the edge of the belly, and counting clear across to the other side is represented by the path in light gray. Counting using a V-shaped path is represented by the pale green.

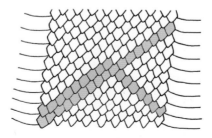

Illustration of scale-counting process.

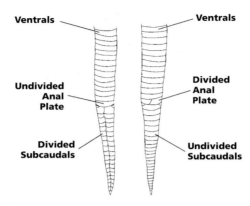

This illustration shows a comparison of divided and undivided anal plates, as well as divided and undivided subcaudal scales.

All snakes possess a scale covering the cloaca, called the anal plate. In some species, this is a single scale and so it is described as "undivided." In other species, it is covered by two scales that closely overlap, and so it is described as "divided."

Snake Sheds

The shed skins of snakes are sometimes common in the field, and at times can be encountered more frequently than the snakes themselves. Here is a variety of relatively common and widespread species whose sheds you may encounter. Note that although the patterns are mentioned for each, these may not be useful in all conditions, especially where they have been bleached by ultraviolet light. When visible, they can aid in the identification process. It is more reliable, however, to use the details of the scales to identify species accurately.

Rough Green Snake (*Opheodrys aestivus*)

General description: Sheds are small, narrow, and tapered. The tail is long and tapered. This species does not have any dorsal patterns, but is solid green with a paler yellow ventral surface. Sheds are often found in shrubs or trees.

Dorsal head scales: There is 1 nasal scale, 1 (sometimes 2) preocular scales, 2 (sometimes 1 or 3) postocular scales, and 1 loreal scale. There are 3 temporal scales. Generally there are 7 (sometimes 6 or 8) supralabials. The spectacle is proportionately large in size. The front scale is broad where it touches prefrontals, and narrow where it touches the parietals.

Ventral head scales: There are generally 8 (sometimes 7, occasionally up to 10) infralabials. The posterior chin shields are long, and arc outward away from each other for half their length.

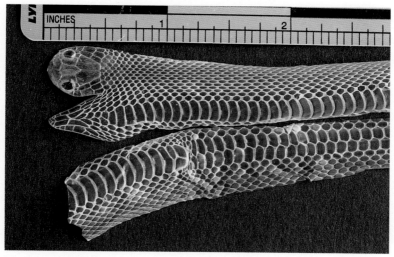

Rough green snake shed.

Vent scales: The anal plate is divided (rarely undivided).

Scale count: There are 17 rows of scales at midbody. Dorsal scales are long, narrow, strongly keeled, and have 2 apical pits.

Adult length: 20–45 inches in total length.

Racer (*Coluber constrictor*)

General description: Sheds are long, moderate or narrow in thickness, and tapered. The tail is long and tapered. This species varies some-what in color, but generally adult racers have a gray, bluish, or black dorsal area and a whitish or yellowish ventral surface. Juvenile racers have a series of brown blotches down the middle of the back and small dark spots on either side. These fade as the snake matures to adulthood. Sheds are usually found on the ground among rocks, at the bases of shrubs, or under debris.

Dorsal head scales: There are 2 nasal scales (partially divided by nostril), 2 (sometimes 1 or 3) preoculars, 2 (rarely 1 or 3) postoculars, and 1 (occasionally 2) loreal scale. There are 3 rows of 2 temporal scales (rarely 1 or 3). Generally there are 7 or 8 (rarely 6) supralabials. The spectacle is proportionately large in size. The frontal scale is narrow, slightly broader where it touches prefrontals, and narrower than the supraoculars where it touches the parietals.

Ventral head scales: There are generally 8 or 9 (sometimes between 7 and 11) infralabials.

Vent scales: The anal plate is divided. Subcaudal scales are typically divided. Notice several undivided scales in the shed of the western

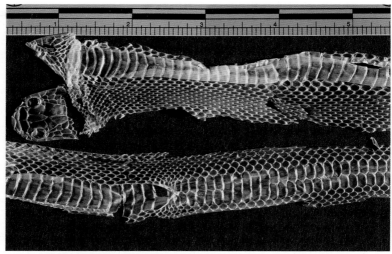

Western yellow-bellied racer shed.

Northern black racer shed. BETHANY AVILLA

yellow-bellied racer (*Coluber constrictor mormon*) pictured on page 372. The photo above is of the shed of a northern black racer (*Coluber constrictor constrictor*).

Scale count: There are 17 (occasionally between 15 and 19) rows of scales at midbody. Dorsal scales are smooth and with 2 apical pits.

Adult length: 36–77 inches in total length.

Coachwhip (*Coluber flagellum*)

General description: Sheds are long to very long, moderate in thickness, and tapered. The tail is long and tapered. The patterns of this species vary with the subspecies. The shed of the red coachwhip (*Coluber flagellum piceus*) is pictured on page 374. This subspecies tends to have black bands on the neck that fade to tan or pinkish-red bands on the body. As these bands progress, they often fade to a solid color from the midbody onward. Some populations of this species may be mostly or completely black. Sheds are usually found on the ground among rocks, at the bases of shrubs, or under debris.

Dorsal head scales: There are 2 nasal scales (separated partially or fully by the nostril), 2 (rarely 1, 3, or 4) preoculars, 2 (sometimes 1 or 3) postoculars, and 1 (sometimes 2 or 3) loreals. There are 3 rows of temporal scales, including the following number of scales in each: 2 or 3 (rarely 1), 3 (rarely 2), and 3 scales. Generally there are 8 (occasionally 7 or 9) supralabials. The spectacle is proportionately large in size. The frontal scale is broad where it touches frontals, and narrows to a line between the supraoculars.

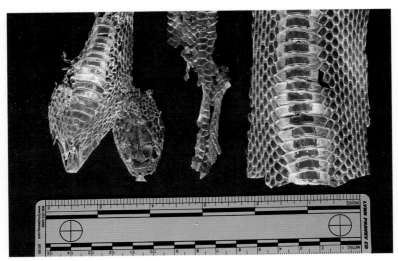

Red coachwhip shed.

Ventral head scales: There are generally 10 (sometimes 8–13) infralabials.
Vent scales: The anal plate is divided. Subcaudal scales are typically divided.
Scale count: There are 17 (sometimes 15) rows of scales at midbody. Dorsal scales are smooth and with 2 apical pits.
Adult length: 36–102 inches in total length.

Common Kingsnake (*Lampropeltis getula*)

General description: Sheds are moderately long and moderate to robust in thickness. The tail is medium in length. The patterns of this species vary considerably between subspecies, and even show considerable variation within each subspecies. The shed pictured on page 375 is a striped individual of the subspecies California kingsnake (*Lampropeltis getula californiae*). Two common patterns for this subpecies include a banded form, which has wide dark bands with pale bands in between that encircle the body and run from the neck down the length of the animal, and a striped form, where the animal has a long, pale middorsal stripe running most of its length, sometimes with a similar lateral stripe on either side. Sheds are usually found on the ground or under debris, such as bark or boards.
Dorsal head scales: There are 2 nasal scales, 2 (rarely 1, 3, or 4) preoculars, 2 (sometimes 1 or 3) postoculars, and 1 (rarely 0) loreal scale. There are 2 rows of temporal scales including 2 (rarely 1 or 3) and 3 (sometimes 2 or 4) scales. Generally there are 7 (sometimes 6 or 8)

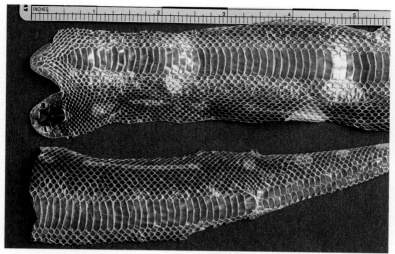

California kingsnake shed.

supralabials. The spectacle is moderate in size. The frontal scale is broad overall, as broad or broader than the supraoculars.

Ventral head scales: There are generally 9 or 10 (occasionally 7 or 11) infralabials.

Vent scales: The anal plate is undivided. Subcaudal scales are typically divided.

Scale count: There are 21–23 (sometimes between 19 and 25) rows of scales at midbody. Dorsal scales are smooth and there are two apical pits.

Adult length: 36–82 inches in total length.

Arizona Mountain Kingsnake (*Lampropeltis pyromelana*)

General description: Sheds are small to moderately long, moderate in thickness, and with a tail of medium length. This species has a series of alternating bands of red, black, and white. These can vary somewhat in width. Generally the black bands are narrow, and act as borders for the edges of the wider red bands. In between each pair of black bands is the white to yellowish band. The snout is usually white or cream colored, with a wide black hood on top of the head from in front of the eyes to the parietals and down onto the supralabials. Sheds are usually found among or under rocks.

Dorsal head scales: There is 1 nasal scale, 1 preocular, 2 (sometimes 3) postoculars, and 1 loreal. There are 3 rows of temporal scales, including the following number of scales in each: 2 (sometimes 3), 3 (sometimes

Arizona mountain kingsnake shed.

4) and 4 (sometimes 5) scales. Generally there are 7 (sometimes 8) supralabials. The spectacle is moderate in size. The frontal scale is broader than the supraoculars.

Ventral head scales: There are generally 9 or 10 (sometimes between 8 and 12) infralabials. The posterior chin shields are shorter and narrower than the anterior chin shields.

Vent scales: The anal plate is undivided.

Scale count: There are 23 (sometimes 25) rows of scales at midbody. Dorsal scales are smooth and with two apical pits.

Adult length: 20–41 inches in total length.

Milk Snake (*Lampropeltis triangulum*)

General description: Sheds are small to moderately long, narrow to moderate in thickness, and have medium-length tails. The patterns on the sheds of milk snakes vary by the subspecies. Generally, they have a series of dark bands bordered in black with pale bands in between. The sheds of this species are usually found on the ground under cover objects such as rocks or boards.

Dorsal head scales: There are 2 nasal scales (separated by the nostril), 1 (rarely 0 or 2) preoculars, 2 (rarely 1 or 3) postoculars, and 1 (rarely 0 or 2) loreal. There are 2 rows of temporal scales including the following number in each row: 1 or 2 and 2 (sometimes 3 or 4) scales per row. Generally there are 7 (sometimes between 6 and 9) supralabials. The spectacle is moderate in size. The frontal scale is broader than the supraoculars.

Ventral head scales: There are generally 8 or 9 (sometimes between 7 and 11) infralabials.

Vent scales: The anal plate is undivided. Subcaudal scales are typically divided.

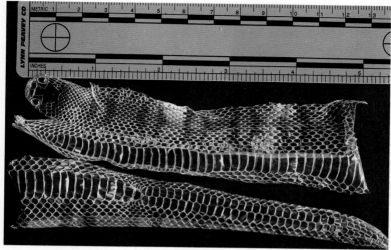

Milk snake shed.

Scale count: There are 19 (sometimes between 17 and 23) rows of scales at midbody. Dorsal scales are smooth with 2 apical pits.

Adult length: 14–52 inches in total length.

Long-nosed Snake (*Rhinocheilus lecontei*)

General description: Sheds are small to moderately long, narrow to moderate in thickness, and have medium-length tails. The pattern varies from individual to individual, but generally includes a series of medium to very wide saddle-shaped bands running down the middle of the dorsal surface. These bands have some white scales that often become more abundant laterally as the bands reach toward the ventral scales. The area between the dark bands may be uniformly pale, or may have some reddish coloring and scattered black dots. Sheds are usually found on the ground under cover objects such as rocks, fallen yuccas, or boards.

Dorsal head scales: The rostral scale protrudes and is slightly upturned. There are 2 nasal scales (separated by the nostril), 1 or 2 preoculars, 2 postoculars, and 1 loreal. There are 2 rows of temporal scales including 2 and 3 scales per row. Generally there are 8 (sometimes between 7 and 10) supralabials. The spectacle is moderate in size. The frontal scale is broader than the supraoculars.

Ventral head scales: There are generally 9 (sometimes between 8 and 11) infralabials.

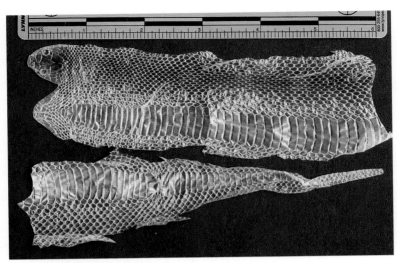

Long-nosed snake shed.

Vent scales: The anal plate is undivided. Subcaudal scales are mostly undivided.

Scale count: There are 23 to 25 rows of scales at midbody. Dorsal scales are smooth with between 0 and 2 apical pits.

Adult length: 22–41 inches in total length.

Common Garter Snake
(*Thamnophis sirtalis*)

General description: Sheds are small to moderately long, narrow to moderate in thickness, and with medium-length tails. There are several subspecies and a diversity of patterns and colors. Many have a longitudinal, middorsal stripe, and a lateral stripe on either side. Sheds are usually found on the ground under cover objects, such as bark, rocks, or boards. The shed pictured on page 379 was collected from an area east of Seattle along the intergrade zone between two subspecies: the Puget Sound garter snake (*Thamnophis sirtalis pickeringii*) and valley garter snake (*Thamnophis sirtalis fitchi*).

Dorsal head scales: There are 2 nasal scales (partially separated by the nostril), 1 (sometimes 2) preoculars, 3 (sometimes 2 or 4) postoculars, and 1 loreal. There are 2 rows of temporal scales, including the following number of scales in each: 1 and 2 (sometimes 1 or 3) scales per row. Generally there are 7 (sometimes between 5 and 9) supralabials. The spectacle is moderate in size. The frontal scale is broader than the supraoculars.

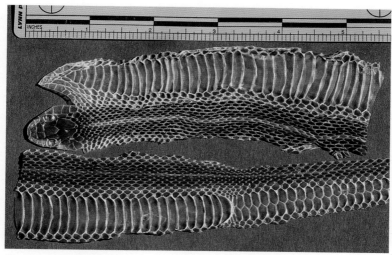

Common garter snake shed.

Ventral head scales: There are generally 10 (sometimes between 8 and 11) infralabials. The posterior chin shields are typically longer than the anterior chin shields.

Vent scales: The anal plate is undivided.

Scale count: There are 19 (occasionally 21) rows of scales at midbody. Dorsal scales are keeled and without apical pits.

Adult length: 18–51 inches in total length.

Northwestern Garter Snake (*Thamnophis ordinoides*)

General description: Sheds are small to moderately long, narrow in thickness, and have medium-length tails. This species also has a wide range of possible patterns, but generally has a longitudinal, middorsal stripe, and a lateral stripe on either side. It has a proportionately small head compared to other garter snakes that share its range. Sheds are usually found on the ground under cover objects, such as bark, rocks, or boards.

Dorsal head scales: There is 1 nasal scale (sometimes separated entirely by the nostril), 1 (sometimes 2) preocular, 3 (sometimes 2 or 4) postoculars, and 1 loreal. There are 2 or 3 rows of temporal scales, including the following number of scales in each: 1, 2, and 2 scales per row. Generally there are 7 (sometimes 6 or 8) supralabials. The spectacle is moderate in size.

Ventral head scales: There are generally 8 (sometimes between 7 and 10) infralabials.

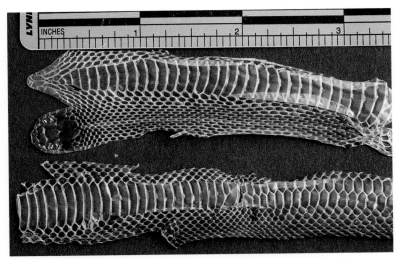

Northwestern garter snake shed.

Vent scales: The anal plate is undivided.

Scale count: There are 17 (sometimes 19) rows of scales at midbody. Dorsal scales are keeled and without apical pits.

Adult length: 15–36 inches in total length.

Mexican Garter Snake (*Thamnophis eques*)

General description: Sheds are small to moderately long, moderate in thickness, and have medium-length tails. Typical pattern is a yellowish, longitudinal middorsal stripe, and a lateral stripe on either side. It has a pair of black blotches behind the parietal scales, and a light, crescent-shaped pale marking just behind the corner of the mouth on either

Mexican garter snake shed.

side. Sheds are usually found on the ground under cover objects, such as bark, rocks, or boards.

Dorsal head scales: There are 2 nasal scales, 1 preocular, 3 (sometimes 4) postoculars, and 1 loreal. There are 2 rows of temporal scales, including the following number of scales in each: 1, and 1 to 4 scales per row. Generally there are 8 (sometimes 7 or 9) supralabials. The spectacle is moderate in size.

Ventral head scales: There are generally 10 (sometimes 9 or 11) infralabials.

Vent scales: The anal plate is undivided. Subcaudal scales are typically divided.

Scale count: There are 19 (sometimes between 17 and 21) rows of scales at midbody. Dorsal scales are keeled and without pits.

Adult length: 18–40 inches in total length.

Western Hog-nosed Snake (*Heterodon nasicus*)

General description: Sheds are small to medium in length, proportionately robust in thickness, and with relatively short tails. The typical pattern has a tan, brown, olive, or grayish base color with a series of brown middorsal blotches, and two rows of smaller dark brown blotches on each side. There is a dark transverse bar lying between the eyes; it extends downward behind each eye with a shorter middorsal blotch and two elongated blotches reaching from either side of the neck and touching the back of the parietal scales. Sheds are typically found on the ground under cover objects. The shed pictured

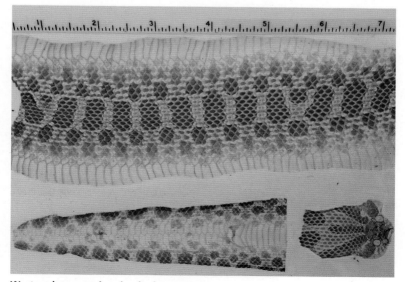

Western hog-nosed snake shed. MARK TOVAR

on page 381 is from the subspecies known as the Plains hog-nosed snake (*Heterodon nasicus nasicus*).

Dorsal head scales: The rostral scale is keeled, only sharply upturned, dorsally concave, and triangular in overall shape. There are 2 nasal scales, an ocular ring of 7 to 13 scales below supraocular scales, and 2 (sometimes between 0 and 4) loreals. There are 2 rows of temporal scales, including the following number of scales in each: 4 (sometimes between 2 and 5) and 5 (sometimes between 3 and 7) scales. Generally there are 8 (sometimes 9) supralabials. The spectacle is moderate in size.

Ventral head scales: There are generally 10 or 11 (sometimes between 9 and 13) infralabials.

Vent scales: The anal plate is divided. Subcaudal scales are typically divided.

Scale count: There are between 21 and 23 (sometimes between 19 and 26) rows of scales at midbody. Dorsal scales are keeled and with apical pits.

Adult length: 16–35 inches in total length.

Eastern Hog-nosed Snake (*Heterodon platirhinos*)

General description: Sheds are small to medium in length, proportionately robust in thickness, and with relatively short tails. Typical pattern is yellow, olive, brown, gray, or black base color with transverse

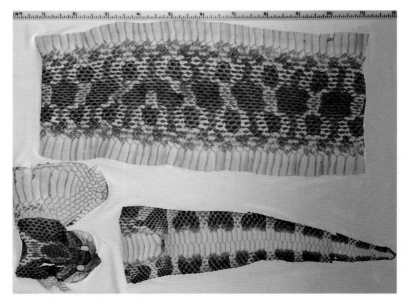

Eastern hog-nosed snake shed. MARK TOVAR

dorsal blotches and often two rows of smaller dark brown blotches on each side. There is a dark, V-shaped mark reaching from the neck onto the parietal scales. There is a transverse dark bar between the eyes that extends downward behind them to the corner of the mouth. Sheds are typically found on the ground under cover objects or at the base of shrubs.

Dorsal head scales: The rostral scale is unkeeled, only slightly upturned, and triangular in overall shape. There are 2 nasals, an orbital ring of 8 to 13 scales below the supraocular scale, and 1 (sometimes between 2 and 4) loreal. There are 2 rows of temporal scales, including the following number of scales in each: 3 or 4, and 4 or 5 (occasionally 3) scales per row. Generally there are 8 (sometimes 7 or 9) supralabials. The spectacle is moderate in size.

Ventral head scales: There are generally 10 or 11 (sometimes between 9 and 14) infralabials.

Vent scales: The anal plate is divided. Subcaudal scales are typically divided.

Scale count: There are typically 25 (sometimes between 21 and 27) rows of scales at midbody. Dorsal scales are keeled and have apical pits.

Adult length: 20–45 inches in total length.

Cornsnake (*Pantherophis guttatus*)

General description: Sheds are small to long, moderate in thickness, and with relatively long tails. Typical pattern is orange ground color with dark orange or reddish blotches with black edges. Usually a black-edged, spear-shaped mark reaches forward from the neck between the eyes. A black-edged stripe also reaches back from each eye beyond the corner of the mouth and onto the neck. This species shows some variability in color and pattern, with certain populations having a more or less orange color with clearly defined markings on the head and body. Sheds may be found on the ground, in trees, and even inside barns and other structures.

Dorsal head scales: There are 2 nasals, 1 (rarely 2) preocular, 2 postoculars, and 1 loreal. There are 2 rows of temporal scales, including the following number of scales in each: 2 (sometimes 3) and 3 (sometimes 2 or 4) scales per row. Generally there are 8 (occasionally between 6 and 9) supralabials. The spectacle is moderate in size.

Ventral head scales: There are generally 11 or 12 (sometimes between 10 and 15) infralabials.

Vent scales: The anal plate is divided. Subcaudal scales are typically divided.

Scale count: There are typically 27 (sometimes between 23 and 29) rows of scales midbody. Dorsal scales are subtly keeled and have apical pits.

Adult length: 24–72 inches in total length.

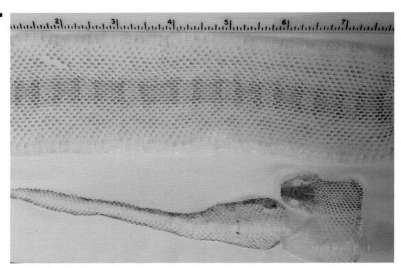

Cornsnake shed. MARK TOVAR

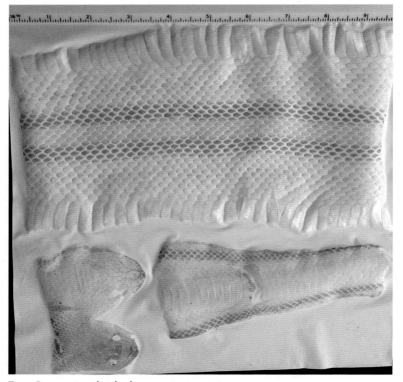

Trans-Pecos rat snake shed. MARK TOVAR

Trans-Pecos Rat Snake (*Bogertophis subocularis*)

General description: Sheds are medium to long, moderate in thickness, and have relatively long tails. Typical pattern is tan or yellow ground color with 2 dark longitudinal stripes on the neck, and black or dark brown, H-shaped blotches down the back. Blotches further back are often broader than those closer to the neck. A dark blotch is found on each side, below each H-shaped middorsal blotch. The shed pictured on page 384 is from the subspecies found in the United States (*Bogertophis subocularis subocularis*). Sheds are found on the ground or in shrubs.

Dorsal head scales: There is 1 nasal divided by the nostril, 1 large preocular, 2 or 3 postoculars, and 1 loreal. There are 5 rows of temporal scales with 3 to 5 scales in each row. Generally there are between 10 and 12 supralabials. The spectacle is moderate in size.

Ventral head scales: There are generally between 13 and 16 (sometimes 17) infralabials.

Vent scales: The anal plate is divided. Subcaudal scales are typically divided.

Scale count: There are typically between 31 and 36 rows of scales at midbody. Dorsal scales are weakly keeled and have 2 apical pits.

Adult length: 34–66 inches in total length.

Green Rat Snake (*Senticolis triaspis*)

General description: Sheds are small to moderately long, moderate in thickness, and have relatively long tails. Adults of this species typically have no pattern, but are solid green or olive dorsally with a pale

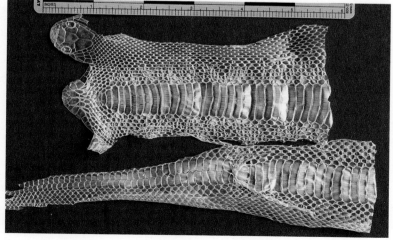

Green rat snake shed.

cream to white ventral surface. Juveniles have a series of blotches running the length of their dorsal surface, and also smaller ones on top of their heads. Sheds are usually found on the ground under cover objects, such as bark, rocks, or boards. They may also be found in shrubs or trees.

Dorsal head scales: There are 2 nasal scales, 1 (sometimes 0 or 2) preocular scales, 2 (sometimes 1 or 3) postoculars, and 1 loreal. There are 2 rows of temporal scales, including the following number of scales in each: 3, and 4 (sometimes 5 scales) per row. Generally there are 9 (sometimes 7 or 9) supralabials. The spectacle is moderate in size.

Ventral head scales: There are generally between 9 and 12 infralabials.

Vent scales: The anal plate is divided.

Scale count: There are 33 (sometimes between 30 and 39) rows of scales at midbody. Dorsal scales are keeled with apical pits.

Adult length: 24–50 inches in total length.

Gopher Snake (*Pituophis catenifer*)

General description: Sheds are moderate to very long, moderate to robust in thickness, and have tails of medium length. The pattern of this species varies with the subspecies, but generally includes a series of large, saddle-like blotches running the length of the dorsal surface from the neck down and becoming bands on the tail. There are usually several rows of dots down the length of the dorsal surface on either side of the body. Sheds are usually found on the ground under cover objects, such as bark, rocks, logs, or boards.

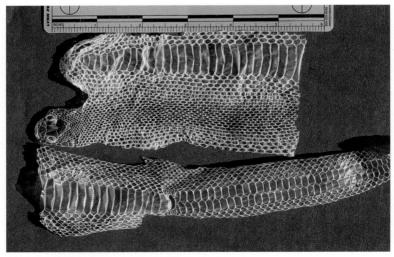

Gopher snake shed.

Dorsal head scales: The rostral scale is as long as it is broad, or longer than broad. In the bullsnake (*Pituophis catenifer sayi*) it projects forward. There are 2 nasals (partially divided by the nostril), 1 or 2 (rarely 3) preoculars, 3 or 4 (sometimes between 1 and 6) postoculars, and 1 (rarely 0 or 2) loreal. There are 3 rows of temporal scales, including the following number of scales in each: 1, 2 or 3, and between 3 and 5 scales per row. Generally there are 8 or 9 (sometimes 7 or 10) supralabials. The spectacle is moderate in size. There are typically 4 prefrontals in this species.

Ventral head scales: There are generally 12 or 13 (sometimes between 10 and 15) infralabials.

Vent scales: The anal plate is undivided.

Scale count: There are 31 or 33 (sometimes between 27 and 37) rows of scales at midbody. Dorsal scales are keeled with two apical pits.

Adult length: 36–100 inches in total length.

Glossy Snake (*Arizona elegans*)

General description: Sheds are moderate to long, narrow to moderate in thickness, and have medium-length tails. The pattern varies somewhat between the different subspecies, but all include a series of saddle-shaped blotches running the length of the dorsal surface from the neck down. There are often one or multiple rows of smaller dots running along the dorsal surface on either side. Sheds are usually found on the ground under cover objects, such as rocks, fallen yuccas, or boards.

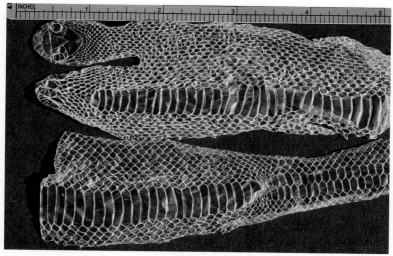

Glossy snake shed.

Dorsal head scales: There are 2 nasals (partially divided by the nostril), 1 (sometimes 2) preoculars, 2 (rarely 3) postoculars, and 1 (sometimes 2) loreal. There are 2 rows of temporal scales, including the following number of scales in each: 2 (sometimes 1), and 3 or 4 (sometimes 5 or 6) scales per row. Generally there are 8 (sometimes 7 or 9) supralabials. The spectacle is moderate in size.

Ventral head scales: There are generally 12 or 13 (sometimes between 11 and 15) infralabials.

Vent scales: The anal plate is undivided.

Scale count: There are 27 to 31 (sometimes between 25 and 35) rows of scales at midbody. Dorsal scales are smooth and with one apical pit.

Adult length: 26–70 inches in total length.

California Lyre Snake (*Trimorphodon lyrophanes*)

General description: Sheds are small to moderately long, narrow in thickness, and have relatively long tails. Sheds are usually found on the ground under cover objects, such as rocks, fallen yuccas, or boards. They may also be found tucked high in cliff faces or rock crevices, where this snake frequently hunts. This species is named for the lyre- or V-shaped mark on top of its head, which is only faintly visible in very fresh sheds.

Dorsal head scales: There are 2 nasals, 3 (sometimes 4) preoculars, 3 or 4 postoculars, and between 2 and 4 loreals. Also, there is a small "lorilabial" scale between the loreals and the supralabials. There are 2 rows of temporal scales, including the following number of scales in each:

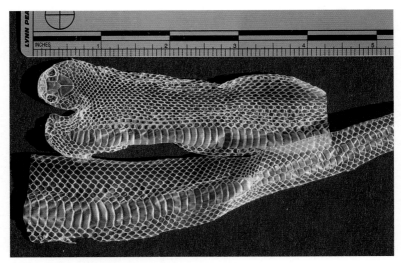

California lyre snake shed.

2 or 3, and 3 (sometimes 4 or 5) scales per row. Generally there are 9 (sometimes 8 or 10) supralabials. The spectacle is proportionately large in size.

Ventral head scales: There are generally between 11 and 13 infralabials.

Vent scales: The anal plate is divided or undivided.

Scale count: There are between 21 and 24 rows of scales at midbody. Dorsal scales are smooth and have 2 apical pits.

Adult length: 18–46 inches in total length.

Western Shovel-nosed Snake (*Chionactis occipitalis*)

General description: Sheds are very small to small in length, narrow in thickness, and have medium-length tails. Typically adults are light yellow dorsally with a series of black saddles or bands spaced well apart down the back. Some subspecies also have a series of reddish bands between the black bands. They typically also have a wide black crescent spanning the back of the head and touching or running through each eye. Sheds are usually found on the ground, underneath shrubs or rocks.

Dorsal head scales: The rostral scale is much wider than long and does not reach back far enough to separate the internasal scales. There is 1 nasal, 1 (sometimes 2) preoculars, 2 (sometimes 1 or 3) postoculars, and 1 (sometimes 0 or 2) loreals. There are 2 rows of temporal scales, including the following number of scales in each: 1 (sometimes 2), and 2 (sometimes 1 or 3) scales per row. Generally there are 7 (sometimes 8

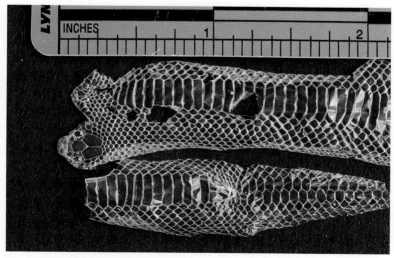

Western shovel-nosed snake shed.

or 9) supralabials. The spectacle is moderately small in size. The frontal scale is very broad and about twice the width of the supraoculars.

Ventral head scales: There are generally 7 (sometimes 6 or 8) infralabials.

Vent scales: The anal plate is divided.

Scale count: There are 15 (sometimes 14 or 16) rows of scales at mid-body. Dorsal scales are smooth and have an apical pit.

Adult length: 10–17 inches in total length.

Brown Vinesnake (*Oxybelis aeneus*)

General description: Sheds are moderate to very long, very narrow proportionately, and with a very long, tapered tail. This species typically is uniform gray or brown dorsally. A thin, dark stripe runs from the tip of the snout through the eyes. The region below this dark stripe is pale cream or white. Sheds are usually found up in shrubs or small trees where this snake spends most of its time.

Dorsal head scales: The rostral scale is small, and wider than long. Both the prefrontals and internasals are very long. There are 2 nasals, 1 preocular, 1 or 2 postoculars, and 0 loreals. There are 2 rows of temporal scales, including the following number of scales in each: 2 (sometimes 1 or 3), and 2 (sometimes between 1 and 4) scales per row. Generally there are 8 or 9 supralabials. The spectacle is moderate in size. The frontal is long and narrower than the supraoculars. The supraoculars project slightly over the spectacle.

Ventral head scales: There are generally 9 (sometimes between 8 and 11) infralabials.

Brown vinesnake shed.

Vent scales: The anal plate is divided.

Scale count: There are 17 (sometimes 15 or 16) rows of scales at mid-body. Dorsal scales are smooth or weakly keeled, and may or may not have apical pits.

Adult length: 36–60 inches in total length.

Ring-necked Snake (*Diadophis punctatus*)

General description: Sheds are very small to moderately small, relatively narrow, and have tails of medium length. The dorsal surface is bluish to charcoal gray. Most subspecies show a thin, pale ring on the neck. The ventral surface is brilliant yellow or orange, becoming darker orange toward the tail. Sheds are usually found under cover objects, such as boards, logs, or rocks. These sheds are known for the narrow neck band found on most, but not all subspecies. Note the presence of the neck band in the photo of this specimen, which is from the subspecies known as the northwestern ring-necked snake (*Diadophis punctatus occidentalis*).

Dorsal head scales: The rostral scale is small, and wider than long. There are 2 nasals, 2 (sometimes 1 or 3, rarely 0) preoculars, 2 (sometimes between 1 and 4) postoculars, and 1 loreal. There are 3 rows of temporal scales, including the following number of scales in each: 1 (rarely 2), 1 (rarely 2), and 2 scales per row. Generally there are 7 or 8 (sometimes 6 or 9) supralabials. The spectacle is proportionately small in size. The frontal scale is about twice as broad as the supraoculars. The parietals are very long and stretch to the back of the head.

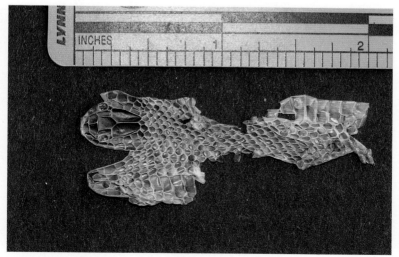

Ring-necked snake shed.

Ventral head scales: There are generally 7 or 8 (sometimes 6 or 9) infra-labials.

Vent scales: The anal plate is divided.

Scale count: There are between 14 and 17 rows of scales at midbody. Dorsal scales are smooth and have an apical pit.

Adult length: 10–30 inches in total length.

Rubber Boa (*Charina bottae*)

General description: Sheds are small to moderate in length, proportionately robust, and have relatively short tails that end in blunt, rounded tips. Typically the dorsal color is uniform tan, olive, or grayish. The ventral surface is usually paler, often yellowish. The sheds of this species are usually found under logs, bark, or boards.

Dorsal head scales: The rostral scale is large, and is wider than long. There are 2 nasals, 1 preocular, 3 or 4 postoculars, and 1 (sometimes 2) loreal. The temporal scales are the same size and shape as the body scales behind them. Generally there are 10 (sometimes 9 or 11) supralabials. The spectacle is proportionately very small. The frontal scale is very broad, and is generally the largest scale on the head. Some of the supralabials may be tall and others might be short. This variation can be used to identify individuals (Hoyer and Hoyer).

Ventral head scales: There are generally 10 reduced infralabials. The chin shields are small in size.

Vent scales: The anal plate is undivided. The ventral scales are proportionately narrow along the entire length of the snake.

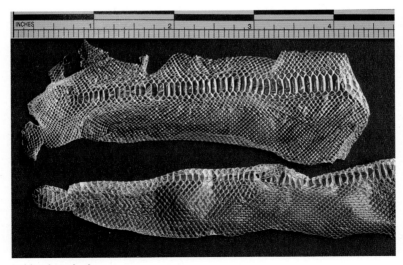

Rubber boa shed.

Scale count: There are between 32 and 53 rows of scales at midbody. Dorsal scales are smooth, very small, and do not have an apical pit.

Adult length: 14–33 inches in total length.

Rosy Boa (*Charina trivirgata*)

General description: Sheds are small to moderate in length, proportionately robust, and have relatively short tails that end in blunt, rounded tips. The pattern varies between subspecies and even specific populations, but generally includes 3 broad, even-edged or irregular longitudinal stripes on the dorsal surface starting at the tip of the snout or the back of the head. Sheds of this species are usually found under logs, bark, or boards. The shed in the photograph below is from the species now known as the northern three-lined boa (*Charina orcutti*), which inhabits the mountains of southern California. Unlike the rubber boa, this species does not have large scales on top of the head.

Dorsal head scales: The rostral scale is large, and is wider than long. It projects slightly in some individuals. There are 2 nasals, a ring of between 7 and 11 ocular scales around the eye, and 3 (sometimes 2 or 4) loreals. There are several rows of between 1 and 3 temporal scales. Generally there are between 13 and 15 (sometimes 12) supralabials. The spectacle is proportionately small in size. Often, the first 2 or 3 supralabials will be tall.

Ventral head scales: There are generally between 14 and 16 (sometimes between 11 and 17) infralabials. The chin shields are absent.

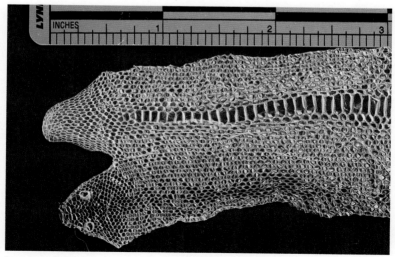

Rosy boa shed.

Vent scales: The anal plate is undivided. The ventral scales are proportionately narrow along the entire length of the snake.

Scale count: There are between 33 and 49 rows of scales at midbody. Dorsal scales are smooth, small, and do not have an apical pit.

Adult length: 24–42 inches in total length.

Western Rattlesnake (*Crotalus oreganus*)

General description: Sheds are moderate to long, proportionately robust in thickness, and with a relatively short tail. The tail ends of sheds are abrupt and open, due to the presence of rattles. A wide variety of patterns exist in this species and its subspecies. Commonly there is a dark line or band bordered by a light one, running through the eye and arcing back toward the rear supralabials. Some subspecies have large faint blotches, while others have distinct blotches that are bordered by dark. The sheds of this species are generally found under boards, bark, or rocks. The shed in the photograph is from a specimen in northern California, and from the subspecies northern Pacific rattlesnake (*Crotalus oreganus oreganus*). Fresh rattlesnake sheds have a strong odor.

Dorsal head scales: The rostral scale is moderate in size, and usually longer than wide. There are 3 or 4 (sometimes between 1 and 8) internasals; behind these are between 4 and 6 (sometimes between 1 and 9) intersupraoculars (the prefrontals are absent), 2 preoculars, several suboculars, 2 (sometimes 1 or 3) postoculars, and 1 (sometimes between 1 and 3) loreal. The 2 supraoculars are proportionately large

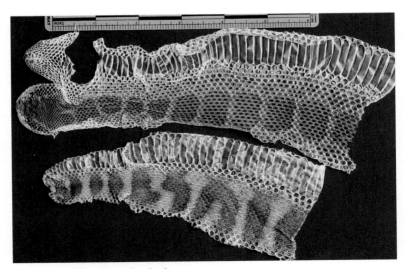

Northern Pacific rattlesnake shed.

and project partially over the spectacle. Generally there are between 15 or 16 (sometimes between 11 and 20) supralabials. The spectacle is moderate in size. Like all rattlesnake species, this snake possesses loreal pits in the region between the nostrils and the spectacle.

Ventral head scales: There are generally between 15 and 16 (sometimes between 11 and 20) infralabials. Posterior chin shields are large and slightly wider than the anterior chin shields.

Vent scales: The anal plate is undivided. The ventral scales in general are proportionately wide. Subcaudal scales are not divided.

Scale count: There are between 23 and 27 (sometimes between 21 and 29) rows of scales at midbody. Dorsal scales are strongly keeled and have apical pits.

Adult length: 16–64 inches in total length.

Tiger Rattlesnake (*Crotalus tigris*)

General description: Sheds are moderate in length, proportionately robust in thickness, and have relatively short tails. The tail ends of sheds are abrupt and open, due to the presence of rattles. This species has a series of faint, irregularly shaped bands crossing the back from the neck all the way to the rattle. The band through the eye of this species is typically very faint. The sheds of this species are generally found at the base of, or underneath, rocks or shrubs. Compared to other rattlesnake species, the head of this snake is relatively small in proportion to its body, and not dramatically wider than the neck. Fresh rattlesnake sheds have a strong odor.

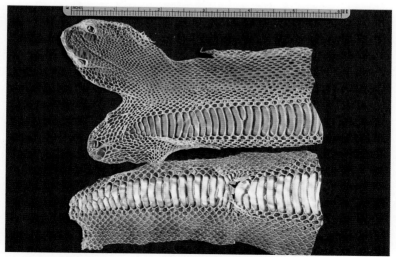

Tiger rattlesnake shed.

Dorsal head scales: The rostral scale is moderate, and wider than long. There are 2 small internasals (with the prenasal scales usually touching the first supralabial, but the postnasal scales rarely touch the upper preocular scales), 2 or 3 preoculars, 1 (sometimes 2) postoculars, and 1 or 2 loreals. The 2 supraoculars are proportionately large and project partially over the spectacle. Generally there are between 12 and 14 (sometimes between 11 and 16) supralabials. The spectacle is moderate in size. Like all rattlesnake species, this snake possesses loreal pits in the region between the nostrils and the spectacle.

Ventral head scales: There are generally between 13 and 15 (sometimes between 11 and 16) infralabials. Posterior chin shields are large, and much wider than the anterior chin shields.

Vent scales: The anal plate is undivided. The ventral scales in general are proportionately wide. Subcaudal scales are not divided.

Scale count: There are 23 (sometime between 20 and 28) rows of scales at midbody. The scales are keeled and have apical pits.

Adult length: 20–36 inches in total length.

Speckled Rattlesnake (*Crotalus mitchelli*)

General description: Sheds are moderate to long, proportionately robust in thickness, and have relatively short tails. The tail ends of sheds are abrupt and open, due to the presence of rattles. This species has a wide variety of patterns, typically some degree of dark bands or blotches crossing the back and running the length of the snake from the neck and onto the tail. Each dorsal scale may have several colors

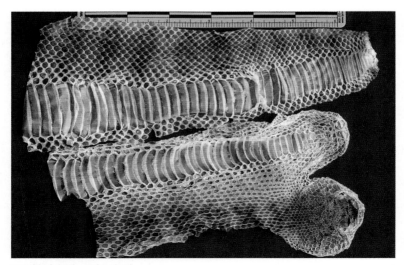

Speckled rattlesnake shed.

on it, which gives this species incredible camouflage. The sheds of this species are generally found at the base of, or underneath, rocks or shrubs. Though similar in pattern to the tiger rattlesnake (*Crotalus tigris*), the head of this species is proportionately much larger. Fresh rattlesnake sheds have a strong odor.

Dorsal head scales: The rostral scale is wider than long. There are up to as many as 50 scales in the internasal and prefrontal region, no prefrontal scales, 2 nasals (the prenasal often touches the first supralabial, but the postnasal does not touch the upper preocular), 2 (sometimes 3) preoculars, 2 (sometimes 3) postoculars, several suboculars, and 2 (sometimes between 0 and 5) loreals. The 2 supraoculars are proportionately large and project partially over the spectacle. Generally there are between 14 or 16 (sometimes between 12 and 19) supralabials. The spectacle is moderate in size. Like all rattlesnake species, this snake possesses loreal pits in the region between the nostrils and the spectacle.

Ventral head scales: There are generally between 15 or 16 (sometimes between 12 and 19) infralabials. Posterior chin shields are very large, and wider than the anterior chin shields.

Vent scales: The anal plate is undivided. The ventral scales in general are proportionately very wide. Subcaudal scales are not divided.

Scale count: There are between 23 and 25 (sometimes between 21 and 27) rows of scales at midbody. Dorsal scales are keeled and have apical pits.

Adult length: 23–52 inches in total length.

Timber Rattlesnake (*Crotalus horridus*)

General description: Sheds are moderate to very long, proportionately very robust in thickness, and have relatively short tails. Tail ends of sheds are abrupt and open, due to the presence of rattles. The tail pattern fades from dark banding to solid black. A typical pattern is yellow to gray (light morph) or dark brown or some mostly black (dark morph). Body has V-shaped bands of black with pale borders and light color in between the bands. A wide, dark stripe reaches from the eye to beyond the corner of the mouth. The shed pictured on page 398 is from the southern part of this species' range, where the snakes tend to be paler and have a reddish vertebral stripe running from the back of the head down the length of the body. Fresh rattlesnake sheds have a strong odor.

Dorsal head scales: The rostral scale is longer than wide. There are 2 nasals, 2 preoculars, 4 (sometimes between 2 and 6) postoculars, and 2 (sometimes 1 or 3) loreals. There are several suboculars and temporal scales. Generally there are between 13 and 15 (sometimes between 10 and 17) supralabials. The spectacle is moderate in size. The 2 supraoculars are proportionately large and project partially over the

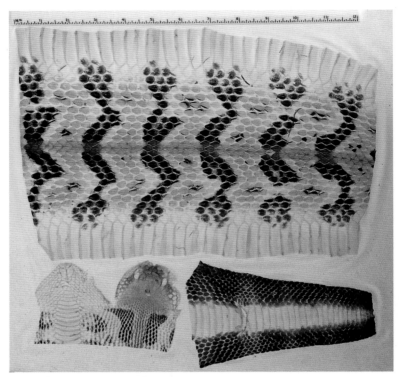

Timber rattlesnake shed. MARK TOVAR

spectacle, with no single large frontal scale in between them. Like all rattlesnake species, this snake possesses loreal pits in the region between the nostrils and the spectacle.

Ventral head scales: There are generally between 14 and 16 (sometimes between 11 and 19) infralabials.

Vent scales: The anal plate is undivided. The ventral scales in general are proportionately wide. Subcaudal scales are not divided.

Scale count: There are typically 23 or 25 (sometimes between 21 and 26) rows of scales at midbody. Dorsal scales are strongly keeled with apical pits.

Adult length: 35–74 inches in total length.

Western Diamond-backed Rattlesnake (*Crotalus atrox*)

General description: Sheds are moderate to very long, proportionately very robust in thickness, and have relatively short tails. Tail ends of sheds are abrupt and open, due to the presence of rattles. The tail pattern is usually an even, alternating series of thick bands of black and white. A broad, dark band runs below the eye, arcing back to the rear supralabials, and is bordered on either side by a paler line. The typical pattern includes large, diamond-shaped marks that are dark brown with broken darker brown boundaries bordered on the outside with yellow or whitish scales. These diamond-shaped marks run down the middle of the back. The shed of this species is generally found at the base of, or underneath, rocks or shrubs. This is the largest western rattlesnake species, and the sheds of large specimens are very impressive. The sheds of red diamond rattlesnakes (*Crotalus ruber*) are very similar, but typically have reddish markings, and are only found in a limited range in southwestern California. Fresh rattlesnake sheds have a strong odor.

Dorsal head scales: The rostral scale is longer than wide. There are 2 small internasals. There are no prefrontals, 2 nasals (with the prenasal touching the supralabials, and the postnasal usually touching the upper preocular), 2 (occasionally 3) preoculars, 3 or 4 (sometimes 2) suboculars, and 1 (occasionally 2) loreal. The 2 supraoculars are proportionately large and project partially over the spectacle, and there are between 4 and 5 (occasionally between 3 and 8) inter-supraocular scales. Generally there are between 15 and 16 (sometimes between 12 and 18) supralabials. The spectacle is moderate in size. Like all rattlesnake species, this snake possesses loreal pits in the region between the nostrils and the spectacle.

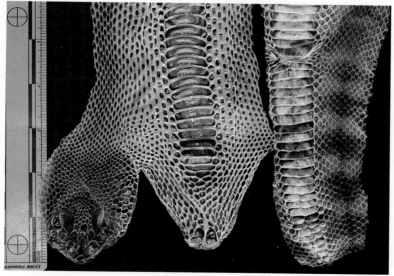

Western diamond-backed rattlesnake.

Ventral head scales: There are generally between 16 and 17 (occasionally between 12 and 18) infralabials. Posterior chin shields are very large and significantly wider than the anterior chin shields.

Vent scales: The anal plate is undivided. The ventral scales in general are proportionately very wide. Subcaudal scales are not divided.

Scale count: There are between 25 and 27 (sometimes between 23 and 29) rows of scales at midbody. The scales are strongly keeled and have apical pits.

Adult length: 34–83 inches in total length.

Eastern Diamond-backed Rattlesnake (*Crotalus adamanteus*)

General description: Sheds are long to very long, proportionately very robust in thickness, and have relatively short tails. Tail ends of sheds are abrupt and open, due to the presence of rattles. The typical pattern

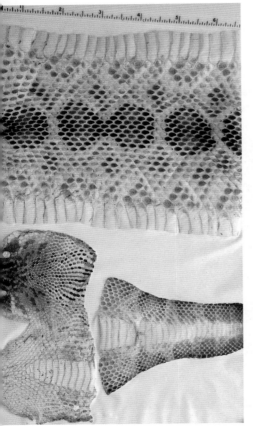

includes large, diamond-shaped marks that are dark brown with black boundaries, bordered on the outside with yellow or whitish scales. These diamond-shaped marks run down the middle of the back. A broad, dark band runs below the eye, arcing back to the rear supralabials, and is bordered on either side by a paler line. The sheds of this species are generally found at the base of, or underneath, shrubs, palmettos, or in gopher tortoise burrows, where this species spends a lot of time. This is the largest rattlesnake species in the world, and the sheds of large specimens are very impressive. Fresh rattlesnake sheds have a strong odor.

Dorsal head scales: The rostral scale is longer than wide. There are 2 internasals. There are no prefrontals but there are between 6 and 7 intersupraoculars, 2 nasals,

Eastern diamond-backed rattlesnake shed. MARK TOVAR

2 preoculars, several suboculars, 2 (occasionally 3) postoculars, and 2 (sometimes between 1 and 3) loreals. The 2 supraoculars are proportionately large and project partially over the spectacle, and there are between 4 and 5 (occasionally between 3 and 8) inter-supraocular scales. Generally there are 14 (sometimes between 12 and 17) supralabials. The spectacle is moderate in size. Like all rattlesnake species, this snake possesses loreal pits in the region between the nostrils and the spectacle.

Ventral head scales: There are generally between 17 and 18 (occasionally between 15 and 21) infralabials. Posterior chin shields are very large and significantly wider than the anterior chin shields.

Vent scales: The anal plate is undivided. The ventral scales in general are proportionately very wide. Subcaudal scales are not divided.

Scale count: There are between 27 and 29 (Sometimes between 25 and 31) rows of scales at midbody. The scales are strongly keeled and have apical pits.

Adult length: 36–96 inches in total length.

Black-tailed Rattlesnake (*Crotalus molossus*)

General Description: Sheds are moderate to long, proportionately very robust in thickness, and have relatively short tails. Tail ends of sheds are abrupt and open, due to the presence of the rattles. The tail pattern is mostly solid and dark, hence the name. The pattern of this species varies somewhat, but it commonly has a series of large diamond-like markings in the form of blotches or wide bands that become more

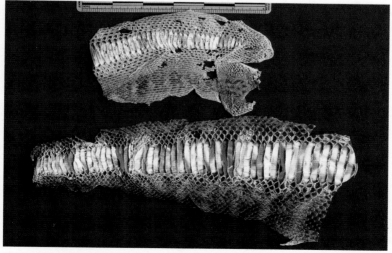

Black-tailed rattlesnake shed.

narrow past the midbody. The diamond-like markings are often more elongated and narrow (often reaching down to the ventral scales) than those of western diamond-backed rattlesnakes, with pale scales scattered both on the inside and on the borders of these bands. A dark band covers the area from the rostrum to the rear of the supraoculars, and narrows into a dark band running through each eye and arcing back to the rear supralabials. Sheds of this species are generally found at the base of, or underneath, rocks or shrubs. Fresh rattlesnake sheds have a strong odor.

Dorsal head scales: The rostral scales are slightly longer than wide. There are 2 large internasals that touch the rostral scale. There are 2 prefrontals, 2 nasals (with the prenasal usually not touching the supralabials, and the postnasal usually not touching the upper preocular), 2 or 3 preoculars, between 3 and 5 (sometimes 6 or 7) postoculars, and 2 or 3 (sometimes up to 9) loreals. The 2 supraoculars are proportionately large and project partially over the spectacle. Generally there are 17 or 18 (sometimes between 13 and 20) supralabials. The spectacle is moderate in size. Like all rattlesnake species, this snake possesses loreal pits in the region between the nostrils and the spectacle.

Ventral head scales: There are generally 17 or 18 (sometimes between 13 and 20) infralabials. The posterior chin shields are very large, and somewhat wider than the anterior chin shields.

Vent scales: The anal plate is undivided. The ventral scales in general are proportionately very wide. Subcaudal scales are not divided.

Scale count: There are 27 (sometimes between 23 and 31) rows of scales at midbody. The scales are keeled and have apical pits.

Adult length: 28–51 inches in total length.

Massasauga (*Sistrurus catenatus*)

General description: Sheds are small to moderate in length, proportionately robust in thickness, and have relatively short tails. Tail ends of sheds are abrupt and open, due to the presence of the rattles. A dark, light-bordered line runs back from the eye and on to the neck. Another similar line is found on either side of the top of the head, and reaches back onto the neck. A series of large, dark brown blotches appears down the middle of the back, and there are 3 rows of smaller dark blotches and spots on either side. Sheds of this species are generally found at the base of, or underneath, rocks or shrubs. Compared to other rattlesnake species, the head of this snake is relatively small in proportion to its body, and not dramatically wider than the neck. Fresh rattlesnake sheds have a strong odor.

Dorsal head scales: The rostral scale is longer than wide. Scales on top of the head are large, unlike in *Crotalus* species. There are 2 internasals, 2 prefrontals, 2 nasals, 2 preoculars (with the upper preocular touching the postnasal), 3 or 4 (sometimes 2 or 5) postoculars, 1 or 2 suboculars, and 1 loreal. The supraoculars are proportionately large and project partially over the spectacle. Generally there are between 11 and 12 (sometimes between 9 and 14) supralabials. The spectacle is moderate in size. A large frontal scale spans the distance between the supraoculars. Parietals are present, and are about the same size as the supraoculars. Like all rattlesnake species, this snake possesses loreal pits in the region between the nostrils and the spectacle.

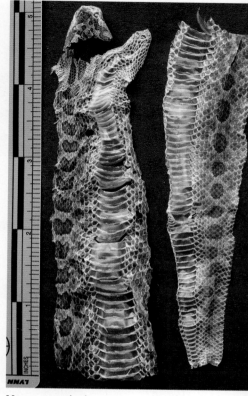

Massasauga shed.

Ventral head scales: There are generally between 11 and 13 (sometimes 10 and 16) infralabials. The posterior chin shields are moderately large.

Vent scales: The anal plate is undivided. The ventral scales in general are proportionately wide. Subcaudal scales are not divided.

Scale count: There are between 23 and 25 (sometimes between 21 and 27) rows of scales at midbody. Scales are strongly keeled and have apical pits.

Adult length: 18–39 inches in total length.

Pygmy Rattlesnake (*Sistrurus miliarius*)

General description: Sheds are small in length, proportionately robust in thickness, and have relatively short tails. Tail ends of sheds are abrupt and open, due to the presence of rattles. The typical pattern is gray ground color with black transverse blotches disrupting a continuous red-brown line running down the middle of the back. Dark spots on the sides tend to line up with the large, dark blotches. This particular shed is from the subspecies known as the dusky pygmy rattlesnake (*Sistrurus miliarius barbouri*). Sheds of this species are generally found at the base of, or underneath, rocks or shrubs. Compared to other rattlesnake species, the head of this snake is relatively small in

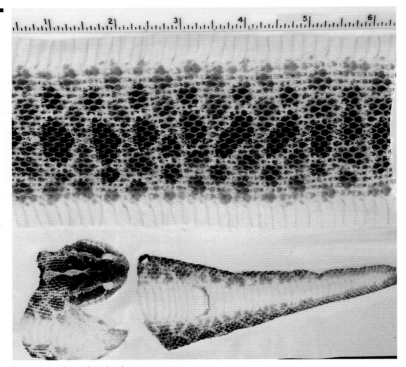

Pygmy rattlesnake shed. MARK TOVAR

proportion to its body, and not dramatically wider than the neck. Fresh rattlesnake sheds have a strong odor.

Dorsal head scales: There are 9 enlarged scales on top of the head, unlike in *Crotalus* species. There are 2 nasals, 2 preoculars, 3 or 4 (sometimes 5 or 6) postoculars, and 1 loreal between the postnasal and upper preocular. The supraoculars are proportionately large and project partially over the spectacle. Generally there are between 10 and 11 (sometimes between 8 and 13) supralabials. The spectacle is moderate in size. A large frontal scale spans the distance between the supraoculars. Parietals are present, and about the same size as the supraoculars. Like all rattlesnake species, this snake possesses loreal pits in the region between the nostrils and the spectacle.

Ventral head scales: There are generally between 11 and 12 (sometimes between 9 and 14) infralabials.

Vent scales: The anal plate is undivided. The ventral scales in general are proportionately wide. Subcaudal scales are not divided.

Scale count: There are typically between 21 and 23 (sometimes between 19 and 25) rows of scales at midbody. Dorsal scales are strongly keeled with apical pits.

Adult length: 15–30 inches in total length.

Lizard Sheds

The shed skins of lizards are found much less frequently, likely due to their small size and fragmented nature. Here are the sheds of several species that you'll commonly encounter in the field.

Chuckwalla (*Sauromalus ater*)

General description: This large, robust lizard species often leaves fragments of sheds in and near rock crevices in boulder piles. Pattern varies somewhat with population, but young and female adult animals show wide bands on the tail. Body has scattered spots of different colors, often pale, black, and reddish. Limbs and head are typically darker, sometimes black. This species grows to between 11 and 16 inches in total length.

Scales: The scales are generally granular, with patches of enlarged scales toward the back of the jaw and especially just ahead of the ear opening. The shed in this photo is from the patch of scales found on the lower portion of the hind limbs. Note the scales' slightly keeled, pointed appearance. This particular piece was photographed in a lava bed in southern California.

Fragment of chuckwalla shed.

Desert Horned Lizard (*Phrynosoma platyrhinos*)

General description: This medium-sized, robust lizard has a crown of large cranial horns and a wide, flat body. This species of horned lizard grows to between 3 and 5 inches in total length. It has a single row of fringe scales on the sides of its body. This species is most likely to leave fragments of its shed near rocks, in open areas, and on the edges of sandy flats.

Scales: The scales are mostly granular, and interspersed with mucronate scales of varying sizes across the dorsal surface, limbs, and tail. The

Desert horned lizard shedding. CAMERON ROGNAN

ventral scales are small and granular. The dorsal pattern varies, but often includes a pair of large, dark blotches near the neck region and a repeating pattern of wavy dark blotches across the back. These blotches usually form alternating dark bands across the tail.

The shed in the photo above is conveniently still coming off of the lizard's back. Notice that many of the large, pointed mucronate scales are darker than the smaller scales. Also note the small fragments of granular ventral scales hanging off in the piece of shed below the lizard's chin.

Desert Iguana (*Dipsosaurus dorsalis*)

General description: This medium to large lizard has a relatively long body of medium build, a short, blunt head, and a long, tapered tail. This species can grow to between 9 and 16 inches in total length. The shed fragments of this species are usually found in association with creosote bushes on hillsides or on flats.

Scales: The scales are very small and granular. A distinct line of slightly larger, keeled scales runs down the length of the spine from the back of the head and well onto the top of the tail. The pattern of this species generally shows prominent, dark, broken lines running laterally down the back. Areas of paler, grayish regions interspersed with pale white spots run across the back. These spots diffuse into continuous pale lines that traverse the back close to the rear limbs.

Notice that the photo of the shed fragments shows the very small, granular scales, and the distinct vertebral line of enlarged, keeled

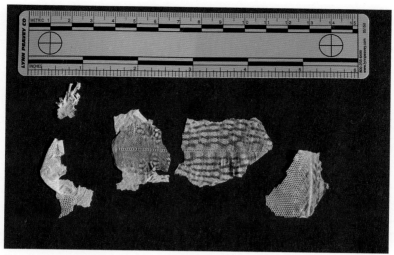

Desert iguana shed fragments.

scales. The two fragments of shed in the middle are both from the back. The piece on the far left is a nearly complete shed of a hind foot and hind leg. The piece on the far right is from the ventral region of the chest, near the forelimbs. Notice the larger, more translucent ventral scales.

Gila Monster (*Heloderma suspectum*)

General description: This is the largest native lizard, with a robust build, large limbs, and a long, thick tail. This species can grow to between 16 and 24 inches in total length. The shed fragments of this species are usually found in and around rock crevices, or under dense shrubs.

Scales: The scales are granular, but large, and are underlaid with osteoderms. The scales are arranged in consistent and even rows. The ventral scales are shaped like rectangles or squares. The dorsal pattern is a very bold series of wide, black bands, or reticulations, and scattered blotches. In between the dark bands, the body is pinkish-orange or yellow-orange. The shed fragments of this species are likely to be confused only with the shed of its larger relative, the beaded lizard (*Heloderma horridum*), which is found much farther south in coastal Mexico.

Notice that the photo of the shed fragments shows the large, rounded granular scales. The large fragment at the top is from the ventral area of the chest, with the right side of it also showing part of the ventral surface of the throat. The piece of shed on the far left is the round, blunted tip of the tail. The shed fragments in the middle, and on the right, are both from the back of the animal.

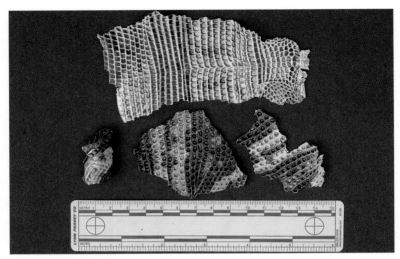

Gila monster shed fragments.

Great Basin Collared Lizard (*Crotaphytus bicinctores*)

General description: This medium-sized lizard has a large head, medium-sized body, and long, tapered tail. This species can grow to between 6 and 13 inches in total length. The shed fragments of this species are usually found among rocks and boulders on rocky hillsides.

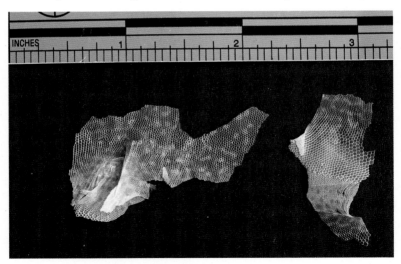

Great Basin collared lizard shed fragments.

Scales: The scales are small and granular, with the largest found near the eyes and around the mouth. The dorsal pattern is greenish-olive to grayish, with scattered pale spots that coalesce together on the hind legs and tail, creating a pale background with brownish or greenish spots. The darker dorsal surface breaks up into dark spots and small blotches on the sides, and fades as you near the belly.

Notice the small to very small granular scales in the photo of the shed fragments. The piece on the left is from the midbody, and shows the dark dorsal color broken up with pale spots. These coalesce and the dark dorsal color breaks up into progressively smaller dots as you near the ventral surface. Notice the scales are smallest where the dark dorsal color forms small dots. Also notice the larger, more translucent scales just visible on the bottom of this shed fragment. These same scales are more obvious and numerous in the piece of shed in the right of the photo. This piece is from the ventral region of the chest and shows the dorsal coloration and scalation at its upper and lower edges. The most distinctive part of the lizard, the double black band forming the namesake collar, is not visible here.

Long-nosed Leopard Lizard
(*Gambelia wislizenii*)

General description: This medium-sized lizard has a large head, medium-sized body, and a long, tapered tail. This species can grow to between 8 and 15 inches in total length. The shed fragments of this species are usually found at the base of shrubs or rocks in open, flat areas.

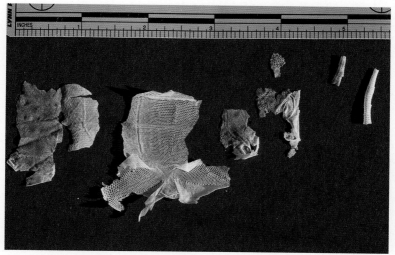

Long-nosed leopard lizard shed fragments.

Scales: The scales are small and granular, with the largest scales found near the eyes and around the mouth. The dorsal pattern is whitish tan to gray overall, and on top of that are darker spots and pale lines that cross in transverse orientation across the body, legs, and tail. The color and pattern fades to pure white on the ventral region.

Notice in the photo on page 409 the small to tiny granular scales. The piece on the far left is a fragment from the dorsal surface near the midbody, and shows the pattern described above. The piece second from the left is the ventral region, including most of the belly, the groin, and upper thighs. Note the larger, more translucent ventral scales on the majority of this shed piece. Look closely for the line of slightly enlarged femoral scales running down either thigh. Also, if you look closely just below the thighs, there is a semicircular region of small scales. This is the opening to the cloaca. The three pieces in the center are from the dorsal surface of the head. The piece at left center includes the scales and skin that once surrounded and covered the ear. The piece at right center includes the scales and skin that once surrounded and covered the other ear, as well as the incredibly delicate skin on the eyelid. Just above this is a tiny fragment that shows the slightly enlarged scales near the nostrils. The two small fragments on the far right are both from the tail, near the tip. The farthest right is the ventral surface, while the piece second from the right is the dorsal surface.

Peninsular Leaf-toed Gecko (*Phyllodactylus nocticolus*)

General description: This very small lizard has a proportionately large head, medium-sized body, and a long tail that is thick at the base. This species can grow to between 4 and 5 inches in total length. The shed fragments of this species are usually found among boulder piles on hillsides.

Peninsular leaf-toed gecko shed fragment.

Scales: The scales are small and granular and interspersed dorsally with slightly larger, rounded tubercles. The tubercles found on the dorsal surface of Mediterranean house geckos are much larger. The largest scales found on the body of this species are present on the front part of the head, and along the edge of the mouth. This species' dorsal pattern is a series of uneven, dark blotches. The ventral surface is unmarked.

Notice in the photo the enlarged scales along the edge of the mouth. The largest scale is the mental scale, at the tip of the mouth. This fragment of shed is from the entire ventral surface of the head. The most distinctive feature of this lizard species—its leaf-like toe tips—is not visible here.

Southern Alligator Lizard (*Elgaria multicarinata*)

General description: This medium-sized lizard has a proportionately large head, relatively long body, and a very long, thick tail. This species can grow to between 10 and 16 inches in total length. The sheds of this species are usually found underneath rocks, logs, or boards.

Scales: The scales come in several types, and are underlaid with osteoderms. The dorsal scales are large and keeled. The ventral scales are also large, but not keeled. The scales found on the limbs, and especially along the lower sides of the body, are small and granular. This species has faint dark lines crossing through the middle of each ventral scale.

Notice in the photo that the shed is in one piece. This is typical of all alligator lizard species found in the United States. The head is

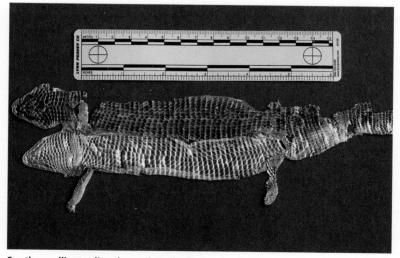

Southern alligator lizard complete shed.

Two complete northern alligator lizard sheds.

wide and triangular, and all the scales on it are relatively large when compared to other lizard species. Notice the tail region at its base is nearly as big around as the body. This particular shed is from a specimen in southern California belonging to the subspecies known as the San Diego alligator lizard (*Elgaria multicarinata webbii*).

Northern Alligator Lizard (*Elgaria coerulea*)

General description: This medium-sized lizard has a proportionately large head, relatively long body, and a long, thick tail. This species is a smaller relative of the southern alligator lizard, and grows between 7 and 13 inches in total length. The sheds of this species are usually found underneath logs, large slabs of bark, and boards.

Scales: The scales come in several types and are underlaid with osteoderms. The dorsal scales are large and keeled. The ventral scales are also large, but not keeled. The scales found on the limbs, and especially along the lower sides of the body, are small and granular. This species has faint dark lines crossing through the middle of each ventral scale.

Notice in the photo the two sheds in the image are in a single piece. Like the southern alligator lizard, this is typical. The head is more elongated and less triangular in this particular species, especially in the subspecies whose shed is seen in the photo, the northwestern alligator lizard (*Elgaria coerulea principis*). Look closely where the ventral and dorsal portions of the lower shed connect and notice the tiny granular scales.

Scat Identification  **10**

Crocodilians

The scat of a very young alligator found along the edge of a lake in Florida. Notice the uniform texture and chalky appearance.

The large, old scat of an adult alligator found along a trail in Florida. Note the lack of hair, bones, or seeds commonly found in similar-sized mammalian scats. The scats of adult American alligators measure 4–8 inches long.

Scat Identification

Turtles

The scat of a captive Pacific pond turtle. This scat shows a taper and a rather uniform texture, which might be a reflection of the diet in captivity. The scats of this species measure 0.4–0.7 inches long. KIM CABRERA

The scat of a three-toed box turtle from central Arkansas. The scats of this species measure 1–2 inches long. STEVE FORTIN

The tapered-looking scat of an adult desert tortoise from Nevada. The scats of this species measure 1.5–3.5 inches long.

A more rounded-looking scat of an adult desert tortoise from California. The scats of this species measure 1.5–3.5 inches long.

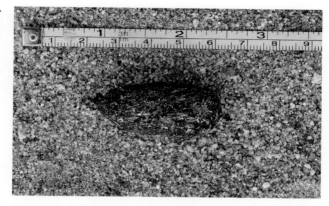

A gopher tortoise scat found on a trail in Florida. The scats of this species measure 1.5–3.5 in long. GEORGE L. HEINRICH

Lizards

The scat of an adult regal horned lizard from Arizona. The scats of this species measure 1–1.75 inches long.

The small scat of an adult Texas horned lizard from New Mexico. The scats of this species measure 0.75–1.75 inches long.

The large scat of an adult desert horned lizard from Nevada. The scats of this species measure 0.75–1.75 inches long.

Scat Identification

The scat of an adult pygmy short-horned lizard from Washington State. The scats of this species measure 0.3–0.6 inches long.

The scat of a small adult flat-tailed horned lizard from California. The scats of this species measure 0.4–1.5 inches long.

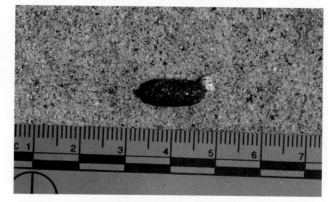

The scat of an adult Coachella Valley fringe-toed lizard. The scats of this species measure 0.4–0.7 inches long.

Scat Identification

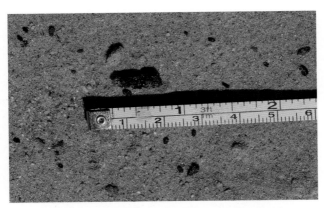

The scat of an adult Colorado Desert fringe-toed lizard. The scats of this species measure 0.4–0.8 inches long.

The scat of an adult Mojave fringe-toed lizard. The scats of this species measure 0.4–0.8 inches long.

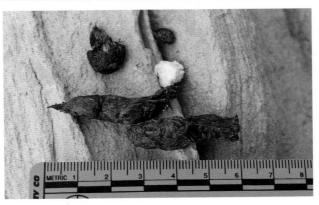

The scat pile of an adult chuck-walla. The single scat pellets of this species measure 0.7–1.5 inches long.

The scat of an adult desert iguana. The single scat pellets of this species measure 0.8–1.5 inches long.

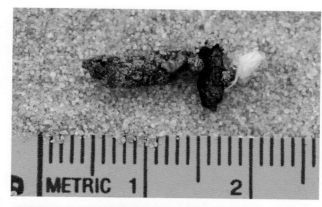

The fresh scat of an adult desert banded gecko. The scats of this species measure 0.3–0.5 inches long.

The scat of an adult eastern collared lizard. The single scat pellets of this species measure 0.6–1 inches long.

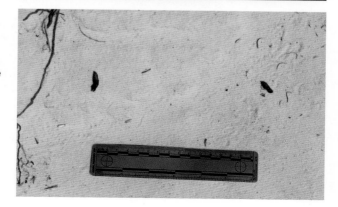

Scat Identification

The scat of an adult long-nosed leopard lizard. The scats of this species measure 1–1.5 inches long.

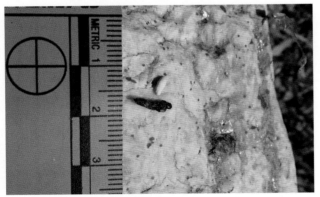

The scat of an adult greater earless lizard. The scats of this species measure 0.3–0.4 inches long.

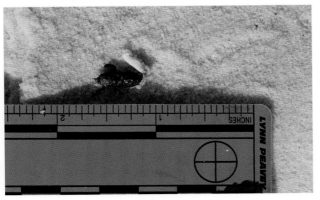

The scat of an adult bleached earless lizard. The scats of this species measure 0.3–0.4 inches long.

The scat of an adult side-blotched lizard. The scats of this species measure 0.19–0.3 inches long.

The scat of an adult southern alligator lizard. The scats of this species measure 0.5–1 inches long.

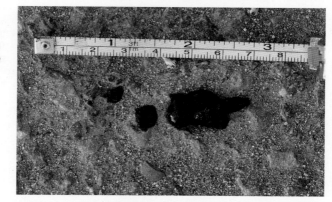

The scat of an adult sagebrush lizard. The scats of this species measure 0.4–0.7 inches long.

The scat of a tiger whiptail lizard. The scats of this species measure 0.5–0.75 inches long.

The scat of an adult western fence lizard. The scats of this species measure 0.3–0.6 inches long.

The scat of an adult yarrow spiny lizard. The scats of this species measure 0.5–1.25 inches long.

Snakes

The scat of an adult, captive speckled rattlesnake.
JASON KNIGHT

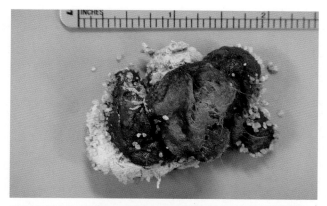

The scat of a small adult spotted leaf-nosed snake.

The scat of an adult, captive gray-banded kingsnake.
JASON KNIGHT

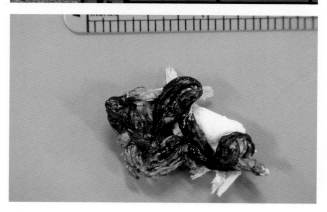

Frogs and Toads

The scat of an adult boreal toad. The scats of this species measure 0.75–1.75 inches long.
KIM CABRERA

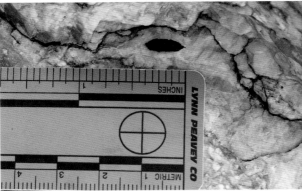

The scat of an adult canyon tree frog. The scats of this species measure 0.25–0.5 inches long.

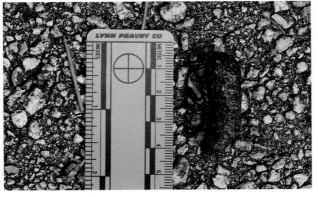

The scat of an adult southern toad. The scats of this species measure 1–1.75 inches long.

Salamanders

The scat of an adult rough-skinned newt. The scats of this species measure 0.5–0.75 inches long.

Scent Marking, Scats, Urea

The scats of reptiles and amphibians are another type of sign that can be encountered in the field. Amphibian scats are must less frequently encountered, as many deposit their scats under cover of rocks, bark, or dense vegetation, or in the water. You'll more frequently encounter reptile scats, especially those of lizards that typically deposit scats where they bask. It is not clear whether any herps use their scats deliberately for the purpose of scent marking, but scats will certainly accumulate in heavily used areas. You can occasionally observe lizards investigating the scats and uric pellets of other lizards.

Many lizard species possess femoral pores that spread the scent of the animal as it travels over surfaces. It is known that at least one lizard species—the desert iguana (Dipsosaurus dorsalis)—can actually see secretions deposited by the femoral pores of other members of its own species (Jones and Lovich 2009).

Scats can be used to further study reptiles and amphibians, giving a window into their diet and showing observers where the animals spend time. Fresh scats will often appear wet with a thin layer of mucus, but will dry quickly in the right conditions. Fresh scats tend to be darker, and will dry to a paler color.

Scats can be used to help act as supportive evidence to help better understand other sign, such as tracks. Sometimes, scats alone can be used to identify the animal down to genus or even species level. The following section will discuss scat identification in more detail. Please remember these are descriptions of typical scats, and some degree of variation in scat shape, size, and content exists even for individual animals depending on diet, health, and other conditions.

Crocodilians

You'll commonly find the scats of crocodilians along the edges of waterways, especially in areas where the animals repeatedly bask. Crocodilians do not possess a urinary bladder, but they do store urine in the kidney ducts. Their scats appear dark brown when fresh with a whitish portion on one end, and are oblong. As they age, they usually bleach to a white color and crumble. The scats tend to be either a single, long piece or several small segments of similar diameter. Crocodilians have powerful digestive systems and so the scats appear to be mostly composed of a uniform material, looking almost like clay when they are very dry. The size of the scat is relative to the size of the animal.

The scats of very young alligators might be confused with those of large lizards when fresh. As they dry, the chalky, uniform texture is distinct. Those of large adults can be impressive, ranging in size from 4 to 8 inches in length and between 1 and 2 inches in diameter.

The scat of a spiny softshell turtle in the shallow water along the shore of a lake. This scat likely started out as a single pellet, but has broken down into a soft mass, and contains crayfish parts and the remains of blueberries. STEVE FORTIN

Turtles

The scats of semiaquatic turtles are rarely encountered, as they typically deposit them in water. When observed, they are elongated pellets that vary in consistency depending on the diet of that particular turtle.

Terrestrial turtles often leave one or two elongated pellets that tend to be tan, dark brown, dark olive green, or nearly black in color. Generally, they are thicker on one end and come to a taper on the other end. Terrestrial turtle scales often include undigested plant materials such as grasses, fragments of leaves, and the seeds from wild fruits. The outer surface of these scats may be relatively uniform in appearance or may show the coarse materials that are present throughout the scat.

The most likely locations to find scats are in and around the entrances of burrows, or in areas of heavy feeding.

Lizards

All scats of the lizard species covered in this book are composed of two main parts: the fecal pellet(s) and the uric pellet. The fecal portion of lizard scats varies in content and shape depending on the species' dietary preferences and physiology. Uric pellets are composed largely of concentrated urea wastes. Generally, lizards excrete the uric pellets first, and immediately follow with the fecal pellets. These two may or may not be attached to each other.

The fecal and uric pellets come in a wide variety of shapes, sizes, and colors. Observed in combination, the fecal and uric pellets can be used in many cases to determine lizard identity down to genus or even to species. Observing context and knowing what species live in the area will also aid you in identifying the lizard. Be mindful that sometimes due to a variety of factors, the scat may have been disturbed and may be missing the uric pellet. For positive identification, it is useful to find multiple examples of deposited scats whenever possible.

Some species, such as the common chuckwalla, leave multiple fecal pellets each time they defecate. These scat piles can be obvious and at times, impressive. Other species, such as side-blotched lizards, leave single small fecal and uric pellets that are scattered more sparsely in a landscape and may be much more difficult to notice.

Many lizard species will deposit their feces and uric materials in the morning after first emerging and becoming active. This is especially common along the trails of species that rest at night under the surface of sand or other fine, loose soils.

The scat and round uric pellet of a regal horned lizard. Look closely to see the ant exoskeletons.

Horned Lizards

These squat, short-limbed lizards have distinct scats that are relatively easy to identify. Their scats tend to be pill shaped, generally thick, evenly round, and often with rounded ends. Horned lizards are predominantly ant eaters, and so their scats tend to be packed full of the exoskeletons of their ant prey.

The uric pellets of all the horned lizard species presented here are small relative to the fecal pellets. They also tend to be shaped like squashed white, or half white and half yellow, spheres. This may be true for all horned lizards, but more research is needed. The uric pellets of the pygmy short-horned lizard (*Phrynosoma douglasii*) are the smallest in proportion to their scats. This species also includes beetles as a large portion of its diet, so you may commonly find beetle parts in their scats.

The presence of scat pellets of horned lizards can help you detect these cryptic species. I talked to research ecologist Dr. Cameron Barrows, and he told me that researchers have used scats to locate the very elusive flat-tailed horned lizard and the endangered Coachella Valley fringe-toed lizard.

Fringe-toed Lizards

This group of sand-dwelling lizards tends to leave scats that are somewhat similar to horned lizard scats, as they also have a diet consisting largely of ants. Fringe-toed lizard scats are generally tapered on one or both ends and are narrow to moderately thick.

The scat and uric pellet of a Coachella Valley fringe-toed lizard.

The uric pellets of fringe-toed lizards tend to be long and narrow relative to the fecal pellets. The uric pellets may be white or white and yellow in color.

Chuckwallas

This large, herbivorous species leaves sizable piles of scat on favored basking rocks. These scats are composed of multiple, long pellets that are often tapered on one or both ends and which show lines of partial segmentation. Coarse fibers of the vegetative matter they consume can be seen in the pellets. These scats can be somewhat similar to the scats of some small mammals.

The uric pellets are soft and often deposited in a small, loose, lumpy pile that tends to end up underneath the fecal pellets.

The fresh scat of an adult Chuckwalla. It will become very pale as it dries.

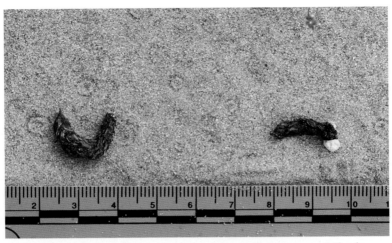

The fresh scat of a desert iguana. Look closely to see the coarse vegetation in the scat.

Desert Iguana (*Dipsosaurus dorsalis*)

This large, blunt-nosed and long-tailed desert-dwelling species is mostly an herbivore. It is especially fond of creosote (*Larrea tridentata*) and is not generally found far from these plants. It will occasionally eat carrion, insects, and fecal pellets.

Desert iguanas generally deposit one or two fecal pellets and a single uric pellet. The fecal pellets are long, dark, and may be partially segmented. The uric pellets are white to yellowish, and often exhibit a fold somewhere along their length. These lizards usually deposit them on the ground instead of on rocks and boulders.

Collared Lizards and Leopard Lizards

These large lizards deposit scats that are oblong and typically dark brown or black in color. They usually leave two pellets, one pellet that is strongly segmented, or a single pellet. The ends of the fecal pellets may be rounded or somewhat tapered. The uric pellets are typically shaped like thick pills or elongated spheres, mostly white with a little bit of yellow color on one end. The uric and fecal pellets are usually deposited together, frequently on top of a rock or similar high perch, but sometimes also on the ground.

Leopard lizard scats are very long, and are typically deposited as a single pellet. These scats are often rounded on one or both ends. The uric pellets of leopard lizards tend to also be shaped like thick, whitish pills with a little bit of yellow coloring on one end. The fecal and uric pellets are usually deposited together on the ground.

The typical, elongated scat of a long-nosed leopard lizard.

Earless Lizards

These medium-sized lizards of the genera *Holbrookia* and *Cophosaurus* have similar scats. Generally, they are relatively small compared to the size of the lizard. The fecal pellets are wider on one end and typically have a taper on the other end. The uric pellets of the greater earless lizard (*Cophosaurus texanus*) are often shaped like pills, and are mostly white. While the uric pellets of the common earless lizard (*Holbrookia maculata*) and northern keeled earless lizard (*Holbrookia propinqua*) are small, they often show a fold and are semispherical.

These lizards deposit their scats on the ground or on low rocks where they bask.

Geckos

Desert banded geckos (genus *Coleonyx*) are small, largely nocturnal ground-dwelling lizards found in arid and semiarid landscapes. The scats of this group of lizards are short, thick, and often have one round end and one end that comes to a finely tapered point. The uric pellets are typically very small and stuck to one end of the fecal pellet. These geckos typically deposit their scats on the ground. The scats of the Mediterranean house gecko (*Hemidactylus turcicus*) are very similar in appearance. This nonnative, nocturnal species of gecko is found in urban and suburban areas throughout the warmer parts of the southern United States. (See scat photo on next page.)

The short, squat scat of a Mediterranean gecko.

Spiny Lizards (Genus *Sceloporus*)

This large group of active, diurnal lizards are found across much of the United States. Typically, their scats are long, narrow, pill shaped or cylindrical, and are often accompanied by semispherical or pill-shaped uric pellets that usually measure about one-fifth to one-fourth the length the scat pellet. The ends of the scat pellets may be rounded or taper slightly. Typically, this species deposits scats on or near their basking sites such as on top of boulders, walls, and fence posts. Larger *Sceloporus* species, such as desert spiny lizard (*Sceloporus magister*) and crevice spiny lizard (*Sceloporus poinsettii*) leave proportionately larger and longer scats.

An example of a typical spiny lizard scat, in this case from a twin-spotted spiny lizard.

Snakes

The scats of snakes are infrequently encountered in the field, but can be found in places where the snake has rested, such as under cover objects or in or beside lays, and they can also be found inside of a shed skin. Snake scats are not very consistent in their form when compared to lizards, and generally appear as a combination of tapered and twisted pellets with rather amorphous blobs that include both feces and white uric deposit mixed in. When the pellets are more firm in consistency, they can appear more like the twisted, tapered scats of small carnivorous mammals. They may include some undigested parts of their prey, such as hair, bones, or teeth. Snake scats sometimes include bits of substrate or small pieces of vegetation the animal incidentally swallowed while consuming prey. The uric deposits of snakes are rather soft and fluid, not like the uric pellets of lizards.

Snake scats are much more difficult to identify down to species or even genus level. You can, however, generally gauge the snake's size by looking at the scat. The size of the scat is relative to the size of the snake that deposited it, with larger snakes generally making larger scats and smaller snakes depositing smaller scats.

The locations of scats can also be a clue to identifying the snake that left it. Many snakes deposit their scat on the ground in the open, under shrubs, or other cover objects. Highly arboreal snakes commonly deposit their scats in trees and shrubs. One U.S. snake that does this is

Look closely for the dark, soft scat hanging from the edge of a sassafras leaf directly below this rough green snake. JOE LETSCHE

the rough green snake (*Opheodrys aestivus*), which often hunts in shrubs and trees for prey such as spiders, cicadas, and grasshoppers. These snakes will also sleep near the tops of tall shrubs and small trees. Look carefully for their scat stuck to or running partially down the upper surface of large leaves.

A scat in the lay of a southern Pacific rattlesnake.

The scat of a juvenile coachwhip, which broke up into smaller pieces when deposited.

The scats of rattlesnakes are commonly found in or next to the lays they create in the substrate. The scats of rattlesnakes are often relatively thick and large, but may shrivel and dry when exposed to very dry conditions and high temperatures.

Very fresh snake scats can be confused with those of birds. Usually snake scats are thicker and more paste-like when compared with the liquid, runny scats of birds. As snake scats age, they will shrink somewhat and often become brittle.

The wet, fresh scat of a northwestern garter snake.

The old scat of a northwestern garter snake.

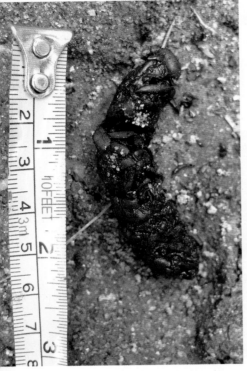

Frogs and Toads

Frog and toad scats are typically tubular, and may have rounded or slightly tapered ends. They do not deposit concentrated urea the way that reptiles do, and so do not excrete uric acid in the form of white liquid or uric pellets. Frogs generally deposit their scats in or near water, while toads may deposit them much farther from the water as they are more tolerant of drier conditions. Tree frogs most often deposit their scat where they rest, such as on bark, large leaves, or in rock crevices. The scats of both often contain the hard parts of arthropods, and in some of the large species, sometimes even the remains of vertebrate prey.

The very large scat of an American toad.
DAN GARDOQUI

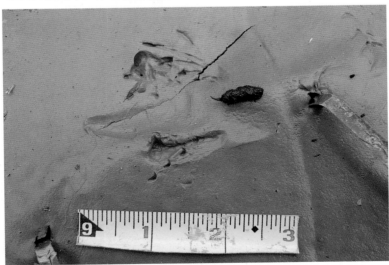

The track group and scat of a bullfrog. GEORGE LEONIAK

The scat of a large, adult rough-skinned newt.

The scats of frogs and toads will often be lightweight when they are dry. Toad scats can sometimes appear like the elongated, cylindrical scats of the northern flicker (*Colaptes auratus*) when they are full of ant exoskeletons, except that toad scats do not have the white uric outer coating typical of bird scats and tend to be less compacted than bird scats.

Salamanders

Salamander scats are not commonly encountered in the field, as most salamander species either spend significant amounts of time in water or under cover. Therefore, their scats are most likely to be found under objects such as boards, bark, or rocks, or in leaf litter. Occasionally, where salamanders travel long distances across open mud or sand, their scats might be found in their trails. Salamander scats are similar to those of frogs, although they tend to be somewhat longer overall. Just like frogs and toads, salamanders do not deposit concentrated urea as a white coating or uric pellet.

PERSONAL STORIES

200 Alligators

It was December on the Gulf Coast of northern Florida, and my wife and I were walking on a raised dike trail through a large wildlife refuge area. The trail acted both as an access point to and a dividing line between a saltwater marsh on one side and a freshwater wetland on the other. The previous night, as we were driving east toward Florida on a long journey from Seattle, we had spotted our first wild alligator.

It was not the most picturesque experience: we spotted the animal loafing on the shoulder of the freeway as we zipped along at seventy miles an hour. Still, for both of us, it was an exciting moment. The sudden dragon-like form looming out of the darkness made an impression. Now, as we found ourselves walking a trail through prime habitat for the American alligator (*Alligator mississippiensis*), we noticed first one, then another. As we continued down the trail, we counted more and more. We counted thirty, forty, fifty, and we kept spotting more. Some were impressively large individuals, as long as 11 feet.

As we walked on, we crossed patches of sand on the otherwise mostly gravel pathway. Here and there were the tracks of various animals, including river otter, grackle, and bobcat. Then we spotted a wide sand opening in the dense grasses of the salt marsh leading up and over the dike to the freshwater ponds on the other side. As we approached and looked more closely, we noticed that the vegetation had been well trampled and the path well worn by use. A thick dragging line led from the water's edge of the salt marsh and up onto the dike. On either side were very large tracks. Some showed five toes, while the larger webbed tracks showed four toes. The trail was nearly two feet wide. We quickly realized this had to be the fresh trail of an alligator. My wife placed her hand next to the track, and it was amazing to see that they were the same size.

A large adult alligator warming in the morning sun.

The well-worn trail of American alligators crossing a dike trail, from a salt marsh to a freshwater wetland.

We were used to seeing small lizard tracks in the southwestern part of the United States, but seeing reptile tracks of this size was mind-blowing. It was as if we had been transported back to prehistoric times, when reptiles ruled the land. Indeed, the alligators here were the top predators of these wetlands.

We counted a total of 199 animals on our walk. We then drove out to a point where we could overlook the Gulf of Mexico. The richly layered clouds and sinking sun came ablaze with color. The sunset of flame-orange clouds and neon pink sun is to this day still one of the most memorable in our shared experience. It was a perfect end to an absolutely astounding day of wildlife watching.

Bird Language

I walked slowly through the towering cypress trees, peeking around their buttresses and looking under pieces of bark and fallen limbs on the ground. December in northern Florida isn't the most ideal time to seek reptiles, but there I was. The swampy forest was surprisingly quiet, with only distant birdsong reaching my ears. I mostly heard the soft crunching of dry leaves and debris underfoot. As I flipped debris and looked around closely, I felt eyes on me.

It took me a moment to figure out what it was. As I raised my head, I saw it at eye level. A little lizard with a beautifully green body, pointed head, and long tail, perched upside down, looked at me without moving. It was familiar to me—a green anole (*Anolis carolinensis*). I wanted more tracks from this species, so I slowly positioned myself and made a swift grab.

It wasn't swift enough. The lizard dodged out of reach at the last second. It stopped and looked at me inquisitively, as if daring me to try again. I did, and missed again. I chased it around the buttresses and knees of one cypress tree and down onto the forest floor. It ran under a large piece of fallen branch. I gave it a moment to fully conceal itself, then I rolled the branch carefully over. I saw not one, but two lizards there. After a couple of lunges, I ended up catching both of them.

I walked them back to a sandy spot and let them move around at their own pace. The trails they left were rather indistinct. They were good enough to glean some measurements, but not clear enough for good track photos. I gathered up the two lizards, and walked them back to release them where I found them. As I walked back from the release site, some vocalizations caught my attention.

I was new to many of the birds in Florida, but I recognized the urgency and tone of the birds as indicating alarm in the presence of a predator. I knew they weren't focused on me, as the birds had plenty of opportunity to focus on me as I slowly wandered about among the trees. To add to that, I could see several birds and they were definitely focused in the canopy of a nearby tree.

I pulled binoculars and a field guide out of my pack and settled onto the forest floor to observe. I identified a red-bellied woodpecker (*Melanerpes carolinus*), a blue jay (*Cyanocitta cristata*), and some Carolina chickadees (*Poecile carolinensis*) spread in a ragged sphere around one particular area of the tree. I scanned slowly, expecting to see an owl perched in the center of the group of birds. But I saw nothing.

I scanned carefully again and again. Then I used the positions of the alarming birds to help triangulate the exact source of this commotion. I saw the woodpecker land on a branch near the center of this area, and carefully approach something that seemed invisible to me. I then noticed a small cavity just barely visible from my position. The woodpecker became more insistent with its vocalizations. Then it suddenly lunged at something and quickly jumped back.

I kept my binoculars fixed on the spot where it had just been. I held as still as I could and hoped the mysterious creature would come into view. I noticed the branch appeared to be slowly developing an odd shape directly below the cavity on the other side. I stood up, keeping the binoculars locked on the target. I finally made out the snake as it slowly moved its head down and then stretched to reach another part of the branch. I never considered I would be watching a snake up in the tree canopy through binoculars. Nevertheless, there it was, emerging from the cavity. The snake appeared to be a large gray rat snake (*Pantherophis spiloides*), a species I was very excited to see.

Once the snake was clearly in view, the birds slowly dispersed and stopped sounding the alarm. The last to leave and return to normal behavior was the woodpecker. I got up, smiling gratefully at the experience. I never got tracks from that rat snake, but the memory of the experience was perhaps a better prize.

Black as a Shadow

On a hot summer morning, my wife and I wandered slowly through sandy riparian wash in southeastern Arizona. I walked off along a sandy trail as she explored closer the flowing water of the stream. I was hoping to find some fresh lizard tracks to photograph for the book. I spotted up ahead the upright form of a sunning greater earless lizard (*Cophosaurus texanus*). As I approached closer, I could see by the rainbow colors across its torso that it was an adult male. I tried to circle around so that when the lizard ran, it would cross the sandy trail and leave some nice, clear tracks.

That is not what happened. Before I even got partway off the trail, the lizard made a dash in the opposite direction toward the nearest mesquite thicket. I followed it halfheartedly, knowing it was not going to leave any tracks for me. Just as the lizard neared the thicket, it turned and ran even faster in a new direction. Immediately in hot pursuit was a lightning-fast shadow that lunged out of the dark area of cover under the shrubs. Out of the shrubs, and moving full speed after the lizard was a solid, shiny black snake between five and six feet in total length. As it spotted me, it took off in another direction and disappeared down a burrow.

I decided to follow my intuition and wait for the snake to come out again. Forgetting the lizard, I sat within about six feet of the burrow and got comfortable. I did not move at all for nearly twenty minutes, even when bees buzzed near my face and flies walked around on my skin. Then I noticed about six inches of the snake was sticking out of the burrow, its head lifted well off the ground and its large eyes locked on me. I froze again, and sat as still as I could. Time passed slowly, and I felt sweat bead up on my forehead.

Then, a subtle scraping started to reach my ears. I slowly turned my eyes and looked in the direction of the sound. A very large adult desert spiny lizard (*Sceloporus magister*) was walking toward me and the snake. I noticed a subtle shift in the snake's position as it pointed its head toward the lizard and continued to stay in that alert, upright, frozen pose. With growing excitement, I wondered what would happen next. The lizard made slow progress in our direction, stopping and starting again. Every time it stopped, it looked around carefully, probably watching both for prey and predators.

So many things could have happened then, and I wondered who would notice whom first. Then, the lizard walked within about three feet of the black coachwhip snake (*Coluber flagellum*). In a movement too fast for me to follow, the snake lunged out of the burrow and grabbed the lizard before it had time to run more than a few inches. It had the lizard by the base of the tail, and it threw a part of its body over lizard to help pin it down. This gave me a moment to move and lift my camera up and take several photos while the snake was distracted.

I soon realized the snake had made a successful catch, but it held the lizard awkwardly. It seemed like both the snake and I realized at the same time that it would have to adjust its grasp to maneuver the lizard for eating. The lizard must have realized this too, because it was holding very still. Then, as soon as the snake released the tail to get a better grip further up the body, the lizard threw itself violently sideways, flipped over, and ran for its life. The snake followed in hot pursuit, but quickly ended up losing its prey.

At this moment my wife rejoined me, and we followed the snake as it moved down toward the creek. It flowed down between the rocks like a

A coachwhip holding a large desert spiny lizard it has just captured.

liquid shadow, poking its head underneath various rocks, hoping to surprise another lizard. We followed its subtle trail through sandy patches and studied it closely. The excitement of witnessing the life-and-death drama stayed with me for the rest of the day and the memory has stuck with me ever since.

Coachwhip with Ted and JBob

The three of us got out of the car just after sunrise to explore a set of small dunes in the Colorado Desert of California. It was already 98 degrees, and the brilliant morning sun was, for the time, mercifully at our backs.

With me were my two travel companions, Theodore "Ted" Packard and Jeremy "JBob" Williams. We were here as part of an extended trip to gather data and images for this book. This morning we were focused on finding and following the most recent snake trails we could locate. Aging the trails was relatively easy this time since strong winds during the previous night would have erased any older trails. As we wandered into the large expanse of sand, we split up, remaining within yelling distance.

Each of us followed the fresh tracks we came across as well as our own inner compasses, scanning for any details that we could pick up from the sands. The beautifully precise sine wave patterns left by western shovel-nosed snakes (*Chionactis occipitalis annulata*) were a frequent story written in the sand. Jeremy and I followed several trails that led us into a dense clump of vegetation. In the middle of this small thicket, several fresh trails converged onto a single, inch-wide burrow entrance. The trails showed different measurements, indicating there were several different individuals. There had to be several snakes in there right now. But, it was not our intention to destroy their nighttime shelter just to get them out, so we parted ways again and continued exploring.

At the crest of a small dune, I spotted a distant broken curving line moving away from me toward the northeast. Upon closer approach, I recognized the trail from a snake species we had come across the previous morning. The swiftly moving snake was only touching down in small areas of sand as it traveled forward using lateral undulation. With the sharp angle of the morning sunlight, the shadows in the trail made it pop. Knowing this animal was likely nearby and traveling fast, I quickly took some photographs and measurements and started jogging down the dune alongside the trail.

Excited, I broke into a run. The trail passed over one dune, around a shrub, and then up and over another dune. I was watching it closely, looking for changes in pattern that might indicate slowing or stopping. Then, I saw a change. The snake had slowed, and its trail had become narrower and more complete. It was now entering an area that was covered in a small concentration of waist-high shrubs.

Like an excited dog trying to pick up the scent of a rabbit, I began to loop back and forth through the shrubs trying to anticipate where the animal might be hiding and to spot where the trail traveled between the shrubs. Then the trail stopped, no longer continuing past a group of shrubs. I walked back and followed the trail to where it ended in one particular shrub. Slowly, I lowered myself down and looked in. It took me a moment to distinguish the lithe serpentine form from the shadowy jumble of branches near the ground. Sure enough, as I spotted the snake, **The lateral undulation trail of a coachwhip.**

it raised its head and stared right back at me, its large yellow eyes meeting mine. Its long, slender body was a mixture of sandy tan, with small areas of orange and black near the head. It was a medium-sized adult coachwhip snake (*Coluber flagellum piceus*).

These keen-eyed, lightning-fast snakes are very alert and quick to escape. After some dodging and chasing around the shrub, I had the animal in hand. I had never seen a coachwhip with this particular color, and I wanted to show my companions this beautiful animal. As I headed back, I spotted Ted in the distance and headed his way. He was bent down inspecting something in the sand.

The temperature was getting pretty high at this point, and it was well over 100 degrees Fahrenheit. I forced myself to slow down despite my excitement to a walking pace to avoid overheating. As I neared Ted, I could see he had a grin on his face. He had just witnessed a loggerhead shrike land and take off again from the sand. The three of us had an unofficial goal set for the trip to see the tracks of a shrike. It was a challenge Jeremy Williams had set for himself years ago, and we adopted it as part of our group goals. Ted and I joked and chatted about how we had finally found shrike tracks, and I held up the beautiful coachwhip for him to inspect up close.

Then we started walking in the direction we had last seen Jeremy. At one point, both of us spotted something odd about a patch of sand we were walking through. The sand was a more compacted, and formed a flat basin like the bottom of a bowl surrounded by rising dunes on all sides. In the sand, several oddly shaped yellowish-white objects caught our attention. We looked close and realized they were bones. We started looking for as many pieces as we could locate, sifting through the sand on our hands and knees. We brought our separate collections together, tried to piece them together, and assessed our finds. The long, curving two-inch claws and short leg bones helped us identify the original owner: a badger.

As Ted and I continued to search for more bones, he accidentally punched through the sand and his hand found a hidden cavity underneath. His fingers brushed against something papery to the touch and with scales. He carefully withdrew it and came away with a piece of reptilian skin. With careful excavation, he discovered the mummified body of sidewinder, complete with rattle still attached. What an amazing find!

We theorized about how it might have ended up there and what had happened to it as we walked back toward Jeremy. Once we caught up to him, we swapped stories and all had one last look at the pale coachwhip snake. Then, I walked back to where I had caught the snake and returned it to its shrub.

The temperature was now around 113 degrees Farenheit and still growing hotter. I quickly rejoined my companions and the three of us sought the shortest route back to our vehicle. It was definitely time to get out of the sun and head to town to cool off, and take a midday siesta in some deep shade.

Trailing a Desert Horned Lizard

A giant sand dune loomed over us as we walked in search of reptiles through a vast, sandy portion of the Mojave Desert. Surrounding us was some of the most beautiful desert wilderness in North America. With me were two companions, Joanna Wright and Robert Mellinger. They were both students of the Alderleaf Wilderness Certification Program where I taught, and were very passionate wildlife trackers.

We were on the lookout for fresh tracks and sign of wildlife—especially lizards—and we were not disappointed. In all directions, for miles, the tracks of a great wealth of different animal species appeared. The cast of characters who left us tracks included black-tailed jackrabbits, desert cottontails, burrowing owls, kit fox, coyotes, roadrunners, tarantulas, scorpions, kangaroo rats, pocket gophers, ravens, and many different types of beetles. And of course, there were lizards—many, many lizards.

The most common lizard tracks we crossed where those of the Mojave fringe-toed lizard (*Uma scoparia*). This is a sand specialist with specially fringed toes on both front and hind feet for dealing with the challenges of loose sand.

The fringe-toed lizards left trails in all directions around us, and appeared to be most concentrated in open areas of the dunes that also included some tufts of grass or small desert shrubs. They are the only lizard species that is likely to be found in the more open portions of the dune systems. We stopped here and there to inspect their trails. Here one dug its way out of its nighttime sandy retreat and started walking around slowly. There, two fringe-toed lizards appeared to chase each other. Just a little further, one of us spotted an area where a lizard had defecated and then dragged its cloacal region for several inches. In another spot, a fringe-toed lizard's trail loitered around the entrance to an active ant nest. The sand told the intimate stories of the lives of these lizards and the animals that share their home.

I stopped to photograph and record information on this species whenever we came across any interesting and photogenic tracks or sign. In a place with such abundant reptile tracks, I could afford to be picky.

Our goal was to learn as much as we could about the diversity of lizard species that lived in and around these sand dunes. We agreed the best place to seek a higher diversity of these scaly creatures was at a meeting of two habitats, an ecotone.

In the distance, we could see the sinuous line of a small wash that sliced through the edge of part of the dune field. On the far side of the wash was a gravelly flat covered with creosote bushes. The wash itself created a meeting place of these two different realms, with some parts of it being overwhelmed by the sandy sea to the west. In other parts of the wash, the hard, gravelly pavement made headway into the dune field. This looked like the perfect place.

As we neared it, we started spotting the evenly spaced walking trails of whiptail lizards, and the very small tracks of occasional side-blotched lizards that had made tentative inroads to the edge of the dunes. A few hundred yards from the wash, one particular trail caught my eye—something about this set of tracks was different. On closer inspection, the tracks looked rather squat, with proportionally short toes, and with front and hind tracks of very similar size. Here was a species we had not seen yet. I stuck my thumb down lightly onto the sand next to the trail and compared my track with that of the lizard trail. Both looked almost equally crisp. This was a fresh trail.

We took turns following this mystery trail, with the lead person pointing to each set of tracks as they were spotted. This kind of hand signaling helped those further back understand that the lead tracker still saw the trail. Initially it started out easy, with us being able to walk at a rapid pace and still follow the tracks. As the tracks passed over patches of firmer sand, we were forced to slow considerably. Even the mild breeze was enough to erase portions of the trail.

Several times, the lizard passed through an area where the falling debris from the large desert shrubs near the edge of the dune field made following the trail extremely difficult. Here the person in lead marked the last track they saw with an exaggerated foot drag through the sand. Then, we carefully fanned out around the shrub to see if the trail would emerge into the open again. We had to tread with great care to avoid stepping on the trail before we spotted it again.

Several times we spooked whiptails and fringe-toed lizards in the process. When we compared the tracks left by these fleeing lizards to the trail we were following, it was encouraging. They looked almost as fresh.

The trail of this mystery lizard continued to move through the open and back into the shady cover of shrubs. It would trot when exposed, and slow to a walk when it was near or within the shade. In one spot, at the edge of a small clump of grass, we saw the trail come to a full stop and found the impression of this animal's full body in the sand. It had laid here for a short time, likely basking in the desert sun. The impressions of the plump width of its belly, the thick tail base, and relatively short limbs told us what we were following: a desert horned lizard (*Phrynosoma platyrhinos*). It was exciting to finally know the identity of our lizard, and it spurred our desire to see this beautifully spiky desert dweller.

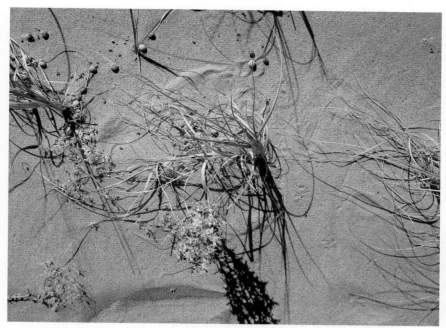

The trail of the desert horned lizard leads from the lower right to the dig above the center of the photo underneath the clump of grass.

From the body impression, the trail continued again through an open area. We noticed now that the horned lizard was slowing down more frequently as it passed through an area of sparse grass clumps. Suddenly, the trail just ended. Holding back disappointment, we got down close to where the last tracks were and started searching on all fours. At about a ninety-degree angle to the trail was a small displaced area of sand. It was not easily seen because of the associated grass. There was a subtle depression in the center of this spot, and a ridge of raised sand on one side.

This must be it, I thought. Joanna poked her finger slowly through the sand right in the center of displaced area. We could tell by her expression and sudden laughter that we had found it. Sure enough, out popped the horned lizard!

We closely admired this prickly lizard's color, pattern, and spines. It was a gorgeous animal, with a wide belly and an impressive crown of cranial horns. It was made even more beautiful by the trail of tracks it left for us, and the story it shared through them.

Life and Death in the Swamp

While traveling in the southern tip of Florida, my wife and I explored the wild, herp-rich landscapes of the Everglades. Though we saw more herp diversity there than almost any other location we had gone to, it was a difficult place to find tracks. Much of the Everglades is a shallow, flooded grassy plain where seeing clear tracks was most likely when the water receded and exposed the sticky gray mud. Even then, it would take some searching to find them.

On this particular day, I was off exploring a trail as my wife rested and took a break from the mosquitoes. I wandered through subtropical swamps, coastal prairies, and mangroves near the shoreline of Florida Bay. Small sandy patches of trail and drying puddles of mangrove were my targets. I scrutinized each potential spot closely, looking for even

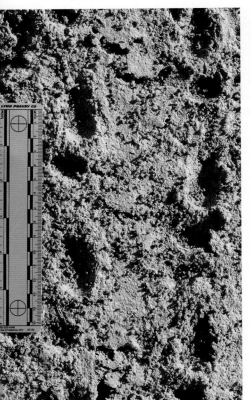

The half understep walking trail of a Florida box turtle.

subtle sign, but other than the occasional old raccoon tracks, I didn't find anything. I continued on until the dense cover of trees opened up onto a wide, flat expanse of exposed soil with patches of low shrubby cover scattered here and there. At the base of the shrubs was a papery layer of algae, and on the surface of the exposed soil was a crust of mineral deposit. This was a dried-up wetland, and it was just the kind of spot I was hoping to come across.

I wandered out on to the open pan, treading carefully in case some of the viscous mud was wetter than it appeared. I noted the tracks of many mammals, including the old tracks of deer, coyotes, gray fox, raccoons, along with those of egrets, ibises, and other wading birds. There were layers upon layers of tracks in some places, indicating that this shallow wetland likely dried slowly. Most of the tracks appeared relatively deep, and almost all of them showed cracks and the rough texture inside them that indicate significant weathering. I found a single

The old, salt-encrusted shell of a box turtle.

front track of an alligator, which must have stepped in the one spot in that area that still held enough moisture to register a track. I followed a fox's trail for a time, and it led to a shallow hole about ten inches deep that the animal had paused to excavate. I peeked inside and found the partial remains of a blue crab. The fox may have been excavating the crab from its last-ditch shelter against the drying of this wetland.

Nor far in the distance, a strange, small lump in the soil caught my eye. As is so often the case when tracking, something about the way this lump seemed to be out of sync with the rest of the patterns drew my attention. It was similar in color to the rest of the soil, a shade of gray and tan. But this was a little darker gray, and was too consistent and rounded to just be a lump of mud. As I neared it, I noticed it had a gaping cavity inside it. As I stooped over it and put my hand on it, I could see and feel a repeating pattern like scales. It was a turtle shell—just the upper portion of it (the carapace) lying upside down in the mud. It was very fragile to the touch and must have been weathering there for some time. I gingerly pulled it out of the muddy crust and turned it over. The shell formed a rather high dome, with two wide, shelf-like flares toward the posterior. The ghostly lines left by the scutes, which were now long gone, helped bring a clearer image in my mind of what this animal's carapace must have looked like before it was reduced to this almost paper-thin husk. I puzzled over the story of what led this animal here and why it came to an end in this now dry, brackish wetland.

Another strange shape in the mud about a hundred feet away drew my attention and I gently placed the shell back in place before I moved on. This next mysterious object revealed more of the story. It was another turtle shell, in a similarly poor state of decomposition, but this one had part of the lower portion of the shell, the plastron. The plastron was very well developed, wide, and provided very little clearance between it and the carapace. What's more, it was broken off perfectly along a seam that must have been a hinge. I carefully extracted this fragile shell out of the mud and ran back to the first shell, and set them side by side. They were a perfect match, with the same high dome and flares.

I ran through my mental list of the turtle species I knew were found in southern Florida, and which had shells that matched these features. Only a few had hinged plastrons like this: mud, musk, and box turtles. I knew that mud turtles have two hinges, one closer to the anterior and the other closer to the posterior of the plastron. This shell clearly only showed one hinge. This left the musk and box turtles. Musk turtles have reduced plastrons and significant gaps between the carapace and plastron, especially around their legs. This shell showed practically no gap between the two major portions of the shell. To add to that, the shells measured 6.5 and 6.75 inches in carapace length. These measurements are too big for even the largest common musk turtle (*Sternotherus odoratus*) ever recorded, by nearly an inch. Those flares on either side of the posterior of the carapace are not a feature found on musk or mud turtles, and in this case, they were present on both shells. This left only one possibility, the box turtle. In this region, the native species is the Florida box turtle (*Terrapene carolina bauri*).

I had figured out the identity of these two turtles by using the evidence at hand and information I had studied in preparation for visiting this subtropical wilderness. I knew that box turtles like to soak in water, especially during the hot periods of the year. But many questions arose that I could not answer with certainty. Why had they died here? Why so near each other? Could the concentration of drying salts in this shallow pan have led to their deaths through desiccation?

I walked as I thought about the deeper mysteries of these turtles. In my path, two parallel lines of what looked like squat exclamation marks stacked on top of each other curved away from me and back toward the higher, more wooded landscape of the main trail. I stared at these marks, puzzled for a moment. Then I realized, at least one of the box turtles had survived after all.

Desert Tortoise at Pinto Dunes

It was an early morning in late March and I was walking across a wide valley in the southern Mojave Desert. A gentle breeze blew from the south and west as I made my way between widely spaced creosote bushes toward a distant area of sand dunes. I scanned the ground and the distant open areas between the shrubs for tracks and for movement of any animals. The clear, ringing song of a black-throated sparrow and the soft hissing of the breeze were the only sounds. Otherwise, the silence was complete. From that spot, you could see for at least ten miles in all directions. Such profound solitude is frightening to some, but it was energizing to me.

Despite the apparently barren landscape, such desert areas can be rich in animal life. As the hard pan under my feet began transitioning to the first patches of sand, tracks popped out to my eyes. Here I spotted the paired, bipedal patterns of desert kangaroo rats. There the continuous lines of the many foraging Great Basin whiptails (*A. tigris tigris*) could be seen winding this way and that. I also saw the zipper-like lines of large beetles meandering across the fine sand.

As the amount of open sand increased, the tracks of more species became visible. The sand was peppered with the tracks of desert cottontails, black-tailed jackrabbits, kit fox, and even a meandering badger trail. The rapid overstep trotting trails of Mojave fringe-toed lizards (*Uma scoparia*) began to appear as I neared a rise in the sand

The first desert tortoise I encountered in the wild.

The half understep walk of a desert tortoise. Direction of travel is from the bottom to the top of the photo.

dune. The faint, smoothly curving Js of an old sidewinder (*Crotalus cerastes*) trail also came into sight. I stopped to photograph each of these finds in turn, and record vital information on measurements, patterns, and location.

These were all wonderful finds, but I was after something bigger and much harder to observe. This animal had eluded me on previous attempts to find it. I noted that there were some small green plants poking through the sand and patches of hard pan near a wash, and my excitement increased. Conditions were good for finding a lumbering desert icon. Letting my eyes and intuition guide me, I wandered down the far-side dune area and walked to the base of a hill. There, I spotted a large, crescent shaped burrow. I almost passed it by, but something about the soil in front of the entrance caught my eye. I saw a set of scuff marks, and as I dropped down to inspect them, the fresh tracks jumped out at me. There were two parallel lines made of circular tracks entering the burrow. I stared hard at what was in front of me, and then saw them: fresher tracks leaving the burrow. Now, the game was on.

A kind of giddiness overtook me as I started to walk next to the line of very fresh tracks headed northwest up the valley. It was the sort of feeling you get when you begin following the fresh tracks of an animal and know that, very likely, you will find it at the end the trail. In this case, it was an animal I had longed to see for years but never had the luck to encounter. The sun was now beating down with the kind of typical desert intensity one has to experience to understand. I took off my shirt in an effort to stay cooler and picked up my pace. Yes, finding these tracks was amazing in and of itself, but now I had another goal: find the animal.

The tracks continued along in a meandering path. The animal had stopped to chomp on some desert dandelion greens. I got down and looked closely at the plant, and could see where the leaves were missing.

What was left of the plant showed jagged edges made by the animal's sharp beak. From here, the trail picked up again in the same general direction. Then, it turned and entered another burrow.

I felt disappointment creeping in as my first impression was that the trail ended there. I didn't just want to find where the animal disappeared, I wanted to see it. I got down and inspected the entrance from all sides. Then, I started circling in a spiral out away from the entrance to see if I could pick up the trail again. Yes! There it was continuing more toward the eastern part of the big valley.

I started walking along the trail again, but now things were getting much more difficult. The sun was much higher in the sky, and the shadows of the morning had receded considerably, now directly under the scattered shrubs. This kind of light makes tracks very difficult to see, as the shadows made by sideways light are what give tracks their greatest definition. The animal had also walked onto an area of mostly hard pan, and the tracks had become very faint. All that I could see were the faint lines left by the large claws on the front feet.

The animal entered two more burrows, and then continued walking out into the most open portion of the valley. I lost the trail several times—the heat and my thirst were starting to affect my ability to maintain my focus. In my excitement to get into the field, I had left my water bottle in the car. To the south and west I could see clouds building behind a distant mountain range and could feel the humidity rising in the air. The trail in front of me was very faint, and led to an area with sparse bunches of desert grass. After nearly four hours and a mile of increasingly difficult trailing in the hot sun, I was feeling spent. I considered turning back, and then I spotted it: a large, adult desert tortoise (*Gopherus agassizii*) crouching close to a small clump of dry grass and trying to use the meager shade it provided.

I approached it slowly, and dropped to my knees to get a closer look. It withdrew its head nervously and uttered a hiss that in the profound silence of the valley was enough to make me jump. I did not wish to disturb it, and so I backed up to a respectful distance and sat still. Slowly, its head emerged again and stared at me with an intensely green eye. Its shell was huge and showed a great deal of wear, and included large gular horns at the leading end of its plastron, telling me this old-timer was a male. The animal was the largest desert tortoise I had seen and was my first wild tortoise. I had no idea how old this animal was, but judging by the size and wear of its shell it must have been very old. To me, it seemed ancient, like some desert rock that had been there for millennia.

I lost track of time looking at the animal, but a twinge of discomfort brought me back to the present. I looked at my bare skin and realized I had sunburn. It was time to put the T-shirt back on and head back to the car. I sat a moment longer and quietly expressed my gratitude to this tortoise and this beautiful, stark desert land for giving me this experience.

As I walked back to the car, the clouds had built up very rapidly and were almost looming overhead. Other changes had taken place around me, including the sudden appearance of bright red, furry alien-like creatures rapidly running across the sand. They were just under half an inch long, and I recognized them as arachnids, specifically a kind called the red velvet mite (genus *Angelothrombium*). At that moment, a common name of this species came to mind: "children of the rain." Moments later a blast of cool air hit my face, carrying with it the first heavy drops of a desert rain.

It was a perfect way to end a day of tracking wildlife in the desert. The arrival of the rains would benefit all the forms of life in that valley and surrounding mountains. Rain in the desert was always worth celebrating. I was nearing my vehicle now, and the rain began to fall hard. I stopped, looked up, and smiled as the squall hit me in earnest.

BIBLIOGRAPHY

Bartlett, R., and Bartlett, P. 2009. *Guide and Reference to the Snakes of Western North America (North of Mexico) and Hawaii.* Gainesville: University Press of Florida.

Brennan, Thomas. *Online Field Guide to the Reptiles and Amphibians of Arizona.* August 2015. reptilesofaz.org.

Brown, H., Bury, R., Darda, D., Diller, L., Peterson, C., Storm, R. 1995. *Reptiles of Washington and Oregon.* Seattle: Seattle Audubon Society.

Carmichael, P., and Williams, W. 2011. *Florida's Fabulous Reptiles and Amphibians.* Hawaiian Gardens, CA: World Publications.

Corkran, C., and Thoms, C. 1996. *Amphibians of Oregon, Washington and British Columbia.* Edmonton, Alberta: Lone Pine Publishing.

Cornett, James. 2003. *Desert Snakes.* Palm Springs, CA: Nature Trails Press.

Cowles, R. 1977. *Desert Journal: Reflections of a Naturalist.* Berkeley: University of California Press.

Davis, Alison R. 2011. *Kin presence drives philopatry and social aggregation in juvenile Desert Night Lizards (Xantusia vigilis).* Oxford University Press.

Dodd, C. Kenneth, Jr. 2013. *Frogs of the United States and Canada: Volumes 1 & 2.* Baltimore: Johns Hopkins University Press.

Dinets, Vladimir, et al. 2013. "Alligators and Crocodiles Use Tools to Hunt." *Live Science.* http://www.livescience.com/41898-alligators-crocodiles-use-tools.html.

Elbroch, M., Kresky, M., and Evans, J. 2012. *Field Guide to Animal Tracks and Scat of California.* Berkeley: University of California Press.

Ernst, C., and Ernst, E. 2003. *Snakes of the United States and Canada.* Washington, DC: Smithsonian Books.

Gannon, Megan. 2013. "Snakes Control Blood Flow to Boost Vision." *Live Science.* http://www.livescience.com/41045-snakes-eyes-blood-flow-control.html.

Gray, Brian. 2005. *The Serpent's Cast: A Guide to the Identification of the Shed Skins from Snakes of the Northeast and Mid-Atlantic States.* Lawrence, KS: The Center for North American Herpetology Monograph Series, Number 1.

Hoyer, R., and Hoyer, R. 2011. *The Rubber Boa (Charina bottae) Information Portal.* www.rubberboas.com.

Hu, David L., and Shelley, Michael. 2012. *Slithering Locomotion.* Spring New York.

Jayne, Dr. Bruce C. "Slow motion of lizards sand swimming." https://www.youtube.com/NDrg6M3Iqb0. Youtube. May 19, 2008.

Jones, L., Leonard, W. P., Olson, D. H., eds. 2005. *Amphibians of the Pacific Northwest.* Seattle: Seattle Audubon Society.

Jones, Lawrence, and Lovich, Robert E., eds. 2009. *Lizards of the American Southwest.* Tucson, AZ: Rio Nuevo Publishers.

Krysko, Kenneth. Florida Museum of Natural History website. August 2015. flmnh.ufl.edu.

Lemm, Jeffrey. 2006. *Field Guide to Amphibians and Reptiles of the San Diego Region.* Berkeley: University of California Press.

Liebenberg, Louis. 1990. *The Art of Tracking: The Origin of Science.* Cape Town, South Africa: David Philip.

Lovich, J., and Ernst, C. 2009. *Turtles of United States and Canada.* Baltimore: Johns Hopkins University Press.

Nafis, Gary. *California Herps: A Guide to the Amphibians and Reptiles of California.* californiaherps.com. August 2015.

Petranka, James. 1998. *Salamanders of the United States and Canada.* Washington, DC: Smithsonian Books.

Powell, Robert, Collins, Joseph T., and Hooper, Errol D. 2012. *Key to the Herpetofauna of the Continental United States and Canada: Second Edition, Revised and Updated.* University Press of Kansas.

St. John, Alan. 2002. *Reptiles of the Northwest: California to Alaska; Rockies to the Coast.* Edmonton, Alberta: Lone Pine Publishing.

Stebbins, R., and McGinnis, S. 2012. *Field Guide to Amphibians and Reptiles of California.* Berkeley: University of California Press.

Stebbins, R., and Miller, A. 1964. *The Lives of Desert Animals in Joshua Tree National Monument.* Berkeley: University of California Press.

INDEX

Page numbers in *italics* indicate photographs or illustrations.